MARINE CONTROL ENGINEERING AND AUTOMATION

MARINE CONTROL ENGINEERING AND AUTOMATION

Editor
Walnes Rego

DPH

DISCOVERY PUBLISHING HOUSE
INDIA

Published by:

DISCOVERY PUBLISHING HOUSE
4383/4B, Ansari Road, Darya Ganj
New Delhi-110 002 (India)
Phone : +91-11-23279245; 23253475; 43596065
Mobile : +91 9811179893 / +91 9871656464
E-mail : discoverybooksindia@gmail.com
orderdphbooks@gmail.com
namitwasan9@gmail.com
web : www.discoverypublishinggroup.com

Marine Control Engineering and Automation
Edited by: **Walnes Rego**

International Standard Book Number: **978-81-19365-84-5** (Hardback)

PREFACE

Marine Control Engineering and Automation have emerged as critical fields within the realm of maritime technology, and this book aims to provide a comprehensive overview of the research, advancements, and practical applications in this dynamic domain. As the maritime industry continues to evolve, the demand for efficient and reliable control systems and automation solutions becomes increasingly paramount.

The primary objective of this book is to serve as a valuable resource for researchers, engineers, and students interested in comprehending the intricacies of marine control engineering and automation. It covers a wide range of topics, including control theory, system modeling, sensor integration, decision-making algorithms, and human-machine interaction in marine environments. By exploring these subjects, readers will gain a deeper insight into the challenges and opportunities presented by the ever-changing marine landscape.

The chapters within this book have been authored by leading experts and researchers from academia and industry who have devoted their efforts to pushing the boundaries of marine control engineering and automation. Their contributions encompass a diverse array of applications, such as ship propulsion systems, maneuvering and navigation, dynamic positioning, power distribution, and safety systems. Each chapter delves into the specific challenges faced in these domains and provides innovative solutions that have been rigorously researched, tested, and validated through practical implementations.

The book also features comprehensive reviews of key technologies and methodologies that are integral to marine control engineering and automation. It addresses advancements in control algorithms, adaptive systems, machine learning, and artificial intelligence, offering readers an in-depth understanding of the cutting-edge techniques driving the field forward. Furthermore, it explores the integration of emerging technologies, such as the Internet of Things (IoT), cyber-physical systems, and autonomous systems, which are revolutionizing the marine industry and paving the way for the vessels of the future.

Throughout the book, emphasis is placed on the practical implications of the presented research. Real-world case studies, simulation results, and experimental findings are provided to illustrate the effectiveness and feasibility of the proposed control strategies and automation solutions. By bridging the gap between theory and application, this book equips readers with the knowledge and tools to tackle the complex challenges encountered in marine environments.

The intended readership of this book includes researchers, engineers, and students in the fields of marine engineering, control systems, robotics, and automation. It also caters to professionals in the maritime industry who seek to enhance their understanding of marine control systems and leverage the latest advancements to improve safety, efficiency, and sustainability in their operations. We hope that readers find this book informative, inspiring, and instrumental in driving innovation within the maritime industry.

The editor has collected reliable and scientifically proven information from various credible sources to provide readers with the latest and most accurate information available. The book is a compendium of contributions from various experts in the field, and the editor expresses gratitude to these contributors for sharing their knowledge and expertise.

Editor

CONTENTS

CHAPTER
1

Machine Learning Methods for Marine Systems

Dr S Thangalakshmi[1] **and Dr K Sivasami**[2]

[1]*Faculty - Electrical, Indian Maritime University, Chennai Campus*
[2]*Associate Professor - Marine Engineering, Indian Maritime University, Chennai Campus*

ABSTRACT

Automation plays a key role in shipping industry and aims towards minimal operating staff. However, the effective automation relies on effective controlling at various levels starting from shipbuilding to navigation. The industry is currently focussing on autonomous shipping which actually requires precise controlling. Although many conventional methods are available for control and automation with regard to automation, Artificial Intelligence Schemes (AIS) are widely attracting the maritime sector because of their benefits. The AIS along with fuzzy logic systems are offering promising results. The emerging use of AIS in a variety of maritime applications can act as a reference wpoint for new researchers. This paper aims to conduct a valid AIS study and to examine the various machine learning approaches used in various maritime applications. It is possible to achieve complete automation in the shipping industry by implementing a related technique.

Keywords: *automation; autonomous shipping; artificial intelligence; machine learning*

1. INTRODUCTION

Presently the digital transformation is one of the inevitable happening in anybody's life and this is quite true after the covid-19 pandemic. While everyone dire ctly or indirectly undergo this digital transformation, a valid and promising technology is needed to accomplish the same. Artificial intelligence systems are one such technique and it applies to almost every industry at a faster rate. Various gadgets and elements incorporate the AIS features in our routine life. These schemes possess altered characteristics based on the device, which adds them or based on a particular application. The approaches and procedures that are followed in seafaring sector are unique and to apply the AIS technology, the distinctiveness of the industry must be appropriately accounted.The scale of the dataset, its authenticity, and the use of recurring key incidents all play a significant role in the success of AIS. This is especially true in the traditional shipping industry, where the transition to full automation is regarded as critical. As a result, a survey was carried out in the most commonly used applications of the maritime industry. The aim of this research is to identify the best area for using AIS and then scale it up to allow full automation. Artificial intelligence can be realized in this field in several facets [1 – 5].

Machine learning is a subset of artificial intelligence and is a probable way to accomplish AI. Machine learning is capable of dealing with minor to major datasets by means of inspecting and associating the data to get mutual arrays and learn the details. Artificial intelligence is a technology where computers / machines can be trained through a relevant learning algorithm to mimic human brain viz. recognition, remembrance, application, etc. Shipping firms these days have understood that financing in quicker communication offers numerous benefits. Most vessels go forward to remote workplaces at sea, which may offer consistent Internet access, simulated / virtual set-ups, e-mail, route organizers and many supplementary schemes and uses. It is now essential for the shipping firms to plan for long-term growth by investing in novel growing technologies to standardize the vessel operation, CTC reduction and business optimization plans. The shipping sector moves towards automation and AI assists in accomplishing the same [16].

Automated / Unmanned systems do not require rest, they effortlessly do repetitive works, and AI development will deliver work with unstable environments. AI is fundamental for automated systems and effectively replaces the manned systems. AI may perhaps an opportunity for modern shippers. Optimization of maritime operations is of prime importance especially in vessel precision operations. There are various means to precisely make ETA / ETD, perfecting container routing & re-routing, fuel saving, etc. by means of a number of geographical data. These data may accurately be processed using suitable machine learning algorithms [17].

This review was inspired by the promise and prospective of the shipping industry necessitating AIS, as well as the popularity of AIS in other areas. This technique uses data, as an instrument for learning from the bygone events and helps in enhancing the future decision-making. Enhanced metrics for strategy making, scheduling, stability, intelligent traffic, and operational productivity are just a few of the exceptional benefits of AI in the freight sector.

This paper is presented under the following headings: literature analysis, developments in maritime transport, AIS summary, Artificial-Intelligence in marine applications, and conclusion. This paper highlights a few applications / case studies pertaining to AI in shipping industry.

2. LITERATURE REVIEW

The use of machine learning in the shipping industry is very constrained since it analyses vast volumes of data to construct a particular logic. Yet, latest scholars have attempted to apply machine-learning practices in data analysis. Various machine-learning techniques viz. neural networks, Bayesian networks and support vector machines are used to explain the problem of maritime anomaly detection. It is pointed out that the AI data has provided a better level of perception in case of anomaly detection [1]. An enhancement to an existing learning-based maritime field knowledge scheme for forecasting potential vessel position using the experience of current vessel behaviour is discussed, and it is shown that such schemes produce better performance [2]. The AI technique is applied for possible prediction of motion of vessel using learning weights [3]. EnmeiTu et al. [4] surveyed AIS data sources and utilized data for safety of seagoing personnel in the areas of traffic irregularity finding, route assessment, collision forecast and path scheduling. Another literature addressed the trick of determining waypoints of marine routes, using the streaming AIS data [5]. Andrea Coraddu et al. have utilized the machine learning approaches for naval propulsion plants [6]. The various sensors used for effective propulsion and for improving the maintenance based on the condition are aptly administered through predictive algorithms for economic reasons. Artificial intelligence in maritime sector is the current survey by many researchers [7].Deep sea connectivity is one of the most difficult issues in the marine industry. Underwater Wireless Sensor Networks (UWSNs) have seen extensive use in marine and military applications in the recent past. However, in order to outspread the network's survivability in UWSN, effective and truthful fault tolerance mechanisms are essential. The Energy-efficient Fault Detection and Recovery Management approach is recommended by A.Prasanth for the UWSN with reasonably enhanced network efficiency. To achieve effective fault detection across the entire network, the Hidden Poisson Markov model has been introduced into the suggested method. The network simulator (NS 3.26) framework is employed to replicate the performance effects. According to the simulator performance the suggested model outshines the existing evidence [27].

3. TRENDS IN MARITIME TRANSPORT

Digitalization is the method of integrating digital devices into structures and procedures in order to upgrade or replace traditional devices with electronic devices. The term automation refers to processes and structures, which are mechanical or electronic machines that are routinely computerised to execute specific tasks without human interference but are frequently supervised by humans. The system is considered to be autonomous if it has control tasks and can use various alternatives to solve a certain class of problems [8].

3.1. Degrees of Autonomy for Ships

There exist a variety of definitions of autonomy levels based on the transport used and perspectives for their own systems. The definition of autonomy levels for merchant ships is listed in Table 1. A degree of autonomy is assigned to the system relative to the amount of labour that must be performed by manned operators in conjunction with an automated system.

Table 1. *Degree of Autonomy for Ships [8]*

Level	Description
Direct control	Management of the ship is exercised directly by the crew on the bridge; no decision assistance is provided.
Decision support	On the bridge, decision support and guidance are given to the crew. The crew makes the final decision.
Automatic bridge	The operation is automated, but the crew is constantly monitoring it.
Periodically unmanned	Shore is constantly monitoring the situation. If required, send in reinforcements.
Remote control	Unmanned, constant surveillance, & command and control from land.
Automatic	Unmanned and under automated supervision, with monitoring from shore.
Constrained autonomous	Unmanned, semi-autonomous, and constantly supervised from shore.
Fully autonomous	It is unmanned and unsupervised.

Lloyd's Register has printed a guide for design of autonomous ships [12]. The guide encompasses a classification of autonomy levels as given in Table 2. Remote monitoring and controlling is not accounted in this taxonomy, except as a choice for shore based decision support [12].

3.2. Infrastructure Operation for Ships

In maritime sector, the infrastructure and control of the vessel operations shift on the way to operation centres that are integrated and interconnected, like fleet service centres, support hubs for marine shipping traffic, etc. These central/integr ated activities - maybe for a geologically limited zone like a harbour or for broader capacities like inland waterways - necessitate a technological framework that gives situational information to the controllers that decide things remotely. Such centralized/integrated controlling schemes progress within a broad framework using modern plans to incorporate automated patterns, operating practises, and conventions to ensure maritime vessel protection, effectiveness, and prediction [9].

3.3. Infrastructure Operation for Ships

In maritime sector, the infrastructure and control of the vessel operations shift on the way to operation centres that are integrated and interconnected, like fleet service centres, support hubs for marine shipping traffic, etc. These central/integr ated activities - maybe for a geologically limited zone like a harbour or for broader capacities like inland waterways - necessitate a technological framework that gives situational information to the controllers that decide things remotely. Such centralized/integrated controlling schemes progress within a broad framework using modern plans to incorporate automated patterns, operating practises, and conventions to ensure maritime vessel protection, effectiveness, and prediction [9].

Table 2. *Classification of Autonomy Levels [12]*

AL	Name	Decision support on board	Decision support on shore	Automatic, human override
0	Manual	No	No	Human
1	On-ship decision support	Yes	No	Human
2	On and off-ship decision support	Yes	Yes	Human
3	'Active' human in the loop	Yes	Yes	Common
4	Human on the loop	Yes	Yes	When needed
5	Fully autonomous	Yes	Yes	Rare
6	Full autonomous	Yes	Yes	Never

3.4. Predictive Maintenance of Vehicles

A significant percentage of servicing tasks in marine shipping could be conducted by personnel who are no longer stationed on-board the vessel, but rather in centrally facilitated support centres. Safety checks, such as software upgrades, may be performed remotely from those centres or through highly trained teams stationed in support sites worldwide and sent to the location where maintenance service is requested. Recent advancements in sensor expertise plus genuine networking help to accomplish the same and permit to observe the condition of apparatus on-board vehicles keenly [10].

Future prognostic maintenance schemes will gradually use an extensive collection of large data points. This knowledge is collected not just from vessels and facilities, but rather from peripheral operational configurations that have an effect upon this particular container such as freight, tides, landscape, and climatic parameters along voyages. Predictions can indeed be improved by creating a visual replica of the vessels and their structures (a "digital twin") [11].

3.5. Automation of Vessels

Fresh and creative forms of robotics and automation can support ship crews, spanning from adaptive vehicle tasks, such as watch keeping, velocity monitoring, and propellant usage to automated driving operation, even without manual assistance. Regulation of vessel activities and facilities is moving toward cohesive and integrated operation centres, such as fleet activity and traffic management hubs for sea vehicles. Independent ships are expected to work in restricted trading areas and parts (Inland Rivers, territorial waters, including surrounding countries). Integrated operation centres will offer increased support and control, with experts moving from ship to shore [16].

For more than a decade, the development of digital, and automation have affected work in the navigational and motor departments of on-board ships, for

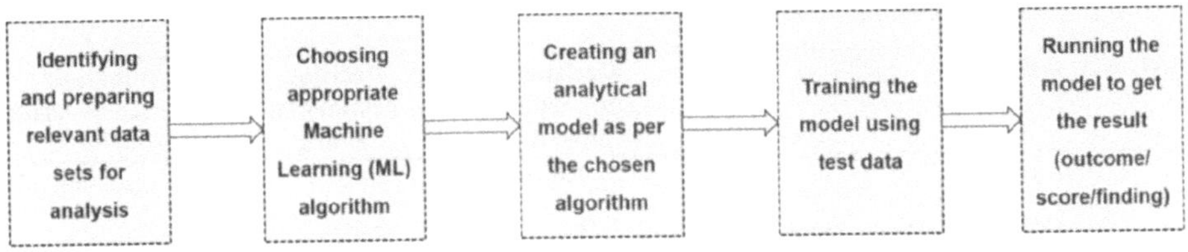

Figure 1. *Flow Diagram of a Machine Learning Programme.*

instance, unmonitored engine areas and maritime "autopilots". The next technological leap would be to reduce the number of crew members on board to provide remote assistance. Experts at shore-based service centres will perform more operations [14].

3.6. User Interfaces

Custom-made services are possible with the help of intelligent interfaces. These interfaces utilize the data mining approaches possibly by means of artificial intelligence. In some cases, robots may be offering the required services to the customers either by digital mode or else physically. The maintenance and repairing jobs onboard vessels may be carried out with the combination of humans and robots. Aerial and subaquatic drones may be employed to examine the parts of the ship that are not reachable are unsafe to reach. The experts may carefully observe the compound engines and other equipment inside the vessel from a distant location (shore). On-request maintenance greatly reduces the transport schedule of the vessel.

4. OVERVIEW OF ARTIFICIAL INTELLIGENCE

Machine training is a tier of algorithm that enables code algorithms to generate more reliable performance predictions instead of being physically coded. As all marine transports entail smart technology, implementing machine-learning tools can have the greatest benefit in terms of performance, resilience, and cost-cutting. Data is perhaps the major aspect of AI that will help to unlock the inconsistency and accelerate shipping. Currently, data collection on maritime transport is minimal. As a result, deployment of machine learning methods in sea industry is hampered in comparison to other industries. However, the use of machine learning in the maritime field is needed for autonomous shipping or ship automation [14].

Meanwhile, machine learning is a method of computer programming that uses vast quantities of data to boost output criterion [17]. According to the essence of training, it can be divided into 3 main classes. In this scenario, the reader should use the flow diagram below for applications of automated learning. There are 3 types of device training techniques: supervised learning, unsupervised learning, and reinforcement learning. Figure. 1 depicts the flow diagram of a machine learning programme.

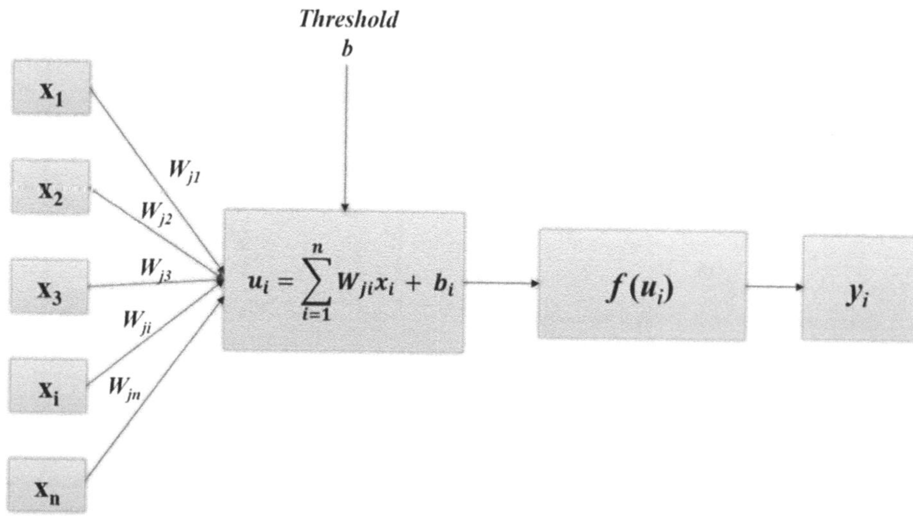

Figure 2. *Elements of ANN.*

4.1. Supervised Learning

Supervised learning involves inputs, and the majority of machine learning approaches involve supervised learnings. The elementary description of methods can indeed be specified as input parameters (x) and target parameters (y) calculating an algorithm to know the function from input to output, as seen in formula below.

$$y = f(x)$$

The meaning is that for a set of input parameters x, the target y can be found. This type can further be divided as (a) regression and (b) classification problems. In (a), the necessary target parameters are continuous. The above learning method is primarily used to achieve point predictions of target variables for given input data. The latter method, on the other hand, necessitates clustering /clumping of target values in a known source database. The widely used methods of supervised learning algorithms include random-forest for classification and regression problems, linear-regressions, support-vector-machines for classification problems, and artificial neural networks for regression problems. Figure. 2 displays the fundamental components of Artificial Neural Network (ANN).

4.2. Unsupervised Learning

This scheme is applicable only when the input data is present. In this situation, none of output variables will be in hand. Unsupervised learning assists in modelling the main structure or allocating the input parameter to boost input learning. In the meanwhile, there is no true response or fix to this sort of problem, because there is no supervision. As a result, it is regarded as unsupervised learning [17].

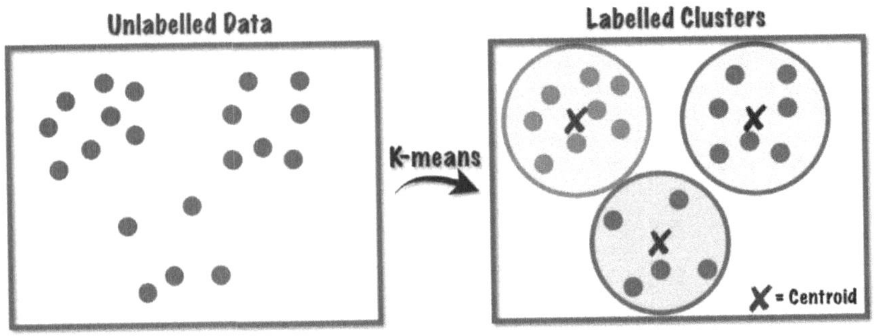

Figure 3. *Unsupervised Learning with K-means.*

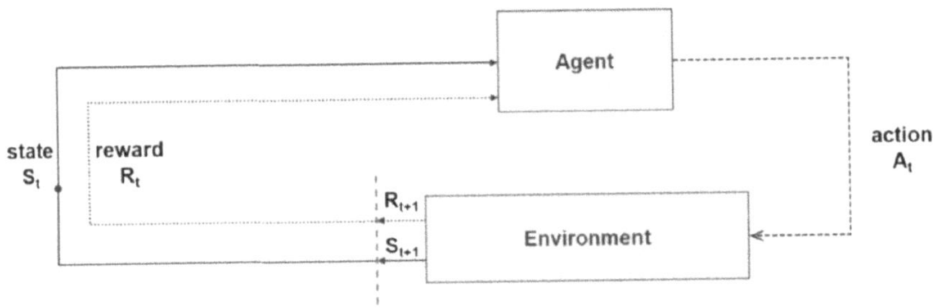

Figure 4. *Reinforced-Learning.*

The machine uses its own algorithm to evaluate and familiarise itself with the original structure in the data in this process. It can also be separated into clustering as well as association problems. The first method can identify inherent classes in data, while the next one is a rule learning problem in which the user is eager to decide distinguishing rules for a larger sample size. The prevalent methods of unsupervised learning rules are k-means for clustering and apriori algorithm for associations rule. One such example for k-clustering is presented in Figure.3.

4.3. Reinforcement Learning

This approach is hardly seen in machine learning algorithms since it incorporates known actions to optimise reward in a given situation. While it is similar to supervised learning, there are significant variations, such as the data collection for instruction and the sequence of contingent resolutions. This learning is a subset of artificial intelligence in which an agent chooses which action to take depending on its current state. Q-learning, which employs the bellman equation, and strategy repetition are the two common types [18]. Figure.4 portrays the reinforcement-learning algorithm.

5. AI IN MARITIME APPLICATIONS

The machine learning application can contribute to enhance shipping because data is one of the chief components in unravelling the uncertainty. It agrees to use

Table 3. *Pros and Cons of AI - Maritime Sector [15]*

Artificial Intelligence – Maritime Sector	
Pros	**Cons**
Cutting-edge Analytics	Reduced data-quality
Computerised Processes/Equipment	Dearth of proper info
Safety/Value-added Security	Panic of occupation replacement
Direction/path optimization	Absence of perfect approach
Performance guessing	Make over into trade is a lengthy course
Cost cutbacks	
Boosting maintenance	

savvy algorithms for data analysis and provides a pragmatic approach to maritime transportation problems. The use of such algorithms is profound mostly in the fields of maritime infrastructure scheduling, flight forecasting, ship planning processes, maintenance procedure, etc. The information/particulars of the marine applications facilitate to focus on proper algorithm. The pros and cons of AI in maritime industry is emphasized in Table 3.

5.1. Autonomous Ships

Any automation starts with the replacement of conventional control by electronic/programmable controllers. Likewise, modern sensors and digital meters exchange the conventional meters. All these circuits are carefully connected throu gh a network. Instead of writing coding that run through pages, a smart algorithm with machine learning technique may be adopted. Independent ships are not predictable ships minus crewmembers; rather, these are mostly a new category of vessel having fewer but highly qualified crewmembers who monitor a variety of organizational activities and services onboard, most likely through direct service stations. Self-directed vessels would not be likely to substitute standard freighters, potentially resulting in work losses for sailors. It is presumed that unmanned ships will be built as component of a constrained transportation infrastructure, creating new nautical routes as a replacement for other modes of transportation. While the daily fleet's "digitalization" accelerates, seafarers' roles will move further toward modern ones, especially in operations testing and system management, and away from functional work. Practicing and preparation will be updated in order to provide seafarers with the requisite latest expertise. Many automated ship techniques can be projected as part of overall shipping, which may possibly eventuate over the near five years. Accordingly, the seafarers may be equipped for data fluency, data interpretation and analysis of huge data. They must be aware of and trained to operate the ships digitally.

5.2. Port Automation

Port operators are normally extra proactive in bringing automation as compared to shipping carriers with respect to the arrangement of autonomous ships. Nonetheless, shipping companies and port operators jointly trust that an integrated

Table 4. *Comparison of Vessel Pattern Tracking With Matching Pursuit-Fletcher Reeves (MPFR) [28]*

Author	Title	Reference	Interpretations
Chen, Zet.al (2018)	Classification of vessel motion pattern in inland waterways based on Automatic identification System	[28]	**Objective:** Creating reliable techniques for automated vessel movement pattern detection in sea routes. **Model Used:** The Lp-minimization-based sparse representation classification (SRC) tool. **Process Created:** Matching Pursuit - Fletcher Reeves (MPFR) **Datasets Employed:** Wuhan-dataset (W) & Taizhou-dataset (T).

Other Details: The Yangtze River sailing in relation to Wuhan and Taizhou Vessels is classified into five groups based on the pattern of their vessel motions. The relevant datasets are listed below.

Class	Training Number		Test Number	
	Wuhan (W)	Taizhou (T)	Wuhan (W)	Taizhou (T)
C1	555	355	200	350
C2	675	265	300	230
C3	730	387	260	300
C4	670	517	200	280
C5	830	233	350	290

The obtained **results** are tabulated and compared to the existing approaches, which are mentioned below.

Class	Suport Vector Machine (%)		k-Nearest Neighbour (%)		SRC-L1 Norm (%)		SRC-L2 Norm (%)		SRC-Lp Norm (%)		MPFR (%)	
	W	T	W	T	W	T	W	T	W	T	W	T
C1	86.5	85.7	88.0	86.2	88.5	88.6	87.0	85.5	89.0	89.1	91.0	91.1
C2	86.7	81.7	88.3	80.8	89.3	83.5	88.7	86.0	90.3	85.2	91.6	88.2
C3	84.6	83.7	85.3	86.0	88.1	87.3	88.8	86.7	89.6	89.7	91.9	90.3
C4	87.0	82.1	87.5	82.9	89.0	85.3	86.0	85.0	88.5	85.3	90.5	86.8
C5	85.4	86.2	86.2	87.2	88.9	89.3	86.6	87.9	89.4	91.3	90.8	92.8
Average	86.0	83.9	87.1	84.6	88.8	86.8	87.4	86.2	89.3	88.1	91.2	89.8

Inference: It can be shown that the precision is determined by the type of dataset as well as the methodology used. On average, the proposed MPFR approach yields slightly higher results for both the Wuhan and Taizhou datasets. This is true with all 5 classes. It also is worth noting that the average training numbers are higher in the Wuhan dataset than the Taizhou dataset, whereas the average test numbers are higher in the Taizhou dataset. It has been discovered that more training/learning fetches more reliable results.

gateway may be built in a controlled region in one country, where it could be protected by the nation's national authority. In contrast, when it comes to programmed ships that traverse the waterways of several countries or travel at sea, all national laws and international treaties must be followed. The technology readiness level (TRL) is an indication of hi-tech maturity. Several market reports examine about port automation. In broad-spectrum, prevailing and future practices of computerization in ports will include levels of implementation, and technological maturity. The Artificial intelligence models are much useful in achieving the above.

5.3. Obstacle Prevention for Under-Actuated Unmanned Marine Vessels

The under-actuated unmanned marine vessel having indefinite dynamics of the environs is protected from obstacle using a deep reinforcement-learning algorithm. The foremost gain of the proposed algorithm is its brevity and extendibility; the usual analytic control rule is not essential for manoeuvring the vessel. The vessel takes actions based on the present observations, together with the detected results for obstacles, and the simple measurements of working states of the vessel by means of the proposed algorithm. Further, this architecture is extendible to other composite/intricate tasks [19].

5.4. Liner Shipping Route for Container Ships

Artificial Intelligence principle is useful in planning the best/ideal route to diminish the overall fuel prices. A cross-machine-learning technique that was first developed aids in the prediction of fuel usage for cargo vessels by learning the practical knowledge of an Asian fleet. The predictive results along with certain assumptions of a liner-shipping path initially utilized. It was joined with the asymmetrical travelling salesperson algorithm to create a prime route to curtail the fuel cost. Deep learning is a data-driven practise that is heavily reliant on the accuracy of the learning dataset [20].

5.5. Fault Detection for Ship Systems Operations

Innovative machine learning based expected behaviour modelling with relevant control charts are applied to ship systems for identifying the fault on system operations. The main objective of the study is to permit for anticipatory actions and scheduling, plummeting interruption and to increase the safety, energy competence and add-on routine watching of vessel progressions. The prompt finding of emerging errors will possibly lessen the suboptimal functions of the supervised system, protecting its energy efficient processes [21].

5.6. Ship Fuel Consumption on a Container Ship

Optimizing the fuel consumption of marine vessel will allow for greater productivity and costeffectiveness in ship management, since fuel price covers majority of running cost. The estimation of vessel fuel usage is a difficult topic since the rate of fuel consumption is directly dependent upon multiple conditions. viz. the state of the primary engine, passenger load, ship draught, sea/ocean weather, meteorological conditions, and so on. Out of an analysis with many predictive models including Multiple-Linear-Regression, Ridge- and LASSO=Regression, Support-Vector-Regression, Tree-Based Algorithms, the Enhancement for a container ship is found to be promising with K-fold cross-validation offering greater accuracy [22]. Another study helped to assess the fuel oil consumption in the main engine using multiple regression algorithms. The following restrictions were taken into account in this study: shipboard information techniques, mid-morning reports, and Automated Data Recording & Tracking schemes [23].

Table 5. *Comparison of Vessel Pattern Tracing by AL and PAL Methodologies [2] & [3]*

Author	Title	Reference	Interpretations
1. Bomberger, N et.al (2006) 2. Rhodes, B at.al (2007)	1. Associative Learning of Vessel Motion Patterns for Maritime Situation Awareness 2. Probabilistic associative learning of vessel motion patterns at multiple spatial scales for maritime situation awareness	[2] [3]	Objectives: Use Associative learning (AL) to recognise the pattern of the Vessl motion [2]. To investigate the motion of a vessel using probabilistic associative learning (PAL) [3].

Other Details: A comparable research has been conducted in both of the preceding literatures by subtly changing the techniques. A standardised square grid is used to discretize vessel position based on vessel velocity over the region of interest surrounding the port of Miami. The zones are split into four westward and eastward zones based on grid positions. The predictions were based on d0 (direct hits/zero distance from predicted cell), d1 (near misses/one cell away from predicted cell), and d2 (two cells away from predicted cell). The output was evaluated using four metrics: recall, coverage, precision and accuracy. The following analyses are tabulated and compared:

Zone Number(s)	Recall		Coverage		Precision		Accuracy	
	AL	PAL	AL	PAL	AL	PAL	AL	PAL
All	0.20	0.20	0.56	0.55	0.20	0.20	0.36	0.36
1 & 2 (d0,d1 & d2)	0.35	0.31	0.83	0.66	0.23	0.28	0.41	0.47
3 & 4 (d2 only)	0.07	0.08	0.30	0.44	0.13	0.09	0.22	0.17
3 (d2 only)	0.12	0.11	0.44	0.36	0.14	0.19	0.27	0.31
4 (d2 only)	0.03	0.06	0.21	0.47	0.11	0.06	0.15	0.13

Inference: The AL approach uses fixed threshold while the PAL approch uses incremental learning. By lookig over the above Table it is seen that the precision and accuracy are improved in zones 1 & 2 with PAL. The performance is almost same when we consider all zones together with regard to both AL and PAL. Other metrics are varying and depend on the learning rate. A multi-scale approach to express spatial position that fits the spatial scale to the track activity in a given area will result in better prediction results. Spatial forecasts would be unnecessarily high if the spatial scale is too coarse to reflect local activities. More training data would be needed if the spatial scale is too wide to achieve reasonable prediction performance. As the spatial scale is properly configured incremental learning can run quite effectively and soon attain productive levels of efficiency as the system tends to track and evaluate vessel activity.

Various other studies on AI schemes include prediction of energy consumption ships in green port [24], Prognostic preservation for ballast pumps on ship-repairing yards [25] and ship collision prevention [26].

While AIS can be used in a variety of shipping applications, the most important application that may yield positive results is vessel location monitoring or detecting a certain vessel's sailing pattern. As a consequence, only the outcome of such an application is elaborated. The same is true for many other applications, depending on the scale of the dataset, the amount of learning and training results, and the data's validity. Table 4 compares vessel pattern tracking with the proposed approach with the other current technologies.

Table 5 compares the results of vessel monitoring patterns produced by the Associative Learning (AL) and Predictive Associative Learning (PAL) methodologies.

6. CONCLUSION

Even if machine learning has previously been included in abundant sections of the online-world, its application in maritime sector is still constrained. To reap the gains for viable conveyance, marine transports need smart software and machine-learning concepts. This paper addressed the different facets of machine learning, as well as their upsides and downsides. The upshot of artificial intelligence on the marine industry's future is also explored. The rigorous investigation reveals that the seafaring experts and investigators ought to focus mainly on apt/suitable algorithm, as the outcomes are different for different algorithms and for different situations. The prospective domains on which the artificial intelligence be applied entails autonomous shipping, port automation and management, route organizing and optimizing, obstacle prevention in underwater environment, fault detection in ship operations, forecasting the fuel usage, energy consumption and management. Rather than focusing on a single application, the approach analysed major and important implementations. In addition, the outcomes of one main application of Vessel path control are elaborated and tabulated.

REFERENCES

1. Obradovic, Ines & Milicevic, Mario & Zubrinic, Krunoslav. (2014). Machine learning approaches to maritime anomaly detection. Nase More. 61. 96-101.
2. Bomberger, Neil & Rhodes, B.J. & Seibert, Michael & Waxman, Allen. (2006). Associative Learning of Vessel Motion Patterns for Maritime Situation Awareness. 1-8. 10.1109/ICIF.2006.301661.
3. M. Zandipour, B. J. Rhodes and N. A. Bomberger, "Probabilistic prediction of vessel motion at multiple spatial scales for maritime situation awareness," 2008 11th International Conference on Information Fusion, Cologne, 2008, pp. 1-6.
4. Tu, E., Zhang, G., Rachmawati, L., Rajabally, E., & Huang, G. (2018). Exploiting AIS Data for Intelligent Maritime Navigation: A Comprehensive Survey from Data to

Methodology. IEEE Transactions on Intelligent Transportation Systems, 19, 1559-1582.

5. Dobrkovic, A., Iacob, M.-E., & van Hillegersberg, J. (2015). Using machine learning for unsupervised maritime waypoint discovery from streaming AIS data. Proceedings of the 15 th International Conference on Knowledge Technologies and Data-Driven Business - i-KNOW '15. doi:10.1145/2809563.2809573.

6. Coraddu A, Oneto L, Ghio A, Savio S, Anguita D, Figari M. Machine learning approaches for improving condition-based maintenance of naval propulsion plants. Proceedings of the Institution of Mechanical Engineers, Part M: Journal of Engineering for the Maritime Environment. 2016;230(1):136-153. doi:10.1177/1475090214540874

7. MacKinnon, Scott & Weber, Reto & Olindersson, Fredrik & Lundh, Monica. (2020). Artificial Intelligence in Maritime Navigation: A Human Factors Perspective. 10.1007/978-3-030-509439_54.

8. Rødseth, Ørnulf & Nordahl, Håvard. (2018). Definition of autonomy levels for merchant ships, Report from NFAS, Norwegian Forum for Autonomous Ships, 2017-08-04. 10.13140/RG.2.2.21069.08163.

9. WMU: Transport 2040: Automation, Technology, Employment - The Future of Work. World Maritime University, Malmö, Sweden (2019)

10. Autor, D. H., Levy, F. and Murnane, R. J., 2003. The Skill Content of Recent Technological Change: An Empirical Exploration, The Quarterly Journal of Economics, 118(4), pp. 12791333. doi: 10.1162/003355303322552801.

11. Bekiaris, E. and Loukea, M., 2017. Future scenarios on skills and competences required by the Transport sector in the short, mid and long-term. D2.1. SKILLFUL Project.

12. Cyber-enabled ships, ShipRight procedure - autonomous ships. First edition, July 2016, A Lloyd's Register guidance document.

13. Benjamin, M., Curcio, J.A.: COLREGS-based navigation of autonomous marine vehicles.In: Proceedings of the IEEE/OES Autonomous Underwater Vehicles (2004).

14. Marine-digital.com. 2021. Artificial Intelligence (AI) And Machine Learning (ML) In Maritime Logistics. [online] Available at: <https://marine-digital.com/article_ai_and_ml > [Accessed 11April 2021].

15. James R. DeChant, V., 2021. How AI Is Influencing The Shipping Industry Today. [online] Adv-polymer.com. Available at: < https://www.adv-polymer.com/blog/artificial-intelligence-inshipping#ch1> [Accessed 11 April 2021].

16. F.R., M., 2021. How AI & Automation Has Overhauled The Shipping Industry. [online] Analytics India Magazine. Available at: https://analyticsindiamag.com/ai-shippingautonomous-drive/ [Accessed 11 April 2021].

17. Alpaydin, Ethem. (2009). "Introduction to machine learning". MIT press.

18. Akyuz, Emre & Cicek, Kadir & Celik, Metin. (2019). A Comparative Research of Machine Learning Impact to Future of Maritime Transportation. Procedia Computer Science. 158. 275 280. 10.1016/j.procs.2019.09.052.
19. Cheng, Y., & Zhang, W. (2018). Concise deep reinforcement learning obstacle avoidance for underactuated unmanned marine vessels. Neurocomputing, 272, 63-73. doi:10.1016/j.neucom.2017.06.066.
20. Bui-Duy, L., & Vu-Thi-Minh, N. (2020). Utilization of a deep learning-based fuel consumption model in choosing a liner shipping route for container ships in Asia. The Asian Journal of Shipping and Logistics. doi:10.1016/j.ajsl.2020.04.003.
21. Cheliotis, M., Lazakis, I., & Theotokatos, G. (2020). Machine learning and data-driven fault detection for ship systems operations. Ocean Engineering, 216, 107968. doi:10.1016/j.oceaneng.2020.107968.
22. Uyanık, Tayfun & Karatuğ, Çağlar & Arslanoğlu, Yasin. (2020). Machine learning approach to ship fuel consumption: A case of container vessel. Transportation Research Part D Transport and Environment. 84. 10.1016/j.trd.2020.102389.
23. Gkerekos, C., Lazakis, I., & Theotokatos, G. (2019). Machine learning models for predicting ship main engine Fuel Oil Consumption: A comparative study. Ocean Engineering, 188, 106282. doi:10.1016/j.oceaneng.2019.106282.
24. Peng, Y., Liu, H., Li, X., Huang, J., & Wang, W. (2020). Machine learning method for energy consumption prediction of ships in port considering green ports. Journal of Cleaner Production, 121564. doi:10.1016/j.jclepro.2020.121564.
25. Kimera, D., & Nangolo, F. N. (2020). Predictive maintenance for ballast pumps on ship repair yards via machine learning. Transportation Engineering, 100020. doi:10.1016/j.treng.2020.100020.
26. Gao, M., & Shi, G.-Y. (2020). Ship collision avoidance anthropomorphic decision-making for structured learning based on AIS with Seq-CGAN. Ocean Engineering, 217, 107922. doi:10.1016/j.oceaneng.2020.107922.
27. A. Prasanth (2020). Certain Investigations on Energy-Efficient Fault Detection and Recovery Management in Underwater Wireless Sensor Networks, Journal of Circuits, Systems and Computers, doi: 10.1142/S0218126621501371.
28. Chen, Z., Xue, J., Wu, C., Qin, L., Liu, L., & Cheng, X. (2018). Classification of vessel motion pattern in inland waterways based on Automatic Identification System. Ocean Engineering, 161, 69-76. doi:10.1016/j.oceaneng.2018.04.072.

CHAPTER

2

DCA-Based Collision Avoidance Path Planning for Marine Vehicles in Presence of the Multi-Ship Encounter Situation

Jiagen Yu[1], Zhengjiang Liu[1,2] and Xianku Zhang[1]

[1]*Navigation College, Dalian Maritime University, Dalian 116026, China*
[2]*Key Laboratory of Navigation Safety Guarantee of Liaoning Province, Dalian Maritime University, Dalian 116026, China*

ABSTRACT

The problem of ship collision avoidance path planning is one of the key problems in the ship motion control field. Aiming at the high computational time problem of path planning in multi-ship encounter situations and the impact of the target ship's action changes on path planning, this paper proposes a dynamic path-planning method based on dynamic cluster analysis (DCA), which is used to dynamically cluster target ships with similar attributes into a group ship, reducing the number of calculated targets and improving the efficiency of path planning. Taking into full consideration the action requirements of the International Regulations for Preventing Collisions at Sea (COLREGs), the course alteration action matrix (CAAM) for collision avoidance is established to limit the space of candidate solutions. On the basis of the rapid optimization capability of the deterministic optimization algorithm (DOA), a dynamic monitoring mechanism is introduced to establish a multi-ship encounter intelligent collision avoidance decision-making model that meets the needs of real-time collision avoidance. The simulation results showed that the method can obtain a dynamic collision avoidance path that is safe and feasible.

Keywords: *multi-ship encounter; collision avoidance; path planning; dynamic cluster analysis*

1. INTRODUCTION

Multi-ship encounter collision avoidance path planning is a hot research topic in the field of ship motion control. Many methods have been used to solve this problem. As early as the 1980s and 1990s, a knowledge-based system was used for multi-ship encounter collision avoidance path planning [1,2]. The problems of decision-making blind spots and low decision-making efficiency are affected by the collision avoidance knowledge base and reasoning mechanism.

Analytical geometry methods are also used for collision avoidance path optimization. In [3], a cooperative path planning algorithm was proposed, and the concept of ship priority was used in the algorithm. However, only three typical encounter situations (the overtaking, head-on, and crossing situations) were considered, and Rule 18 and 19 of the COLREGs were not considered. For the multi-ship encounter situation, the uncertainty of the target ship's action makes this cooperative collision avoidance difficult to achieve in the practice of ship collision avoidance. In [4], a deterministic collision avoidance path optimization algorithm TBA (Trajectory Base Algorithm) was proposed, which effectively shortened the calculation time but did not consider the action changes of the target ship. The CCDWA (COLREG-compliant dynamic window approach) was used to solve the collision avoidance problem in ref. [5]. However, this method only considers three typical encounter situations between the two ships and does not consider the changes in the action of the target ship.

Control theory methods are also used for collision avoidance path optimization [6-8] In [9], a concept of a ship collision avoidance system based on model predictive control was proposed. This algorithm considers the action changes of the target ship and the constraints of the COLREGs. With an increase in the number of target ships, the computational complexity of the algorithm increases, which affects the decision-making time. A combined nonlinear model predictive control (NMPC) method was used for trajectory tracking and collision avoidance [10]. However, this approach only considers three typical encounter situations and does not take into account the collision-avoidance action requirements of the COLREGs.

Fuzzy logic methods are used for collision avoidance path optimization. Aiming at the problem of collision avoidance in complex multi-ship encounter situations, ref. [11] proposed a Fuzzy-Bayesian ship intelligent collision avoidance decision-making model to achieve continuous collision avoidance actions. In addition, this method was upgraded in ref. [12], but the action changes of the target ship were not considered. In [13-15], Fuzzy theory was employed to infer the collision risk index for collision avoidance, but Rules 18 and 19 of the COLREGs were not considered.

Artificial Potential Field (APF) has been used for collision avoidance path optimization in some studies [16-18]; the principle is to guide the target to avoid obstacles through the action rules of Virtual Field Force (VFF). Ref. [19] proposed a multi-vessel cooperative path planning method based on APF, which makes collision avoidance decisions according to the priority of target vessels. Additionally, the uncertainty of the action of the target ship makes it difficult to realize the coordinated path planning scheme. Ref. [20] proposed a COLREGs-constrained real-time path-planning method for autonomous ships using modified artificial potential fields; this method has high decision-making efficiency, but it only considers three typical encounter situations.

Evolutionary algorithms are also used for multi-ship encounter collision avoidance path optimization. In [21], the decision to avoid the four target ships took about 900 s. A multivessel path planning method based on an evolutionary algorithm was proposed [22, 23] called ESoSST (Evolutionary Sets of Safe Ship Trajectories). Subsequently, the method was used in TSS (Traffic Separation Scheme) waters [24], as well as in waters with restricted visibility [25]. ESoSST is a cooperative path planning method, and the computational time is about $10 - 30$ s.

The swarm intelligence optimization algorithm has been used for collision avoidance path optimization. A method of ship collision avoidance path planning based on the ant colony optimization (ACO) algorithm has been proposed [26,27], which assumes that target ships maintain their course and speed. The simulation results show that the decisionmaking time to avoid four target ships and eight target ships can reach up to 29 s and 57 s, respectively. In [28], particle swarm optimization (PSO) was used for the path planning problem of unmanned surface vehicles with currents effects, but the targets were static.

The differential game theory method is also used for collision avoidance path optimization [29,30]. The method takes into account the action changes of the target ship. However, the drawback is that the high computational complexity in multiship encounters reduces the decision-making efficiency.

Distributed decision-making methods are used for collision avoidance path optimization [31,32]. These methods rely on information exchange and negotiation between ship agents to make collision avoidance decisions, but they are not applicable to non-agent ships.

The velocity obstacle (VO) method, based on the relative motion principle in physics, is also used for ship collision avoidance path optimization. In [33], VO was used for avoiding dynamic targets, but only three typical encounter situations were considered. Ref. [34] used VO to avoid target ships with predictable trajectories, but the constraints of the COLREGs were not considered.

With the rapid development of artificial intelligence technology, there has been related research on the application of machine learning in the field of ship collision avoidance [35,36]. In [37], a concise deep reinforcement learning obstacle avoidance method was presented, but the target ships were static. Ref. [38] proposed an automatic collision avoidance of multiple ships method based on deep Q-learning, but only three typical encounter situations were considered. In [39], deep reinforcement learning was used for AUV (autonomous underwater vehicle) obstacle avoidance planning, but the constraints of the COLREGs were not considered. Moreover, when machine learning deals with complex problems, it is often inefficient because of the high amount of computation.

In view of the above literature review, this paper presents a DCA-based collision avoidance path planning method for marine vehicles in presence of a multi-ship encounter situation. The proposed method fully takes into account the constraints of the COLREGs and the changes in the action of the target. The purpose of the method is to give an optimal collision avoidance path quickly and dynamically. The rest of the paper is organized as follows. In Section 2, a simplification of an encounter situation based on DCA is presented. The multi-ship encounter intelligent collision avoidance decision-making method is given in Section 3. Simulation results are shown in Section 4, and conclusions are given in Section 5 .

2. SIMPLIFICATION OF ENCOUNTER SITUATION BASED ON DCA

In practice at sea, in fishing areas, in TSS waters, or in habitual waterways, for a group of ships with a similar course, speed and position are often encountered (e.g., a fleet of fishing vessels). In the process of collision avoidance, the officer on watch (OOW) usually considers such a group of ships as one target. On the basis of this collision avoidance logic, it is possible to simplify the multi-ship encounter situation. In this paper, a dynamic cluster analysis (DCA) method is used to classify the target ships with similar attributes as group ships.

2.1. Cluster Analysis Method

In this paper, the Hierarchical Clustering Method (HCM) is used. The method divides the n samples into n classes, and then, the two classes with the closest attributes are merged into a new class by calculating the distance D_G. After multiple clustering, all samples are merged into one class, and then a dendrogram is drawn. Finally, the number of classes and the sample composition of classes are determined.

2.1.1. Calculation of Distance between Samples

Suppose there are n samples $X_i(i = 1,2,\ldots,n)$, and each sample has m attributes; x_{ij} $(j = 1,2,\ldots,m)$ represents the j-th attribute value of the i-th sample.

d_{ij} is the Euclidean distance between X_i and X_j.

$$d_{ij} = \left[\sum_{k=1}^{m} \left(x_{im} - x_{jm}\right)^2\right]^{1/2} \tag{1}$$

$\bar{x}_j$ and s_j represent the sample mean and standard deviation of the j-th attribute value, respectively. The calculation formula is as follows.

$$\bar{x}_j = \frac{1}{n}\sum_{i=1}^{n} x_{ij} \tag{2}$$

$$s_j = \left[\frac{1}{n-1}\sum_{i=1}^{n} \left(x_{ij} - \bar{x}_j\right)^2\right]^{1/2} \tag{3}$$

xI_{ij} represents the normalized data of each attribute value.

$$x\prime_{ij} = \frac{x_{ij} - \bar{x}_j}{s_j} \tag{4}$$

The standardized Euclidean distance $d\prime_{ij}$ can be obtained by calculation Formula (5).

$$d\prime_{ij} = \left[\sum_{k=1}^{m} \left(x\prime_{im} - x\prime_{jm}\right)^2\right]^{1/2} \tag{5}$$

2.1.2. Calculation of the Distance between Classes

In this paper, the Group Average Method was used to calculate the distance between classes. There are two classes, G_{p1} and G_{p2}, containing q_{p1} and q_{p2} samples, respectively.

The distance D_G between two classes is calculated as follows.

$$D_G\left(p_1, p_2\right) = \frac{1}{q_{p1} \cdot q_{p2}} \sum_{i=1}^{q_{p1}} \sum_{j=1}^{q_{p2}} d_{l_{ij}} \tag{6}$$

2.2. Group Ship Model

Suppose the target ship's course is C_{ts}, the speed is V_{ts}, the distance to own ship (OS) is D_{ts}, and the true bearing is TB_{ts}. Take each target ship (TS) as the sample of cluster analysis, choose C_{ts}, V_{ts}, D_{ts}, and TB_{ts} as characteristic indices, and multiple target ships with a close course, speed, and position are classified into one group ship (GS).

The constraints of the cluster analysis are as follows. If the course difference of the target ships is more than E_c degrees, the speed difference of the target ships is more than E_v knots, or the distance between the target ships is greater than $2SD$ (safe distance), these three cases cannot be classified as one class. E_c, E_v, and SD are not fixed and can be determined according to specific encounter situations.

N target ships $(\mathrm{TS_1} - \mathrm{TS_n})$ are classified into $K(K \leq N)$ group ships $(\mathrm{GS_1} - \mathrm{GS_k})$ on the basis of the HCM. $\mathrm{GS_p}(p = 1, 2, \ldots k)$ contains $q_p(1 \leq q_p \leq N, q_1 + \ldots + q_2 = N)$ target ships. The domain of GS_p is a circle, the central position is (x_{gsp}, y_{gsp}), the radius is R_{gsp}, the course is C_{gsp}, the speed is V_{gsp}, $DCPA_{gsp}$ is the distance to the closest point of approach (DCPA) from the center of GS_p to OS, and the passing distance to OS is SD_{gsp}. The calculation formula of each variable of GS_p is as follows.

$$\begin{cases} x_{gsp} = \left(\max\left(x_{tsqp}\right) + \min\left(x_{tsqp}\right)\right)/2 \\ y_{gsp} = \left(\max\left(y_{tsqp}\right) + \min\left(y_{tsqp}\right)\right)/2 \\ R_{gsp} = \left(\left(\max\left(x_{gsp}\right) - \min\left(x_{gsp}\right)\right)^2 + \left(\max\left(y_{gsp}\right) - \min\left(y_{gsp}\right)\right)^2\right)^{1/2} \\ C_{gsp} = \text{average}\left(C_{tsqp}\right) \\ V_{gsp} = \text{average}\left(V_{tsqp}\right) \\ SD_{gsp} = \text{abs}\left(DCPA_{gsp}\right) - R_{gsp} \end{cases} \tag{7}$$

where x_{tsqp} is the X coordinate value of each target ship in GS S_p; y_{tsqp} is the Y coordinate value of each target ship in GS_p; C_{tsqp} is the course of each target ship in GS_p; V_{tsqp} is the speed of each target ship in GS_p.

Two samples of the group ship are shown in Figure 1.

2.3. Dynamic Cluster Analysis

The cluster analysis method was introduced for the collision avoidance problem [40], but it does not consider the action changes of the target ship. Actually,

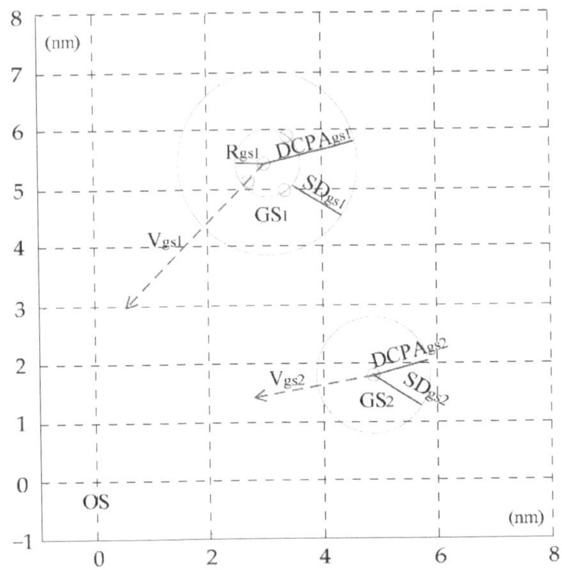

Figure 1. *Samples of the group ship.*

when the action of the target ship changes, it may be necessary to re-cluster the target ships to simplify the encounter situation. In order to realize the dynamic cluster analysis of the target ships, it is necessary to detect the action changes of the target ship, and the detection period is set as T_c. When it is detected that the course difference between the target ships exceeds the threshold E_c or the speed difference exceeds the threshold E_v, the clustering analysis is performed again. A flow chart of the DCA of the target ships is shown in Figure 2.

3. MULTI-SHIP ENCOUNTER INTELLIGENT COLLISION AVOIDANCE DECISION-MAKING

3.1. Course Alteration Action Matrix

In order to determine the course-altering direction of the collision avoidance path, this paper establishes the course alteration action matrix (CAAM) model AM.

$$AM = \begin{bmatrix} a_{11} & a_{12} & \cdots & a_{1n} \\ a_{21} & a_{22} & \cdots & a_{1n} \\ \vdots & \vdots & \vdots & \vdots \\ a_{m1} & a_{m2} & \cdots & a_{mn} \end{bmatrix} \tag{8}$$

where the rows represent the relative bearing (RB) of the target ship, the columns represent the type of the target ship, a_{mn} is action value, and the range of action values is $(-1, 0, 1)$. A value of -1 indicates that OS should change course to the port side, 0 means that either a change to the port side or starboard is possible, and 1 means that OS should change course to starboard.

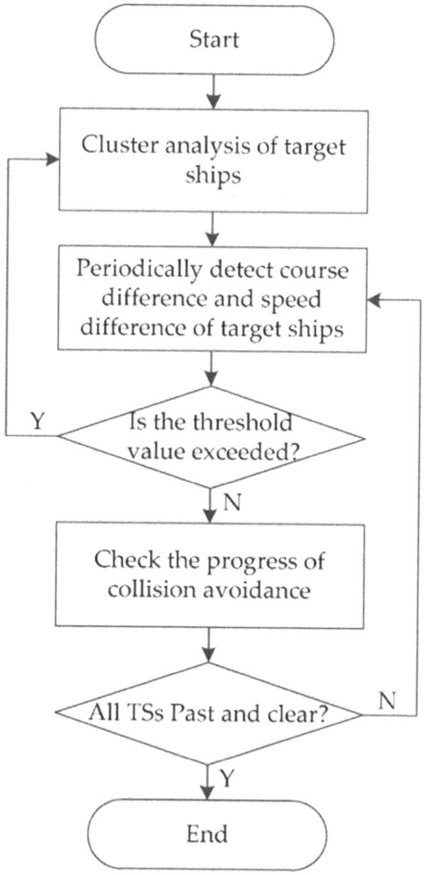

Figure 2. *Flow chart of the DCA of target ships.*

When encountering multiple ships, according to the RB of each target ship and the ship type, the action vector *AT* of the ship's course alteration to avoid collision is constructed.

$$AT = [AM_1, AM_2, \cdots AM_N] \tag{9}$$

The priority of the action value is from high to low: $1, -1, 0$. If a value of 1 appears in *AT*, it is limited to altering the course to starboard. Then, whether there is a -1 value is judged, and if so, it is limited to altering the course to the port side. If the values in *AT* are all 0 , it means that OS can change course to the port side or to starboard.

3.1.1. CAAM in Sight of One Another

In sight of one another, considering the constraints of the COLREGs (Rules 13, 14,15, 17, and 18) and good seamanship, the encounter situation is divided as shown in Figure 3a.

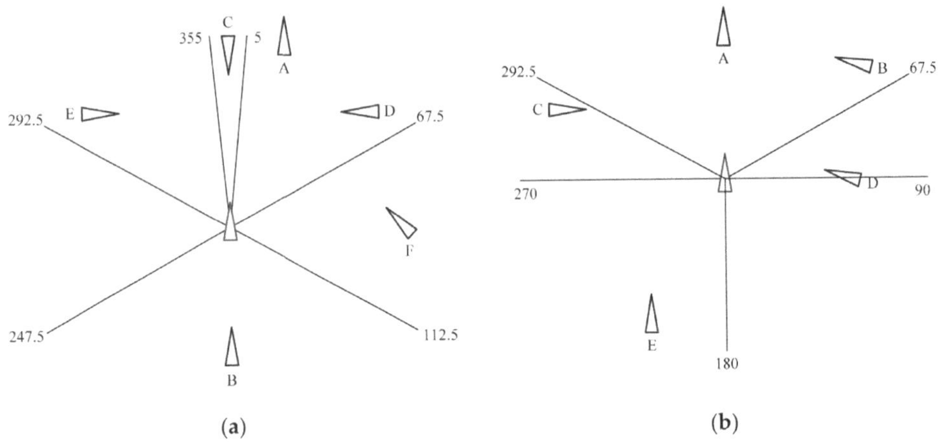

Figure 3. *Encounter situation division (A, B, C, D, E, F represent the target ships): (a) Encounter situation division in sight of one another; (b) encounter situation division not in sight of one another.*

Table 1. *CAAM in sight of one another.*

RB\TS	NUC	RAM	VEF	SV	PDV1	PDV2
[355.0, 005.0]	0	0	0	0	0	1
[005.0, 067.5]	0	0	0	0	0	1
[067.5, 112.5]	−1	−1	−1	−1	0	−1
[112.5, 247.5]	−1	−1	−1	−1	0	−1
[247.5, 292.5]	1	1	1	1	0	1
[292.5, 355.0]	0	0	0	0	0	1

OS is a regular power-driven vessel (PDV). For a vessel being overtaken (A), the action value is 0 . For an overtaken vessel (B), the action value is -1 . For a head-on vessel (C), the action value is 1 . For a crossing vessel from starboard (D), the action value is 1 . For a crossing vessel from the port side (E), the action value is 1 . For the other types of vessels in Rule 18, the action value is 0 . For an abeam vessel (F), OS should avoid altering course toward it. Based on these, the CAAM in sight of one another is shown in Table 1.

NUC is a vessel not under command, RAM is a vessel restricted in her ability to maneuver, VEF is a vessel engaged in fishing, SV is a sailing vessel, PDV1 is overtaken PDV, and PDV2 is not overtaken PDV.

3.1.2. CAAM Not in Sight of One Another

Not in sight of one another, considering constraints of the COLREGs (Rule 19) and good seamanship, the encounter situation is divided as shown in Figure 3 b.

OS is a regular power-driven vessel (PDV). For a vessel being overtaken (A), the action value is 0 . For vessels forward of the beam (B, C), the action value is 1 . For a vessel abeam (D) or abaft the beam (E), OS should avoid altering course towards it. Based on these, the CAAM not in sight of one another is shown in Table 2.

Table 2. *CAAM not in sight of one another.*

RB\TS	Overtaken Vessel	Other Vessel
[292.5, 067.5]	0	1
[067.5, 090.0]	0	1
[090.0, 180.0]	0	−1
[180.0, 270.0]	0	1
[270.0, 292.5]	0	1

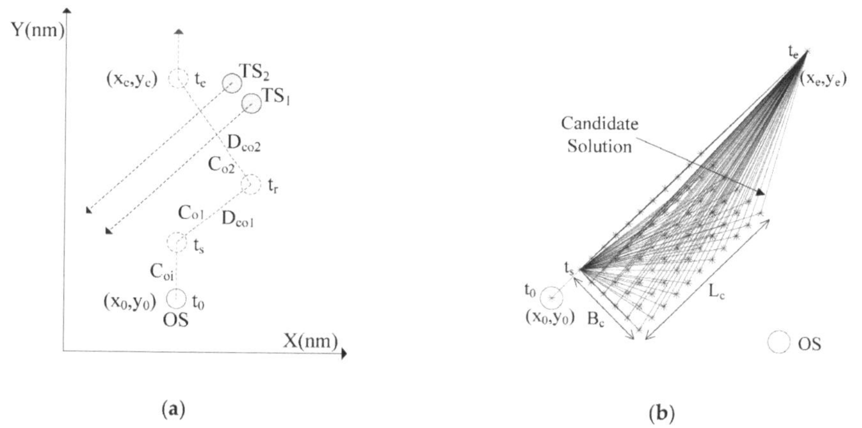

Figure 4. *Deterministic optimization algorithm: (a) Process of path planning for collision avoidance; (b) candidate solutions for collision avoiding path.*

3.2. Deterministic Optimization Algorithm

3.2.1. Problem Description

At time t_0, OS is at position (x_0, y_0) with initial course C_{oi} and speed V_{oi} encountering N target ships ($\text{TS}_1 - \text{TS}_\text{n}$), and a risk of collision (ROC) exists. To pass these target ships at a safe distance (SD), OS intends to take action to avoid a collision at time t_s taking steer course C_{01} to sail D_{c01}n miles and steer course C_{02} at time t_r to sail D_{c02}n miles to the goal position (x_e, y_e) and return to the initial course C_{oi} at time t_e. The values of t_0, (x_0, y_0), C_{oi} and V_{oi} can be obtained from OS. The data for the target ship's C_{ts}, V_{ts}, D_{ts}, and TB_{ts} are available from radar (these data are assumed to be known in this paper). The value of t_s and (x_e, y_e) can be preset according to a specific encounter situation.

The problem of collision avoidance path planning can be presented by an optimization problem, namely searching for the optimal (C_{01}, t_r, C_{02}) among the candidate solutions, which make OS pass target ships at a safe distance, and voyage losses are minimal under the constraints of COLREGs. Figure 4a shows the process of path planning for collision avoidance.

3.2.2. Determination of Candidate Solutions

The side to be steered can be determined on the basis of CAAM. A rectangular area on the side to be steered is defined, and the length (L_c) and width (B_c) of

Algorithm 1: Pseudocode for collision avoidance path planning on the basis of DOA.	
Input:	$C_{oi}, V_{oi}, C_{ts}, V_{ts}, D_{ts}, TB_{ts}, SD_{gsp}$
Process:	
	Initialization: $(x_0, y_0), t_0, t_s, (x_e, y_e), L_c, B_c$
	Calculate the sum of D_{co1} and D_{co2}
	If SD_{gsp} >=SD && course change is at least 30 degrees
	keep the candidate solution
	Else
	abandon the candidate solution
	End if
	Search for a solution in which the sum of D_{co1} and D_{co2} goes to the minima
Output:	C_{o1}, t_r, C_{o2}

the area are determined according to the specific encounter situation. Candidate waypoints are determined at equal intervals along the length and width of the rectangular area. The value of the interval (D_{wp}) is determined according to own ship's length overall (LOA); the larger the LOA, the larger the interval, and vice versa. Each candidate waypoint corresponds to a (C_{01}, t_r, C_{02}), which is a candidate solution for the collision avoidance path, as shown in Figure 4 b.

3.2.3. Objective Function

In the algorithm, each group ship is used as the calculation target, and a global search is performed in the candidate solutions. The optimal collision avoidance path should satisfy the following conditions.

- The SD_{gsp} of each group ship is not to be less than SD
- The course change is at least 30 degrees
- The sum of D_{c01} and D_{c02} takes the minimum value

3.2.4. Steps of the Algorithm

Pseudocode for collision avoidance path planning on the basis of the deterministic optimization algorithm (DOA); see Algorithm 1.

3.3. Collision Avoidance Decision-Making Model

Aiming at the action changes of the target ship, this paper introduces a dynamic monitoring mechanism to realize dynamic collision avoidance path planning. When the action of the target ship leads to a new ROC, the collision avoidance decision needs to be updated, that is, the collision avoidance path of OS needs to be re-planned. If the action change of the target ship does not form a new ROC, the current collision avoidance path will be maintained.

The multi-vessel encounter intelligent collision avoidance decision-making model established in this paper includes three modules: the encounter situation simplification module, the ship collision avoidance decision-making module, and the dynamic monitoring and evaluation module. The model is shown in Figure 5.

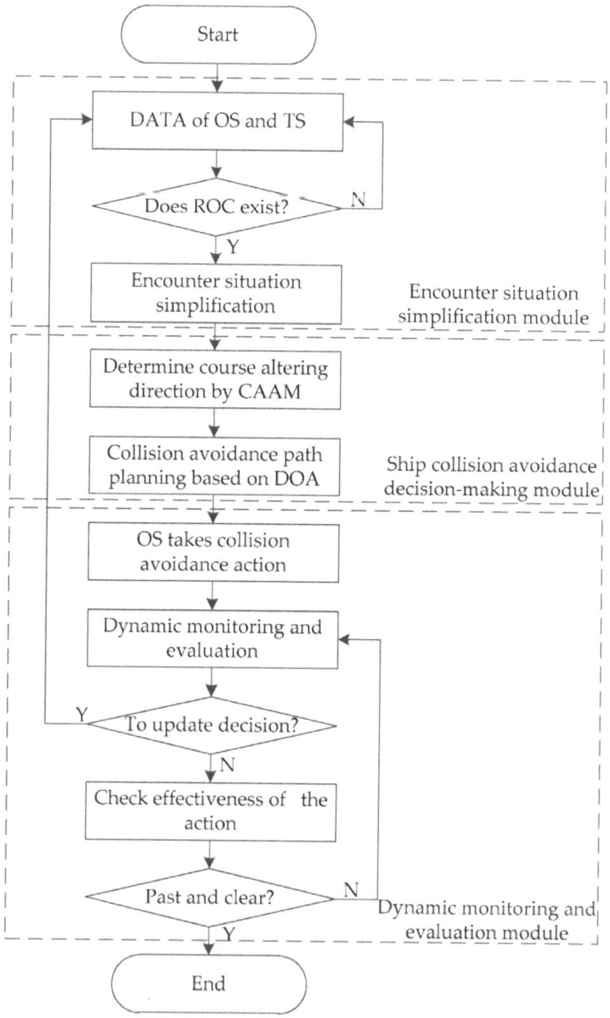

Figure 5. *Model of collision avoidance decision-making.*

4. SIMULATION AND ANALYSIS

To verify the validity of the decision-making model, two simulation cases were implemented using MATLAB software; one case is for in sight of one another, and the other is for not in sight of one another.

4.1. Experimental Conditions and Explanations

The simulation tests were conducted on the basis of a computer with an Intel(R) Core (TM) i7-3770 3.4GHz processor, 4G RAM, and a 32-bit Windows 7 Professional system. The simulation results are presented in the form of data tables and ship trajectory graphs.

Table 3. *Data of target ships at 0000.*

Data	C_{ts} (°)	V_{ts} (kn)	D_{ts} (nm)	TB_{ts} (°)	$DCPA_{ts}$ (nm)	$TCPA_{ts}$ (h)	Type
TS_1	270	7.80	4.20	028	0.1890	0.2737	PDV
TS_2	270	7.80	4.50	026	0.3593	0.2926	PDV
TS_3	270	7.80	4.40	032	−0.1091	0.2869	PDV
TS_4	270	7.80	4.70	030	0.0475	0.3065	PDV
TS_5	270	7.80	4.60	036	−0.4346	0.2987	PDV
TS_6	270	7.80	4.90	034	−0.2924	0.3190	PDV
TS_7	180	4.70	6.10	003	0.3192	0.3403	VEF
TS_8	180	4.70	5.80	001	0.1012	0.3240	VEF
TS_9	160	4.60	4.50	325	−2.2411	0.2218	VEF
TS_{10}	160	4.60	4.50	320	−2.5727	0.2099	VEF

The following parameter values were used in the simulation: $E_c = 1°, E_v = 0.5$kn, $T_c = 5$ s, $SD = 1$ nm, LOA $= 116$ m, $L_c = 4$ nm, $B_c = 2$ nm, and $D_{wp} = 0.2$ nm. In the simulation cases, the initial data of OS and target ships were considered to be known, and the process of acquiring data from the radar is not considered in this paper.

In the simulation cases, DCPA and TCPA (time to the closest point of approach) were used to determine the ROC. When an ROC exists, all target ships with approximate TCPA values were considered to give the optimal collision avoidance path.

4.2. Path Planning in Sight of One Another

4.2.1. Initial Situation of the Encounter

In sight of one another, OS's course is 000 , the initial speed is 13.2kn, the initial position is $(0,0)$, and the initial time is 0000 . OS encounters 10 target ships, $TS_1 - TS_{10}$; $TS_1 - TS_6$ are fishing vessels propelled by machinery but not engaged in fishing. The data of the target ships at 0000 are known, as shown in Table 3. It can be determined that a ROC exists, and the TCPA values for each TS are approximate.

4.2.2. Simulation Process and Analysis

A cluster analysis was performed, and target ships $(TS_1 - TS_{10})$ were classified into three group ships $(GS_1 - GS_3)$. The distribution of target ships and group ships at 0000 is shown in Figure 6.

According to the CAAM in sight of one another, OS should alter course to starboard. The candidate solution area is on OS's starboard. The action time t_s is 0002 , and the goal position is $(0,7)$. The optimal collision avoidance decision is to steer course 032 at 0002 $(C_{0002} = 032)$, 329.2 at 0019 $(C_{0019} = 329.2)$, and 000 at 0037 $(C_{0037} = 000)$.

At 0005, TS_5 and TS_6 change their course to 245 , and their speed remains unchanged. The trajectories of each ship from 0000 to 0005 are shown in Figure 7a.

The course changes of TS_5 and TS_6 lead to a new ROC, so OS should update the collision avoidance decision. The data of the target ships at 0005 are shown in Table 4.

After the re-clustering analysis, the target ships $(TS_1 - TS_{10})$ were classified into four group ships $(GS_1 - GS_4)$. The distribution of target ships and group ships at 0005 is shown in Figure 7*b*.

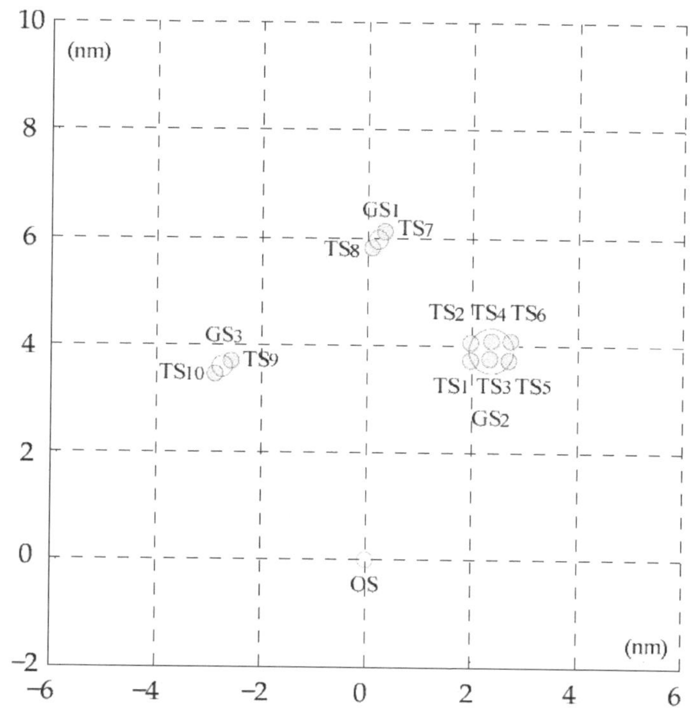

Figure 6. *Distribution of target ships and group ships at 0000.*

Table 4. *Data of target ships at 0005.*

Data	C_{ts} (°)	V_{ts} (kn)	D_{ts} (nm)	TB_{ts} (°)	$DCPA_{ts}$ (nm)	$TCPA_{ts}$ (h)	Type
TS_1	270	7.8	2.8820	19.7558	1.5752	0.1301	PDV
TS_2	270	7.8	3.2007	17.7361	1.8428	0.1411	PDV
TS_3	270	7.8	3.0434	25.9978	1.3765	0.1463	PDV
TS_4	270	7.8	3.3586	23.7444	1.6356	0.1581	PDV
TS_5	245	7.8	3.2155	32.0471	0.6739	0.1557	PDV
TS_6	245	7.8	3.5268	29.6075	0.8852	0.1691	PDV
TS_7	180	4.7	4.7054	359.6280	−1.9233	0.2473	VEF
TS_8	180	4.7	4.4197	356.7759	−2.0050	0.2268	VEF
TS_9	160	4.6	3.6436	309.7770	−3.4124	0.0777	VEF
TS_{10}	160	4.6	3.7496	303.9154	−3.6276	0.0577	VEF

According to the CAAM in sight of one another, OS should alter course to starboard. The candidate solution area is on OS's starboard. The action time t_s is 0007 , and the goal position is $(0, 7)$. The optimal collision avoidance decision is to steer course 068.9 at 0007 $(C_{0007} = 068.9)$, 333.5 at 0016 $(C_{0016} = 333.5)$, and 000 at 0041 $(C_{0041} = 000)$.

At 0010, TS_9 and TS_{10} change their courses to 180 , and their speed remains unchanged. The trajectories of each ship from 0000 to 0010 are shown in Figure 7c.

The course changes of TS_9 and TS_{10} do not lead to a new ROC, so OS should maintain the current collision avoidance decision.

At 0018, TS_7 and TS_8 change their course to 065 , and their speed remains unchanged. The trajectories of each ship from 0000 to 0018 are shown in Figure 7d.

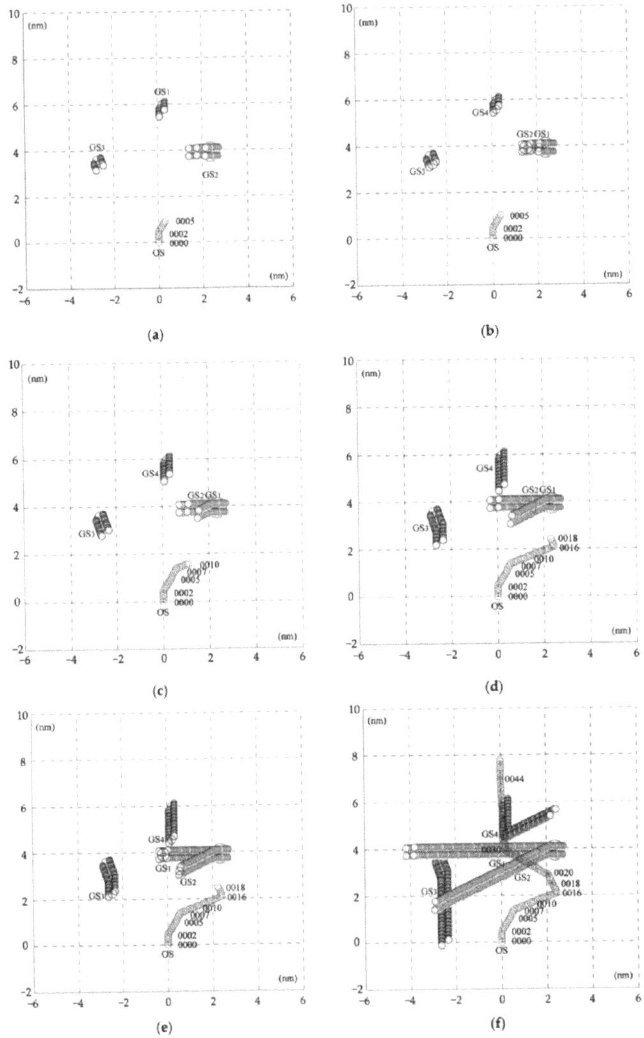

Figure 7. *Simulation results: (a) Ship trajectories (0000-0005); (b) distribution of target ships and group ships at 0005; (c) ship trajectories (0000-0010); (d) ship trajectories (0000-0018); (e) distribution of target ships and group ships at 0018; (f) ship trajectories (0000-0044).*

The course changes of TS_7 and TS_8 lead to a new ROC, so OS should update the collision avoidance decision. The data of the target ships at 0018 are shown in Table 5.

After the re-clustering analysis, target ships $(TS_1 - TS_{10})$ were classified into four group ships $(GS_1 - GS_4)$. The distribution of target ships and group ships at 0018 is shown in Figure 7e.

According to the CAAM in sight of one another, OS altering course to starboard or the port side is acceptable. To pass the stern of TS_7 and TS_8, OS chooses to alter course to the port side, and the candidate solution area is on OS's port

Table 5. *Data of target ships at 0018.*

Data	C_{ts} (°)	V_{ts} (kn)	D_{ts} (nm)	TB_{ts} (°)	$DCPA_{ts}$ (nm)	$TCPA_{ts}$ (h)	Type
TS_1	270	7.80	2.8973	295.4838	2.7809	0.0679	PDV
TS_2	270	7.80	3.0563	301.1898	2.8336	0.0957	PDV
TS_3	270	7.80	2.5883	299.3745	2.4293	0.0746	PDV
TS_4	270	7.80	2.7554	305.7158	2.4653	0.1029	PDV
TS_5	245	7.80	1.8101	287.5982	1.7628	0.0271	PDV
TS_6	245	7.80	1.9084	297.7328	1.7532	0.0497	PDV
TS_7	65	4.70	2.9474	318.9801	0.2520	0.2079	VEF
TS_8	65	4.70	2.8918	311.8991	−0.1098	0.2045	VEF
TS_9	180	4.60	4.5747	268.6760	4.3404	0.0829	VEF
TS_{10}	180	4.60	4.8970	265.9638	4.7142	0.0760	VEF

Table 6. *DCPA of target ships.*

DCPA	C_{0002}	C_{0007}	C_{0016}	C_{0020}	C_{0030}
TS_1	1.5738	2.1656	2.7917	1.9259	2.0922
TS_2	1.8413	2.4926	2.8444	1.7616	2.0825
TS_3	1.3750	2.1052	2.4401	1.6003	1.7320
TS_4	1.6342	2.4308	2.4761	1.4194	1.7078
TS_5	1.1425	−1.5959	1.7736	1.5948	1.8884
TS_6	1.3925	−1.8967	1.7639	1.4232	1.5925
TS_7	−1.9216	−3.4719	1.0840	1.3743	1.3631
TS_8	−2.0033	−3.3727	1.3876	1.0325	1.0443
TS_9	−3.4105	3.4895	3.9984	3.5225	3.5229
TS_{10}	−3.6257	3.6497	4.3814	3.9110	3.9127

side. The action time t_s is 0020 , and the goal position is $(0, 7)$. The optimal collision avoidance decision is to steer course 299.8 at 0020 $(C_{0020} = 299.8)$, 356.6 at 0030 $(C_{0030} = 356.6)$, and 000 at 0044 $(C_{0044} = 000)$.

From that point to time 0044 , the actions of each TS remain unchanged. The trajectories of each ship from 0000 to 0044 are shown in Figure 7f.

During the period from 0000 to 0044, OS changes its course to C_{0002}, C_{0007}, C_{0016}, C_{0020}, and C_{0030} successively. The DCPA values of target ships corresponding to different courses are shown in Table 6. As shown in the table, the absolute value of all DCPA is greater than the SD value, indicating that all target ships can pass OS at a safe distance.

4.3. Path Planning Not in Sight of One Another

4.3.1. Initial Situation of the Encounter

Not in sight of one another, OS's course is 000 , the initial speed is 12.8kn, the initial position is $(0, 0)$, and the initial time is 0000 . OS encounters 10 target ships, $TS_1 - TS_{10}$; $TS_1 - TS_5$ are fishing vessels propelled by machinery but not engaged in fishing. The data of the target ships at 0000 are known, as shown in Table 7. It can be determined that an ROC exists, and the TCPA values for each TS are approximate.

4.3.2. Simulation Process and Analysis

A cluster analysis was performed, and target ships $(TS_1 - TS_{10})$ were classified into three group ships $(GS_1 - GS_3)$. The distribution of the target ships and group ships at 0000 is shown in Figure 8 a.

Table 7. *Data of target ships at 0000.*

Data	C_{ts} (°)	V_{ts} (kn)	D_{ts} (nm)	TB_{ts} (°)	$DCPA_{ts}$ (nm)	$TCPA_{ts}$ (h)	Type
TS_1	249	8.20	4.20	035	−0.6615	0.2370	PDV
TS_2	249	8.20	4.40	037	−0.8442	0.2467	PDV
TS_3	249	8.20	4.10	028	−0.1475	0.2341	PDV
TS_4	249	8.20	4.30	030	−0.3046	0.2451	PDV
TS_5	249	8.20	4.50	032	−0.4752	0.2557	PDV
TS_6	181	4.60	5.40	356	0.4015	0.3095	VEF
TS_7	181	4.60	5.50	359	0.1214	0.3160	VEF
TS_8	181	4.60	5.20	349	1.0157	0.2931	VEF
TS_9	181	4.60	5.00	353	0.6322	0.2851	VEF
TS_{10}	135	14.1	4.40	330	−0.4869	0.1759	PDV

Table 8. *Data of target ships at 0004.*

Data	C_{ts} (°)	V_{ts} (kn)	D_{ts} (nm)	TB_{ts} (°)	$DCPA_{ts}$ (nm)	$TCPA_{ts}$ (h)	Type
TS_1	220	8.20	2.9905	31.9752	−0.6682	0.1391	PDV
TS_2	220	8.20	3.1837	34.9236	−0.5508	0.1497	PDV
TS_3	249	8.20	2.9305	22.0352	1.6450	0.1174	PDV
TS_4	249	8.20	3.1159	25.1585	1.6059	0.1292	PDV
TS_5	249	8.20	3.3037	28.1626	1.5520	0.1411	PDV
TS_6	181	4.60	4.4285	350.916	−3.1409	0.1917	VEF
TS_7	181	4.60	4.5048	354.6700	−2.9802	0.2074	VEF
TS_8	181	4.60	4.2967	342.1831	−3.4720	0.1554	VEF
TS_9	181	4.60	4.0573	346.7227	−3.0791	0.1622	VEF
TS_{10}	135	14.10	3.0653	322.7536	−1.7946	0.1340	PDV

According to the CAAM not in sight of one another, OS should alter course to starboard. The candidate solution area is on OS's starboard. The action time t_s is 0002 , and the goal position is $(0, 7)$. The optimal collision avoidance decision is to steer course 048 at 0002 $(C_{0002} = 048)$, 337.3 at 0015 $(C_{0015} = 337.3)$, and 000 at 0039 $(C_{0039} = 000)$.

At 0004, TS_1 and TS_2 change their course to 220 , and their speed remains unchanged. The trajectories of each ship from 0000 to 0004 are shown in Figure 8b.

The course changes of TS_1 and TS_2 lead to a new ROC, so OS should update the collision avoidance decision. The data of the target ships at 0004 are shown in Table 8.

After the re-clustering analysis, target ships $(TS_1 - TS_{10})$ are classified into four group ships $(GS_1 - GS_4)$. The distribution of target ships and group ships at 0004 is shown in Figure 8c.

According to the CAAM not in sight of one another, OS should alter course to starboard. The candidate solution area is on OS's starboard. The action time t_s is 0006 , and the goal position is $(0, 7)$. The optimal collision avoidance decision is to steer course 083.6 at 0006 $(C_{0006} = 083.6)$, 338.0 at 0014 $(C_{0014} = 338.0)$, and 000 at 0043 $(C_{0043} = 000)$.

At 0016, TS_6 and TS_7 change their course to 121 , their speed increases to 7.8kn, and they are not engaged in fishing; TS_8 and TS_9 change their courses to 186 , and their speeds remain unchanged. The trajectories of each ship from 0000 to 0016 are shown in Figure 8d.

The course changes of TS_8 and TS_9 do not lead to a new ROC, but the course and speed changes of TS_6 and TS_7 lead to a new ROC, so OS should update the

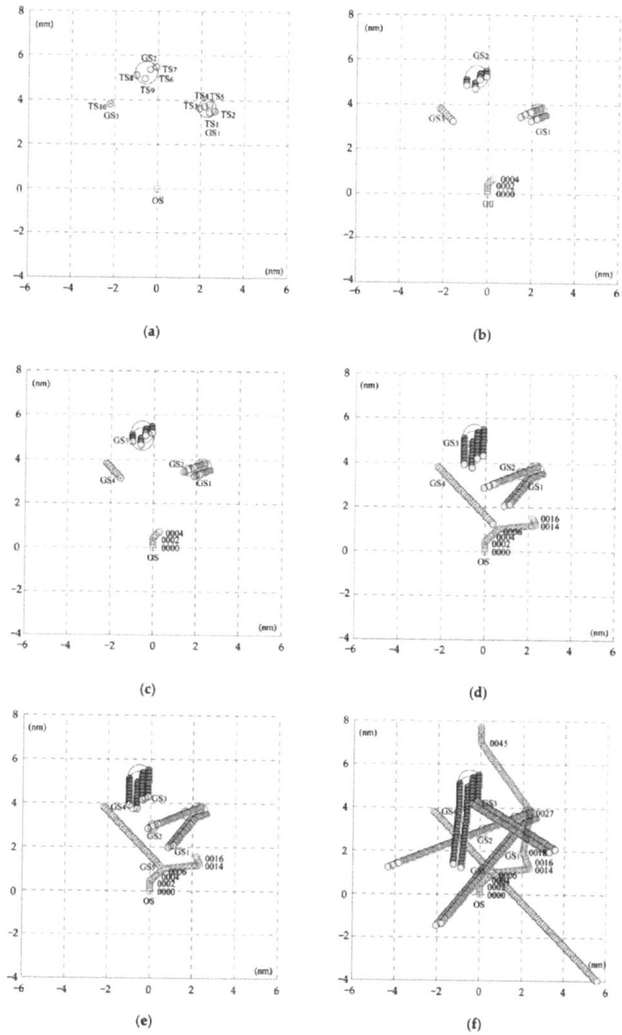

Figure 8. *Simulation results: (a) Distribution of target ships and group ships at 0000; (b) ship trajectories (0000-0004); (c) distribution of target ships and group ships at 0004; (d) ship trajectories (0000-0016); (e) distribution of target ships and group ships at 0016; (f) ship trajectories (0000-0045).*

Table 9. *Data of target ships at 0016.*

Data	C_{ts} (°)	V_{ts} (kn)	D_{ts} (nm)	TB_{ts} (°)	$DCPA_{ts}$ (nm)	$TCPA_{ts}$ (h)	Type
TS_1	220	8.20	1.4046	287.5424	1.3501	0.0214	PDV
TS_2	220	8.20	1.2074	294.3046	1.1132	0.0258	PDV
TS_3	249	8.20	2.6274	298.9221	2.4994	0.0538	PDV
TS_4	249	8.20	2.4886	303.5264	2.2981	0.0634	PDV
TS_5	249	8.20	2.3531	308.5622	2.0853	0.0724	PDV
TS_6	121	7.80	3.6635	315.1114	−0.5767	0.1846	PDV
TS_7	121	7.80	3.5559	319.5975	−0.2834	0.1809	PDV
TS_8	186	4.60	3.9493	305.8532	2.5107	0.1794	VEF
TS_9	186	4.60	3.5576	307.6149	2.1762	0.1656	VEF
TS_{10}	135	14.10	1.7820	256.6961	−1.6666	0.0240	PDV

Table 10. *DCPA of target ships.*

DCPA	C_{0002}	C_{0006}	C_{0014}	C_{0018}	C_{0027}
TS_1	1.2238	−1.4464	1.3535	1.3676	3.8540
TS_2	1.1519	−1.4198	1.1166	1.1217	3.6474
TS_3	1.6425	−2.1703	2.5029	2.5182	4.1397
TS_4	1.6035	−2.2246	2.3015	2.3421	3.8922
TS_5	1.5496	−2.2655	2.0887	2.1664	3.6401
TS_6	−3.1381	−3.9930	1.7759	−1.5933	−1.3650
TS_7	−2.9774	−3.9748	1.4752	−1.2931	−1.0871
TS_8	−3.4691	−4.0002	2.4454	−3.2954	3.4485
TS_9	−3.0763	−3.7086	2.1162	−2.8951	3.1648
TS_{10}	−1.7920	−2.2157	−1.6690	1.6848	4.4668

collision avoidance decision. The data of the target ships at 0016 are shown in Table 9.

After the re-clustering analysis, the target ships $TS_1 - TS_{10}$ were classified into five group ships $(GS_1 - GS_5)$. The distribution of target ships and group ships at 0016 is shown in Figure 8e.

According to CAAM not in sight of one another, OS should alter course to starboard. The candidate solution area is on OS's starboard. The action time t_s is 0018 , and the goal position is $(0,7)$. The optimal collision avoidance decision is to steer course 010 at 0018 $(C_{0018} = 010)$, 323.4 at 0027 $(C_{0027} = 323.4)$, and 000 at 0045 $(C_{0045} = 000)$.

From then to time 0045 , the actions of each TS remained unchanged. The trajectories of each ship from 0000 to 0045 are shown in Figure 8f.

During the period from 0000 to 0045, OS changed its course to C_{0002}, C_{0006}, C_{0014}, C_{0018}, and C_{0027} successively. The DCPA values of the target ships corresponding to different courses are shown in Table 10. As shown in the table, the absolute value of all DCPA is greater than the *SD* value, indicating that all target ships can pass OS at a safe distance.

5. CONCLUSIONS

In this paper, a method to simplify the encounter situation based on dynamic cluster analysis was proposed to improve the efficiency of collision avoidance path planning. Target ships with similar properties were dynamically classified into one group ship, the model of the group ship was given, and the number of computational targets was effectively reduced. Fully considering the constraints of the COLREGs, a course alteration action matrix model (CAAM) was established to determine the side of course changing, which narrows the range of candidate solutions and helps improve the efficiency of the decision-making. In order to realize dynamic collision avoidance path planning, a dynamic monitoring mechanism was introduced, and a multi-ship encounter intelligent collision avoidance decision-making model was established. The simulation results show that the dynamic path planning method in this paper is safe and feasible. However, in the simulation tests, the ship's maneuverability was not considered; this will be taken into account in subsequent studies.

Author Contributions

Methodology, J.Y. and Z.L.; software, J.Y.; validation, J.Y.; writing-original draft preparation, J.Y. and Z.L.; writing-review and editing, J.Y. and X.Z.; supervision, Z.L. and X.Z. All authors have read and agreed to the published version of the manuscript.

REFERENCES

1. Iijima, Y.; Hagiwara, H.; Kasai, H. Results of Collision Avoidance Manoeuvre Experiments Using a Knowledge-Based Autonomous Piloting System. J. Navig. 1991, 44, 194-204.
2. Coenen, F.P.; Smeaton, G.P.; Bole, A.G. Knowledge-based Collision Avoidance. J. Navig. 1989, 42, 107-116.
3. Tam, C.K.; Bucknall, R. Cooperative path planning algorithm for marine surface vessels. Ocean Eng. 2013, 57, 25-33.
4. Lazarowska, A. A new deterministic approach in a decision support system for ship's trajectory planning. Expert Syst. Appl. 2017, 71, 469-478.
5. Kim, H.G.; Yun, S.J.; Choi, Y.H.; Ryu, J.K.; Suh, J.H. Collision Avoidance Algorithm Based on COLREGs for Unmanned Surface Vehicle. J. Mar. Sci. Eng. 2021, 9, 863.
6. Zheng, H.; Negenborn, R.R.; Lodewijks, G. Predictive path following with arrival time awareness for waterborne AGVs. Transp. Res. Part C Emerg. Technol. 2016, 70, 214-237.
7. Hagen, I.B.; Kufoalor, D.K.M.; Brekke, E.F.; Johansen, T.A. MPC-based Collision Avoidance Strategy for Existing Marine Vessel Guidance Systems. In Proceedings of the 2018 IEEE International Conference on Robotics and Automation (ICRA), Brisbane, Australia, 21-25 May 2018.
8. Xie, S.; Garofano, V.; Chu, X.; Negenborn, R.R. Model predictive ship collision avoidance based on Q-learning beetle swarm antenna search and neural networks. Ocean Eng. 2019, 193, 106609.
9. Johansen, T.A.; Perez, T.; Cristofaro, A. Ship Collision Avoidance and COLREGS Compliance Using Simulation-Based Control Behavior Selection With Predictive Hazard Assessment. IEEE Trans. Intell. Transp. Syst. 2016, 17, 3407-3422.
10. Abdelaal, M.; Fränzle, M.; Hahn, A. Nonlinear Model Predictive Control for trajectory tracking and collision avoidance of underactuated vessels with disturbances. Ocean Eng. 2018, 160, 168-180.
11. Perera, L.P.; Carvalho, J.P.; Soares, C.G. Intelligent Ocean Navigation and Fuzzy-Bayesian Decision/ Action Formulation. IEEE J Ocean. Eng. 2012, 37, 204-219.
12. Perera, L.P.; Carvalho, J.P.; Soares, C.G. Solutions to the Failures and Limitations of Mamdani Fuzzy Inference in Ship Navigation. IEEE Trans. Veh. Technol. 2014, 63, 1539-1554.
13. Lee, H.J.; Rhee, K.P. Development of collision avoidance system by using expert system and search algorithm. Int. Shipbuild. Prog. 2001, 48, 197-212.
14. Zhao, Y.; Li, W.; Shi, P. A real-time collision avoidance learning system for Unmanned Surface Vessels. Neurocomputing 2016, 182, 255-266.
15. Namgung, H. Local Route Planning for Collision Avoidance of Maritime Autonomous Surface Ships in Compliance with COLREGs Rules. Sustainability 2022, 14, 198.

16. Zhu, Z.; Lyu, H.; Zhang, J.; Yin, Y. An Efficient Ship Automatic Collision Avoidance Method Based on Modified Artificial Potential Field. J. Mar. Sci. Eng. 2022, 10, 3.
17. Singh, Y.; Sharma, S.; Sutton, R.; Hatton, D. Towards use of Dijkstra Algorithm for Optimal Navigation of an Unmanned Surface Vehicle in a Real-Time Marine Environment with results from Artificial Potential Field. TransNav Int. J. Mar. Navig. Saf. Sea Transp. 2018, 12, 125-131.
18. Mousazadeh, H.; Jafarbiglu, H.; Abdolmaleki, H.; Omrani, E.; Monhaseri, F.; Abdollahzadeh, M.; Mohammadi-Aghdam, A.; Kiapei, A.; Salmani-Zakaria, Y.; Makhsoos, A. Developing a navigation, guidance and obstacle avoidance algorithm for an Unmanned Surface Vehicle (USV) by algorithms fusion. Ocean Eng. 2018, 159, 56-65.
19. Xue, Y.; Clelland, D.; Lee, B.S.; Han, D. Automatic simulation of ship navigation. Ocean Eng. 2011, 38, 2290-2305.
20. Lyu, H.; Yin, Y. COLREGS-Constrained Real-time Path Planning for Autonomous Ships Using Modified Artificial Potential Fields. J. Navig. 2019, 72, 588-608.
21. Tam, C.K.; Bucknall, R. Path-planning algorithm for ships in close-range encounters. J. Mar. Sci. Technol. 2010, 15, 395-407.
22. Szlapczynski, R. Evolutionary Sets of Safe Ship Trajectories: A New Approach to Collision Avoidance. J. Navig. 2011, 64, 169-181.
23. Szlapczynski, R.; Szlapczynska, J. On evolutionary computing in multi-ship trajectory planning. Appl. Intell. 2012, 37, 155-174.
24. Szlapczynski, R. Evolutionary Sets of Safe Ship Trajectories Within Traffic Separation Schemes. J. Navig. 2013, 66, 65-81.
25. Szlapczynski, R. Evolutionary Planning of Safe Ship Tracks in Restricted Visibility. J. Navig. 2015, 68, 39-51.
26. Lazarowska, A. Ship's Trajectory Planning for Collision Avoidance at Sea Based on Ant Colony Optimisation. J. Navig. 2015, 68, 291-307.
27. Lazarowska, A. Multi-criteria ACO-based Algorithm for Ship's Trajectory Planning. TransNav Int. J. Mar. Navig. Saf. Sea Transp. 2017, 11, 31-36.
28. Ma, Y.; Hu, M.; Yan, X. Multi-objective path planning for unmanned surface vehicle with currents effects. ISA Trans. 2018, 75, 137-156.
29. Lisowski, J. The dynamic game theory methods applied to ship control with minimum risk of collision. WIT Trans. Ecol. Environ. 2006, 91, 293-302.
30. Lisowski, J. Computational intelligence methods of a safe ship control. Procedia Comput. Sci. 2014, 35, 634-643.
31. Zhang, J.; Zhang, D.; Yan, X.; Haugen, S.; Soares, C.G. A distributed anti-collision decision support formulation in multi-ship encounter situations under COLREGs. Ocean Eng. 2015, 105, 336-348.
32. Kim, D.; Hirayama, K.; Okimoto, T. Distributed Stochastic Search Algorithm for Multi-ship Encounter Situations. J. Navig. 2017, 70, 699-718.
33. Kuwata, Y.; Wolf, M.T.; Zarzhitsky, D.; Huntsberger, T.L. Safe Maritime Autonomous Navigation with COLREGS, Using Velocity Obstacles. IEEE J. Ocean. Eng. 2014, 39, 110-119.
34. Huang, Y.; van Gelder, P.H.A.J.M.; Wen, Y. Velocity obstacle algorithms for collision prevention at sea. Ocean Eng. 2018, 151, 308-321.

35. Xu, Q.; Yang, Y.; Zhang, C.; Zhang, L. Deep Convolutional Neural Network-Based Autonomous Marine Vehicle Maneuver. Int. J. Fuzzy Syst. 2018, 20, 687-699.

36. Lin, C.; Wang, H.; Yuan, J.; Yu, D.; Li, C. An improved recurrent neural network for unmanned underwater vehicle online obstacle avoidance. Ocean Eng. 2019, 189, 106327.

37. Cheng, Y.; Zhang, W. Concise deep reinforcement learning obstacle avoidance for underactuated unmanned marine vessels. Neurocomputing 2018, 272, 63-73.

38. Shen, H.; Hashimoto, H.; Matsuda, A.; Taniguchi, Y.; Terada, D.; Guo, C. Automatic collision avoidance of multiple ships based on deep Q-learning. Appl. Ocean Res. 2019, 86, 268-288.

39. Yuan, J.; Wang, H.; Zhang, H.; Lin, C.; Yu, D.; Li, C. AUV Obstacle Avoidance Planning Based on Deep Reinforcement Learning. J. Mar. Sci. Eng. 2021, 9, 1166.

40. Yu, J.; Liu, Z.; Bu, R.; Gao, X.; Li, W. A Novel Fast Decision-making Method for Ship's Course Altering for Collision Avoidance. In Proceedings of the 2020 Chinese Automation Congress (CAC), Shanghai, China, 6-8 November 2020.

CHAPTER 3

Ship Steering Adaptive CGS Control Based on EKF Identification Method

Wei Guan, Haowen Peng, Xianku Zhang and Hui Sun

Navigation College, Dalian Maritime University, Dalian 116026, China

ABSTRACT

In recent years, marine autonomous surface vessels (MASS) have grown into a ship research issue to increase the level of autonomy of ship behavior decision-making and control while sailing at sea. This paper focuses on the MASS motion control module design that aims to improve the accuracy and reliability of ship steering control systems. Nevertheless, the stochastic sea and wind environment have led to the extensive use of filters and state observers for estimating the ship-motionrelated parameters, which are important for ship steering control systems. In particular, the ship maneuverability Nomoto index, which primarily determines the designed ship steering controller's performance, cannot be observed directly due to the model errors and the external environment disturbance in the process of sailing. Hence, an adaptive robust ship steering controller based on a closed-loop gain shaping (CGS) scheme and an extended Kalman filter (EKF) on-line identification method is explored in this paper. To verify the effectiveness of the proposed steering controller design scheme, the motor vessel YUKUN was taken as the control plant and a series of simulation experiments were carried out. The results show the advantages of the dynamic response performance of the proposed steering controller compared with the classical PD and traditional CGS controllers. Therefore, the proposed adaptive CGS steering controller would be a good solution for MASS motion control module design.

Keywords: *ship steering control; Marine autonomous surface vessel (MASS); extend Kalman filter (EKF) identification; adaptive closed-loop gain shaping (CGS) controller*

1. INTRODUCTION

Marine surface vessels autonomously sailing at sea had been in the dreams of ship designers for several decades. With the development of artificial intelligent decisionmaking technology, satellite communication systems, high-tech sensors,

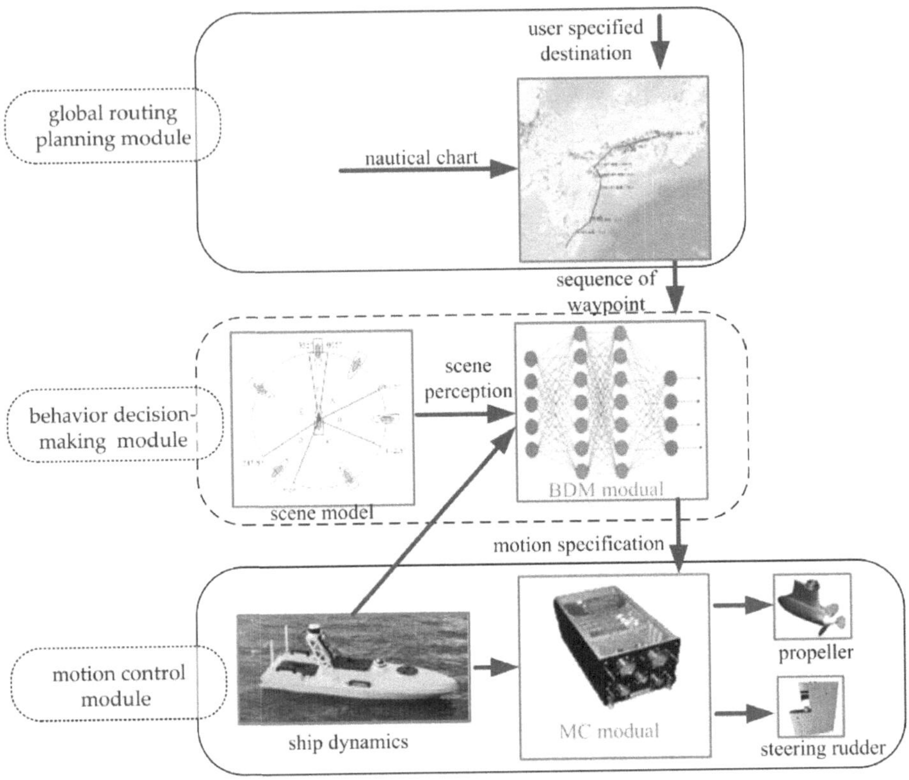

Figure 1. *The concept of the MASS HDMC model.*

and advanced actuators, their dreams might turn into a possible reality. Therefore, the concept of marine autonomous surface vessels (MASS) had been introduced by the International Maritime Organization (IMO) for fueling present ship navigation and control technologies [1]. To unify the MASS theoretical research and project practice, a typical MASS hierarchical decision-making and control (HDMC) model was introduced [2]. The MASS HDMC model can be hierarchically decomposed into three components, as shown in Figure 1. At the highest level, a global routing planning module is designed. This is followed by a behavior decision-making (BDM) module, which decides on a local driving task to progress the MASS towards the destination and to keep it collision-free with obstacles and other ships. At the lowest level, a motion control (MC) module is used to control the ship actuator mechanism to execute the BDM module output commands (such as steering rudder and propeller rotation). These modules are in charge of the different tasks. In particular, the MC module is the foundation module which can determine the performance of MASS autonomous navigation and the sailing process [3]. The MASS steering operation to maintain or change ship course is the key function of the MC module. Hence, ship steering controller design is one of the most important tasks of ship motion control.

With the last few decades of research, researchers have improved and optimized the ship steering controller consistently. A proportional-integral-derivative (PID) ship steering controller was proposed in 1922, which is the milestone of the modern ship steering control system and is still widely used today [4]. An adaptive steering controller is based on the filter and observer, which was proposed to improve the ship steering controller's dynamic response performance [5]. The ship steering adaptive disturbance rejection control (ADRC) solution is a typical adaptive steering controller which can estimate and compensate for both the internal and external disturbance based on the state observer [6, 7]. However, there still exist problems relating to difficult stability analysis and arbitrary tuning parameter settings with this solution.

Optimal control is also a classical ship steering controller design scheme, which can achieve optimum performance in certain control constraints [8]. For instance, the ship steering model predictive control (MPC) is derived from the concept of optimal control [9]. Although there exist limitations relating to poor stability in the open-loop control system, the ship steering MPC is becoming a hot academic issue nowadays.

Furthermore, incorporating the unmodelled uncertainties and external disturbance into the consideration of ship steering controller design, the H_∞ robust controller [10] and the sliding mode controller [11] are both typical robust ship steering controllers and have good practical ship steering control application benefits.

However, linear controllers are widely available in the design of a ship steering controller. The validity of linear controllers is limited to the course-keeping maneuver, since only a small rudder angle action is involved. Once the ship maneuver calls for rapid and large course-changing movements, hydrodynamic nonlinearities need to be taken into consideration in the steering controller design. Hence, the feedback linearization technique was used in ship steering controller design with the cancellation of the system model nonlinear term [12]. The backstepping algorithm is the most successful controller synthesis method in dealing with the nonlinear control problem in the last few decades. Based on the Lyapunov stability theorem, the processes of controller design and controller stability analysis are synthesized. Fossen [13] presented a nonlinear backstepping design for the ship steering controller, and a nonlinear steering controller with "good nonlinearities term" was explored. Furthermore, the novel nonlinear feedback technique [14] and nonlinear decoration technique [15] were used to improve the nonlinear backstepping design methodology and the ship steering steady state performance.

Additionaly, a series of intelligent ship steering controllers, such as the neural network controller [16], quantum neural network controller [2], expert system controller [17], adaptive fuzzy controller [18], adaptive neuro-fuzzy inference controller [19], and the stateof-the-art artificial intelligence (AI) controller based on deep reinforcement learning [20] were proposed to enhance ship steering controllers' adaptability and self-learning ability. However, due to their strong subjectivity and poor theoretical interpretation, the practical applications of the above intelligent ship steering controllers have not been reported.

The above achievements might be a good solution for ship steering control under the concept of the MASS HDMC model, while there are still two limitations which need to be addressed in future studies:

(1) The proposed steering controller might be a hybrid algorithm which has a highcomplexity structure and poor real-time control performance. Hence, these steering controllers might not be suitable for the MC module design under the MASS HDMC concept.

(2) The optimization or intelligent algorithm might be used as the steering controller design, but the tuned parameters of the controllers might have indefinite physical meaning, resulting in instability and poor robustness.

Based on the overlook of the ship steering control scheme in previous works, a ship steering adaptive closed-loop gain shaping (CGS) controller, which combined the CGS controller design approach [21] with the extended Kalman filtering (EKF) on-line identification technique [22,23], was proposed to address the ship steering control problem based on the MASS HDMC concept. To ensure the stability and robustness of the proposed controller, the conventional CGS controller design methodology was used as the basic structure of the proposed ship steering controller, since the CGS controller has a concise controller structure and definite controller design procedures. However, the conservative robustness controller design scheme might undermine the dynamic response performance of the CGS controller. Hence, the on-line EKF identification technique was introduced to improve the controller's dynamic response performance. Consequently, the proposed ship steering adaptive CGS controller would have a concise structure, the definite physical meaning of tuning parameters, and better dynamic response performance of the ship course-keeping and course-changing.

This paper is organized as follows. In Section 2, the nonlinear model of the ship steering control system is described. Section 3 is devoted to the proposed ship steering adaptive CGS controller design procedure. In Section 4, analysis of system stability is carried out. In Section 5, numerical simulations are given. Section 6 contains the conclusions and discussions.

2. MATHEMATICAL MODELS

The nonlinear ship motion mathematic model is generally described as

$$\ddot{\psi} = -\frac{1}{T_R}\dot{\psi} - \frac{a}{T_R}\dot{\psi}^3 + \frac{K_R}{T_R}\delta - K_d \tag{1}$$

where ψ is the yaw angle, δ is the rudder angle, K_R and T_R are donated as the ship maneuverability index, a is the coefficient of the nonlinear term function, and K_d is considered as the bounded external interference.

As shown in Figure 2, only the ship lateral drift velocity v and the yaw angular velocity r are considered, and the ship motion mathematic model can be expressed as

$$\begin{bmatrix} \dot{v} \\ \dot{r} \end{bmatrix} = \begin{bmatrix} a_{11} & a_{12} \\ a_{21} & a_{22} \end{bmatrix} \begin{bmatrix} v \\ r \end{bmatrix} + \begin{bmatrix} b_1 \\ b_2 \end{bmatrix} \delta \tag{2}$$

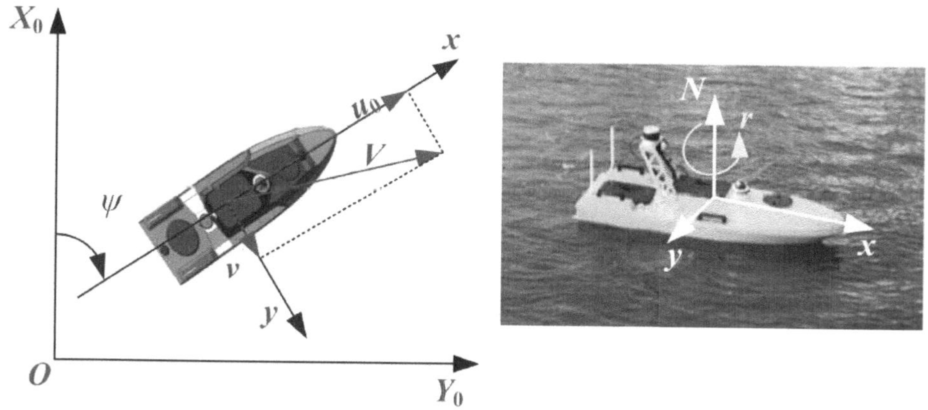

Figure 2. *Ship motion and yaw motion coordination.*

where $a_{11}, a_{12}, a_{21}, a_{22}, b_1$, and b_2 are the ship maneuverability parameters. Ignoring the lateral drift velocity v in Equation (2), the response equation of the ship steering rudder to yaw motion can be written as

$$T_1 T_2 \ddot{r} + (T_1 + T_2)\,\dot{r} + r = K_R \delta + K_R T_3 \dot{\delta} \tag{3}$$

where T_1, T_2, T_3, and K_R are maneuverability indexes. According to [22], their values can be estimated by $T_1 T_2 = 1/(a_{11}a_{22} - a_{12}a_{21}), T_1 + T_2 = (a_{11} + a_{22})/(a_{12}a_{21} - a_{11}a_{22})$, $T_3 = b_2/(a_{21}b_1 - a_{11}b_2), K_R = (a_{21}b_1 - a_{11}b_2)/(a_{11}a_{22} - a_{12}a_{21})$. Then, the Laplace transform of Equation (3) can be carried out to obtain the transfer function of the ship steering control system, as shown in Equation (4)

$$G(s) = \frac{\psi(s)}{\delta(s)} = \frac{K_R\,(1 + T_3 s)}{s\,(1 + T_1 s)\,(1 + T_2 s)} \tag{4}$$

Since the ship has large inertia characteristics, only the low-frequency part of the ship's dynamic characteristics is considered. Let $s = j\omega \to 0$; the transfer function (4) can be rewritten as Equation (5),

$$G(s) = \frac{K_R}{s\,(T_R s + 1)} \tag{5}$$

where $T_R = T_1 + T_2 - T_3$, and the model is a second-order linear Nomoto model [11]. Furthermore, the transfer function Equation (5) can be expressed as Equation (6).

$$T_R \ddot{\psi} + \dot{\psi} = K_R \delta \tag{6}$$

Let the ship yaw velocity term $\dot{\psi}$ be a nonlinear function $H(\dot{\psi}) = \dot{\psi} + a\dot{\psi}^3$. Therefore, Equation (6) can be written as Equation (1). Defining $\alpha = -1/T_R, \beta = -a/T_R$ and $\gamma = K_R/T_R$, Equation (1) can be written as Equation (7) for the following ship steering controller design.

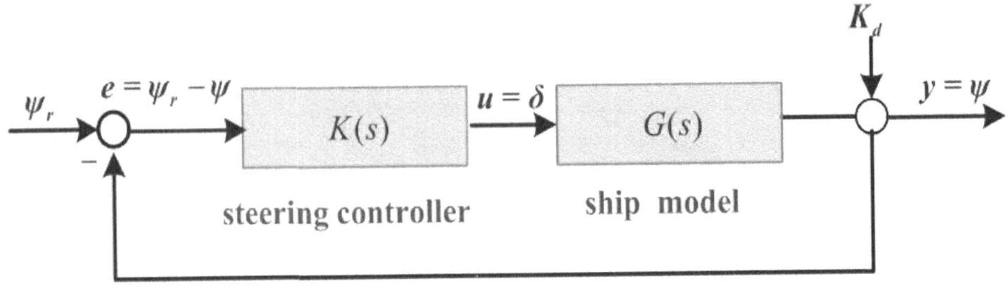

Figure 3. *Typical structure of ship steering control system.*

$$\ddot{\psi} = \alpha\dot{\psi} + \beta\dot{\psi}^3 + \gamma\delta - K_d \tag{7}$$

3. SHIP STEERING ADAPTIVE CGS CONTROL BASED ON EKF

3.1. Ship Steering CGS Controller Design

Considering a typical structure of the ship steering control system as shown in Figure 3, $G(s)$ and $K(s)$ represent the transfer function of the ship mathematical model and steering controller, respectively. y is defined as the steering to yaw motion system output signal, and ψ_r is the demand course reference signal. K_d is the external disturbance of the system. u is defined as the ship steering rudder angle δ. Defining the error of the steering control system as $e = \psi_r - \psi$, the purpose of the steering controller design is to produce the ship steering rudder angle δ to make the output yaw angle y track the input demand course signal ψ_r under the disturbance K_d, which can be expressed as a transfer function as follows:

$$\frac{y(s)}{\psi_r(s)} = \frac{G(s)K(s)}{1 + G(s)K(s)} \tag{8}$$

To design the ship steering CGS controller, the following hypotheses should be made:

Hypothesis 1. The spectrum of the ship steering control system works as a low-frequency pass filter. That is, the main work frequency band of the system is in the low-frequency region, and the high-frequency region is regarded as an interference signal.

Hypothesis 2. The maximum singular value of the ship steering control system is set to 1. That is, the peak value of the system is 0, thus ensuring the system could track the reference signal without any overshoot.

Hypothesis 3. The spectrum slope of the ship steering closed loop control system could be set as -20 dB/ dec to suppress external disturbance.

Based on the above hypotheses, the controlling performance of the entire ship steering to the yaw control system will be primarily determined by the system bandwidth $1/\lambda$, which simplifies the controller design procedures, and causes the controller tuning parameters to have definite physical meanings. Therefore, the transfer function of the ship steering control system can be rewritten as

$$\frac{G(s)K(s)}{1+G(s)K(s)} = \frac{1}{\lambda s+1} \tag{9}$$

Then, the ship steering controller $K(s)$ can be obtained according to Equation (9), which is

$$K(s) = 1/(\lambda G(s)s) \tag{10}$$

Therefore, according Equation (5), the ship steering CGS controller shown in Figure 3 can be realized by Equation (11),

$$\delta = \lambda_1 (\lambda_2 e + \dot{e}) \tag{11}$$

where $\lambda_1 = T_R/(\lambda K_R), \lambda_2 = 1/T_R$. Since λ is the only tuning parameter, the estimation of the ship maneuverability index T_R and K_R would be predominant in the process of ship steering CGS controller design. Hence, an extend Kalman filter (EKF) identification method can be used in the ship steering parameter identification technique to complete the ship steering CGS controller design.

3.2. Extend Kalman Filter Identification

For the estimation of the ship maneuverability index T_R and K_R, the least-square method [24], multi-innovation method [25], and support vector machine method [26] are all good solutions. Additionally, the parameters of the ship steering control system can be estimated by the on-line EKF identification technique, and successful results were reported in [22]. Due to the effectiveness of capturing the dynamic behavior of ship steering to yaw motion, the EKF on-line identification technique might be a good solution to improve the ship steering controller's dynamic performance.

The EKF identification technique for adaptive ship steering controller design can be summarized as follows:

(1) The ship steering control system, described in Equation (1), can be rewritten as

$$\dot{X}(t) = f(X(t)) + W(t) \tag{12}$$

where $\boldsymbol{X}^{\mathrm{T}}(t) = \left[\psi(t)\ \dot{\psi}(t)\ \alpha(t)\ \beta(t)\ \gamma(t)\ \delta(t)\right], \boldsymbol{W}(t)$ is the noise vector, and $f(X(t))$ is represented as

$$f(\boldsymbol{X}(t)) = \begin{bmatrix} \dot{\psi}(t) \\ \alpha(t)\dot{\psi}(t) + \beta(t)\dot{\psi}(t)^3 + \gamma(t)\delta(t) \\ 0 \\ 0 \\ 0 \\ 0 \end{bmatrix} \tag{13}$$

Then, the Jacobian matrix of Equation (13) can be given by

$$\frac{\partial}{\partial \boldsymbol{X}(t)}(\boldsymbol{f}(\boldsymbol{X}(t)))=\begin{bmatrix}0 & 1 & 0 & 0 & 0 & 0\\ 0 & \alpha(t)+3\beta(t)\dot{\psi}^3(t) & \dot{\psi}(t) & \dot{\psi}^3(t) & \delta(t) & \gamma(t)\\ 0 & 0 & 0 & 0 & 0 & 0\\ 0 & 0 & 0 & 0 & 0 & 0\\ 0 & 0 & 0 & 0 & 0 & 0\\ 0 & 0 & 0 & 0 & 0 & 0\end{bmatrix} \tag{14}$$

(2) Define the measured state matrix as

$$z(k)=\boldsymbol{h}(\boldsymbol{X}(k))+\xi(k) \tag{15}$$

where $z^T(k)=[z_\psi(k)\ z_{\dot{\psi}}(k)\ z_\delta(k)]$ and $\boldsymbol{\xi}(k)$ is the measured noise, and $\boldsymbol{h}^T(\boldsymbol{X}(t))=[\psi(t)\ \dot{\psi}(t)\ \delta(t)]$. Then, the Jacobian matrix of $\boldsymbol{h}^T(\boldsymbol{X}(k))$ can be computed as

$$\frac{\partial}{\partial \mathrm{X}(k)}h(\mathrm{X}(k))=\begin{bmatrix}1\,0\,0\,0\,0\,0\\ 0\,1\,0\,0\,0\,0\\ 0\,0\,0\,0\,0\,1\end{bmatrix} \tag{16}$$

(3) Define the extrapolation law of error covariance as

$$\dot{\boldsymbol{P}}(t)=\boldsymbol{F}(\hat{\boldsymbol{X}}(t))\boldsymbol{P}(t)+\boldsymbol{P}(t)\boldsymbol{F}^{\mathrm{T}}(\hat{\boldsymbol{X}}(t))+\boldsymbol{Q}(t) \tag{17}$$

where $\boldsymbol{Q}(t)$ is the covariance of state error, and $\boldsymbol{F}(\hat{\boldsymbol{X}}(k))$ can be achieved by Equation (18).

$$\boldsymbol{F}(\hat{\boldsymbol{X}}(k))=\frac{\partial}{\partial \boldsymbol{X}(t)}f(\boldsymbol{X}(k))\bigg|_{\boldsymbol{X}(k)=\hat{\mathrm{X}}(k)} \tag{18}$$

(4) Define the update law of estimated states as

$$\hat{\boldsymbol{X}}(k^+)=\hat{\boldsymbol{X}}(k^-)+\boldsymbol{K}(k)\left[\boldsymbol{z}(k)-\boldsymbol{h_k}\left(\hat{\boldsymbol{X}}(k^-)\right)\right] \tag{19}$$

where $\hat{\mathbf{X}}(k^+)=\left[\begin{array}{cccccc}\hat{\psi}(k^+) & \hat{\dot{\psi}}(k^+) & \hat{\alpha}(k^+) & \hat{\beta}(k^+) & \hat{\gamma}(k^+) & \hat{\delta}(k^+)\end{array}\right]^{\mathrm{T}}$ is the estimated state vector, $\hat{\mathbf{X}}(k^-)=\left[\begin{array}{cccccc}\hat{\psi}(k^-) & \hat{\dot{\psi}}(k^-) & \hat{\alpha}(k^-) & \hat{\beta}(k^-) & \hat{\gamma}(k^-) & \hat{\delta}(k^-)\end{array}\right]^{\mathrm{T}}$ is the prior state vector, $\mathbf{z}(k)=\left[\begin{array}{ccc}z_\psi(k) & z_{\dot{\psi}}(k) & z_\delta(k)\end{array}\right]^{\mathrm{T}}$ is the measured state vector, and $\boldsymbol{h_k}(\hat{\mathbf{X}}(k^-))=\left[\begin{array}{ccc}\hat{\psi}(k^-) & \hat{\dot{\psi}}(k^-) & \hat{\delta}(k^-)\end{array}\right]^{\mathrm{T}}$. Equation (19) can be rewritten as

vector, $z(k)=[z_\psi(k)\ z_{\dot{\psi}}(k)\quad z_\delta(k)]^{\mathrm{T}}$ is the measured state vector, and $\boldsymbol{h_k}(\hat{\boldsymbol{X}}(k^-))=[\hat{\psi}(k^-)\hat{\dot{\psi}}(k^-)\hat{\delta}(k^-)]^{\mathrm{T}}$. Equation (19) can be rewritten as

$$\begin{bmatrix}\hat{\psi}(k^+)\\ \hat{\dot{\psi}}(k^+)\\ \hat{\alpha}(k^+)\\ \hat{\beta}(k^+)\\ \hat{\gamma}(k^+)\\ \hat{\delta}(k^+)\end{bmatrix}=\begin{bmatrix}\hat{\psi}(k^-)\\ \hat{\dot{\psi}}(k^-)\\ \hat{\alpha}(k^-)\\ \hat{\beta}(k^-)\\ \hat{\gamma}(k^-)\\ \hat{\delta}(k^-)\end{bmatrix}+\boldsymbol{K}(k)\begin{bmatrix}z_\psi(k)-\hat{\psi}(k^-)\\ z_{\dot{\psi}}(k)-\hat{\dot{\psi}}(k^-)\\ z_\delta(k)-\hat{\delta}(k^-)\end{bmatrix} \tag{20}$$

Through the above steps, the values of $\hat{\alpha}, \hat{\beta}$ and $\hat{\gamma}$ can be estimated, and their estimated values are used to design the adaptive steering controller. However, only the Kalman gain needs to be determined.

(5) Define the update law of error covariance as

$$\boldsymbol{P}\left(k^{+}\right)=\left[1-\boldsymbol{K}(k) \boldsymbol{H}_{k}\left(\hat{\boldsymbol{X}}\left(k^{-}\right)\right)\right] \boldsymbol{P}\left(k^{-}\right) \tag{21}$$

where $\boldsymbol{P}\left(k^{+}\right)$is the priori error covariance, $\boldsymbol{P}\left(k^{-}\right)$is the posteriori error covariance, and $H_k\left(\hat{X}\left(k^{-}\right)\right.$ is defined as

$$\boldsymbol{H}_{k}\left(\hat{\boldsymbol{X}}\left(k^{-}\right)\right)=\left.\frac{\partial h(\boldsymbol{X}(k))}{\boldsymbol{X}(k)}\right|_{\boldsymbol{X}(k)=\hat{\boldsymbol{X}}\left(k^{-}\right)} \tag{22}$$

Then, the Kalman gain can be calculated by

$$\boldsymbol{K}(k)=\frac{\boldsymbol{P}\left(k^{-}\right) \boldsymbol{H}\left(\hat{\boldsymbol{X}}\left(k^{-}\right)\right)}{\boldsymbol{H}\left(\hat{\boldsymbol{X}}\left(k^{-}\right)\right) \boldsymbol{P}\left(k^{-}\right) \boldsymbol{H}\left(\hat{\boldsymbol{X}}\left(k^{-}\right)\right)^{T}+\boldsymbol{R}(k)} \tag{23}$$

where $\boldsymbol{R}(k)$ is the measured noise covariance.

Remark 1. To guarantee the convergence of the EKF identification method, the priori and posteriori error covariance $\mathbf{P}\left(k^{+}\right)$and $\mathbf{P}\left(k^{-}\right)$should be positive. The initial value of estimated state $\hat{\boldsymbol{X}}(0)$ and error covariance matrix $\boldsymbol{P}(0)$ should be set suitably.

Therefore, the adaptive steering controller based on the on-line EKF identification method can be expressed as

$$\delta=\hat{\lambda}_{1}\left(\hat{\lambda}_{2} e+\dot{e}\right) \tag{24}$$

where $\hat{\lambda}_1=\hat{T}_R /\left(\lambda \hat{K}_R\right)=1 / \lambda \hat{\gamma}$ and $\hat{\lambda}_2=1 / \hat{T}_R=-\hat{\alpha}$

Remark 2. The ship steering adaptive CGS control concept based on the on-line EKF identification method is described in Figure 4. The final form of the ship steering adaptive CGS controller is described in Equation (24). The significant parameter of the controller $\hat{T}_R$ and $\hat{K}_R$ can be estimated by the EKF on-line identification technique. The whole EKF on-line identification process is an iterative process, which includes the state prediction process (17) and (18), and the measurement state update process (22), (23), (19) and (21). More details of the EKF parameter identification technique can be found in the literature [22].

4. STABILITY ANALYSIS

To facilitate the analysis of the proposed controller, we need the following assumptions.

Assumption 1. According to the conclusions given in [23], the inequality (25) should be satisfied to guarantee the convergence of the ship steering system on-line EKF estimated parameters.

$$\varepsilon^{T}(k) \varepsilon(k) \leq \operatorname{tr}\{\operatorname{var}[\varepsilon(k)]\} \tag{25}$$

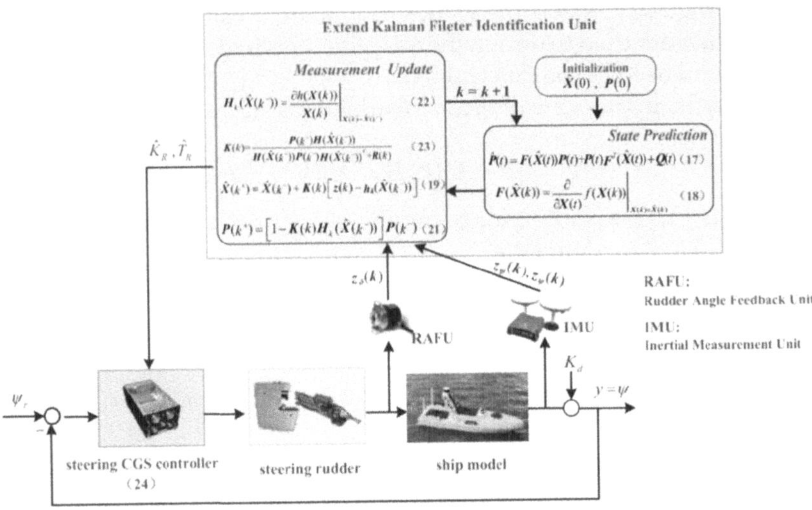

Figure 4. *The concept of ship steering adaptive CGS controller based on the EKF identification method.*

where $\varepsilon(k) = z(k) - \boldsymbol{H}\left(\hat{\boldsymbol{X}}\left(k^{-}\right)\right)$.

Assumption 2. The convergence of EKF estimated values is satisfied with $\|\inf(\frac{K_R}{\hat{K}_R})\|_\infty < \theta_K$, $\|\inf(\frac{\hat{T}_R}{T_R})\|_\infty < \theta_T$, where θ_K, and θ_T are positive constant.

Assumption 3. The interference K_d is a slow time-varying signal, bounded with $\|K_d\|_\infty \leq \rho$, where $\rho > 0$.

Assumption 4. The demand course reference signal ψ_r is a step signal, so $\dot{\psi}_r = \ddot{\psi}_r = 0$.

Hence, the mathematic model of the ship steering control system (1) can be described by Equation (26)

$$\ddot{e} = -\frac{1}{T_R}\dot{e} - \frac{a}{T_R}\dot{e}^3 - \frac{K_R}{T_R}\delta + K_d \tag{26}$$

Define the Lyapunov function (27),

$$V = \frac{1}{2}e^2 + \frac{1}{2}\dot{e}^2 \tag{27}$$

Substituting Equations (24) and (26) into the derivative of the Lyapunov function V, we can obtain Equation (28)

$$\begin{aligned}
\dot{V} &= \dot{e} + \ddot{e}\dot{e} \\
&= \dot{e} + \dot{e}\left(-\frac{1}{T_R}\dot{e} - \frac{a}{T_R}\dot{e}^3 - \frac{K_R}{T_R}\cdot\frac{1}{\hat{K}_R}\cdot\frac{1}{\lambda}\cdot e - \frac{K_R}{T_R}\cdot\frac{\hat{T}_R}{\grave{K}_R}\cdot\frac{1}{\lambda}\cdot\dot{e} + K_d\right) \\
&\leq e\dot{e} - \frac{1}{T_R}\dot{e}^2 - \frac{K_R}{T_R}\cdot\frac{1}{\hat{K}_R}\cdot\frac{1}{\lambda}\cdot e\dot{e} - \frac{K_R}{T_R}\cdot\frac{\hat{T}_R}{\bar{K}_R}\cdot\frac{1}{\lambda}\cdot\dot{e} + K_d\cdot\dot{e}
\end{aligned} \tag{28}$$

Since the inequality $K_d \cdot \dot{e} \leq \eta \dot{e}^2 + \rho^2/(4\eta)$, where $\eta > 0$, then

$$\dot{V} \leq e\dot{e} - \frac{1}{T_R}\dot{e}^2 - \frac{K_R}{T_R} \cdot \frac{1}{\hat{K}_R} \cdot \frac{1}{\lambda} \cdot e\dot{e} - \frac{K_R}{T_R} \cdot \frac{\hat{T}_R}{\hat{K}_R} \cdot \frac{1}{\lambda} \cdot \dot{e}^2 + \eta \dot{e}^2 + \rho^2/(4\eta) \tag{29}$$

According to Assumption 2, we can obtain

$$\dot{V} \leq e\dot{e} - \frac{1}{T_R}\dot{e}^2 - \frac{1}{T_R} \cdot \frac{1}{\lambda} \cdot \theta_K \cdot e\dot{e} - \frac{1}{\lambda} \cdot \theta_T \cdot \theta_K \cdot \dot{e}^2 + \eta \dot{e}^2 + \rho^2/(4\eta)$$
$$\leq -\left(\frac{1}{T_R} \cdot \frac{1}{\lambda} \cdot \theta_K - 1\right) e^2 - \left(\frac{1}{\lambda} \cdot \theta_T \cdot \theta_K + \frac{1}{T_R} + \frac{1}{T_R} \cdot \frac{1}{\lambda} \cdot \theta_K - \eta - 1\right) \dot{e}^2 + \rho^2/(4\eta) \tag{30}$$

Observing the inequality (30), we can define
$a_0 = \min\{(\frac{1}{T_R} \cdot \frac{1}{\lambda} \cdot \theta_K - 1), (\frac{1}{\lambda} \cdot \theta_T \cdot \theta_K + \frac{1}{T_R} + \frac{1}{T_R} \cdot \frac{1}{\lambda} \cdot \theta_K - \eta - 1)\}, \varepsilon_0 = \rho^2/(4\eta)$,
Taking into account the definition of the Lyapunov function (27), the inequality (31) can be obtained:

$$\dot{V} \leq -2a_0 V + \varepsilon_0 \tag{31}$$

By the integration of (31), it follows that

$$V_0(t) \leq \frac{\varepsilon_0}{2a_0} + \left(V(0) - \frac{\varepsilon_0}{2a_0}\right) \exp\left(-2a_0 t\right) \tag{32}$$

Hence, while $a_0 > 0$, we can achieve

$$\lim_{t \to \infty} \left(V(t) - \frac{\varepsilon_0}{2a_0}\right) = 0 \tag{33}$$

Therefore, based on the above analysis, we can give the main result of this paper in the theorem.

Theorem. Consider the system (1), and the control law (24) with the estimated parameters update law (19). If Assumptions 1, 2,3, and 4 are satisfied, then the proposed steering controller can suppress the external disturbance and make the whole ship steering control system semi-globally uniform and ultimately bounded, and the course tracking error can converge to a compact set $\Omega = \{e : e \leq \sqrt{\varepsilon_0/a_0}\}$ gradually.

5. SIMULATIONS AND ANALYSIS

5.1. Ship Model Description

In order to verify the feasibility of the proposed adaptive CGS ship steering controller based on the EKF identification technique, the motor vessel ***YUUU***, a special teaching and training ship belonging to Dalian Maritime University, was taken as the simulation case The value of K_R was 0.28, T_R was 71.84 , the nonlinear coefficient was 0.9 . We then obtained the transfer function of the motor vessel

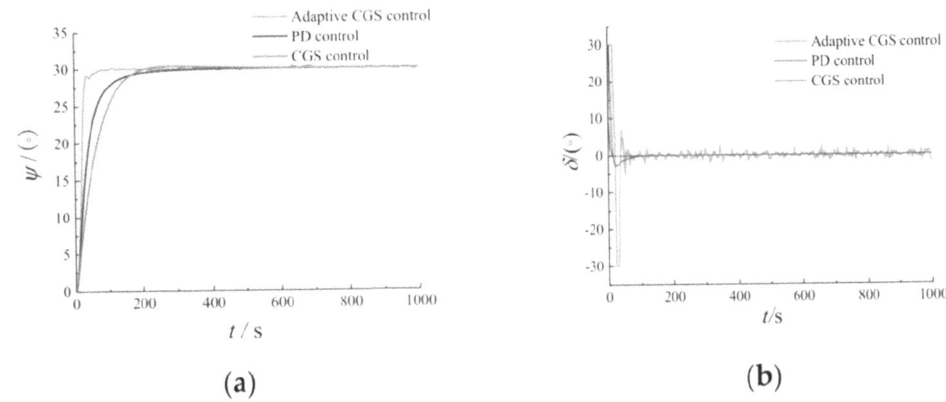

Figure 5. *Comparison of different ship steering controllers. (a) Yaw angle. (b) Steering rudder angle.*

YUKUN steering control system formulated as Equation (34) according to Equation (5).

$$G(s) = \frac{0.28}{s(71.84s + 1)} \tag{34}$$

5.2. Disturbance Description

While the ship is sailing at sea, sea waves are considered as the main disturbance of the ship steering control system. The description of sea waves can be formulated as a second-order transfer function driven by Gaussian white noise. The work frequency is 1.0, and the second-order wave transfer function can be formulated as

$$h(s) = \frac{2\xi\omega_n\sigma_w s}{s^2 + 2\xi\omega_n s + \omega_n^2} \tag{35}$$

where $\sigma_w = \sqrt{0.0185T_w} \cdot h_{1/3}$ is the wave spectrum intensity, ξ is the damping coefficient, $\omega_n = 4.85/T_w$ is the wave frequency, and T_w is the wave period. Then, the second-order wave transfer function under wind scale 6 can be written as

$$K_d(s) = \frac{0.5927s}{s^2 + 0.4157s + 0.48801} \tag{36}$$

5.3. Course-Keeping Control Comparison and Analysis

The proposed adaptive CGS ship steering controller based on the on-line EKF identification technique can achieve better course-keeping performance and corresponding steering rudder angle actuation signals, as shown in Figure 5 (the orange line). The adaptive CGS controller parameter λ was taken as 100, and the initial values of K_R and T_R were taken as 0.28 and 71.84 , respectively. In order to further verify the performance of the proposed controller, the traditional PD controller was used to carry out comparison simulation experiments.

(1) PD controller

The control strategy of the PD controller is $\delta_c = K_p e + K_d \dot{e}$. The MATLAB Simulink Control System Tuner APP was used to automatically tune the PD controller coefficients K_p and K_d from the specified tuning goals, such as rising time and overshoot. Finally, we obtained K_p 's value as 0.0984 , and K_d 's value as 7.5549 .

(2) Traditional CGS controller

The traditional CGS controller described in Equation (11) was also used to carry out comparison experiments. The parameters of the traditional CGS controller were taken as $\lambda = 100, K_R = 0.28$ and $T_R = 71.84$.

The final simulation results are shown in Figure 5. Compared with the ship steering PD controller (the blue line) and the traditional ship steering CGS controller (the red line), the proposed adaptive CGS ship steering controller (the orange line) based on the on-line EKF identification technique can achieve better course-keeping dynamic response performance. Additionally, from the EKF on-line identification results as presented in Figure 6, it can be concluded that accurate and convergent estimated values of the ship steering control system parameters α, β and γ can be observed. Furthermore, the significant parameters T_R and K_R of the proposed controller were obtained, where α is about -0.014, β is about -0.0129 , and γ is about 0.0039 . Furthermore, as presented in Figure 7, the measured and estimated ship states of course angle, course turning rate, and steering rudder angle are similar due to the EKF's accurate estimation capabilities. That means that the estimated ship states converge into the measured ship states. This can also be used to monitor the validity of the EKF on-line identification results during the proposed adaptive CGS ship steering controller's working time.

5.4. Simulation Results Analysis

In order to further evaluate the performance of the controllers, the rising time t_r, the mean absolute error of dynamic (MAED) value, the mean absolute input of dynamic (MAID) value, the mean absolute error of static (MAES) value, and the mean absolute input of static (MAIS) were defined to evaluate the controller performance accurately.

In the simulations, the first 500 s were defined as the dynamic phase of the ship steering control, where the values of MAED and MAID were used to describe the dynamic performance of the ship steering controller. The next 500 s were defined as the static phase of the ship steering control, where the values of MAES and MAIS were used to describe the static performance of the ship steering control. The value of the MAED and MAES can be calculated by Equations (37) and (38),

$$\text{MAED} = \sum_{i=1}^{n} |\psi_i| \, / n \tag{37}$$

$$\text{MAES} = \sum_{i=1}^{n} |\psi_i - \psi_r| \, / n \tag{38}$$

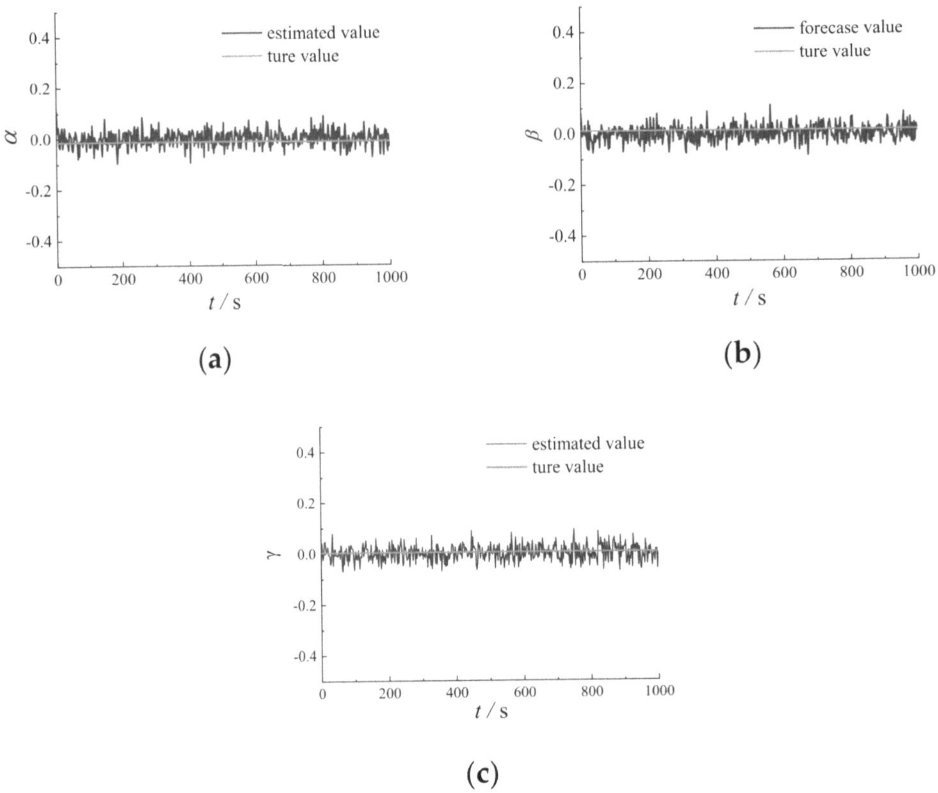

Figure 6. *The estimated parameters of the ship steering adaptive CGS controller. (a) Estimated ship parameter α. (b) Estimated ship parameter β. (c) Estimated ship parameter γ.*

where $|\psi_i|$ is the absolute value of the ship yaw angle in each sampling period. The value of dynamic and static steering input signals can be calculated by Equation (39),

$$\mathrm{MAID} = \mathrm{MAIS} = \sum_{i=1}^{n} |\delta_i| \, / n \tag{39}$$

where $|\delta_i|$ is the absolute value of the steering control rudder angle in each sampling period.

Therefore, the performance of adaptive CGS, PD, and traditional CGS controllers can be summarized in Table 1 , and the advantages and disadvantages of the different controllers can be observed:

(1) The proposed adaptive CGS controller has a faster rising time and better dynamic response performance compared with the PD controller and traditional CGS controller.

(2) In the ship steering control steady state phase, compared with the PD controller and traditional CGS controller, the proposed adaptive CGS controller has

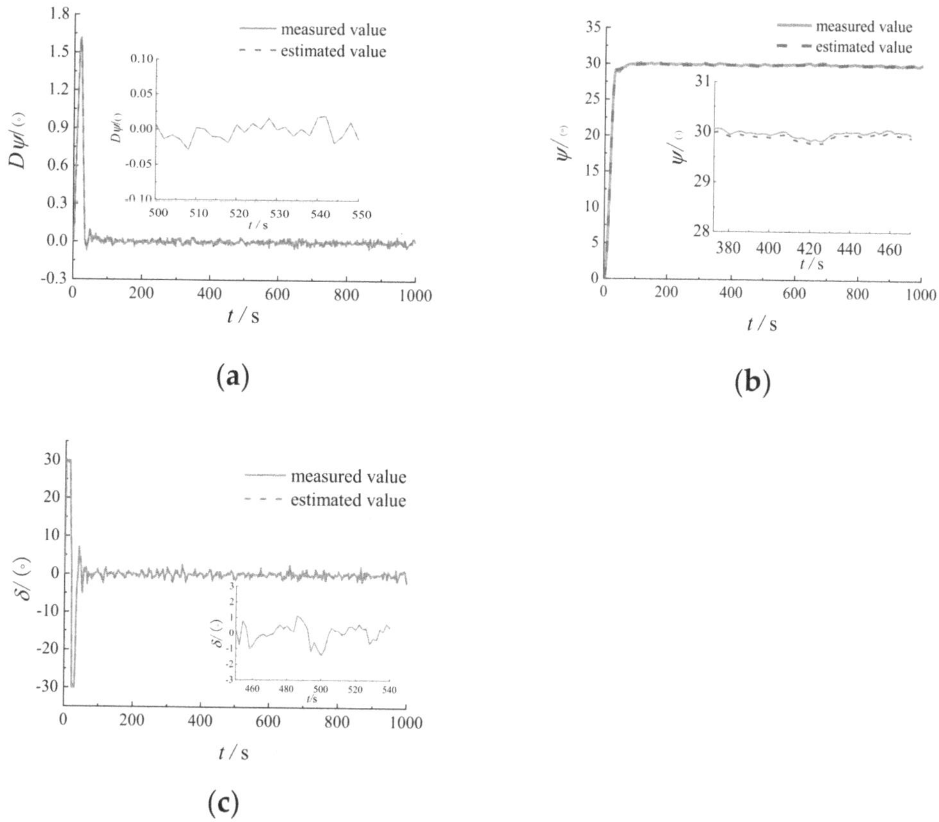

Figure 7. *The ship's measured and estimated ship states based on the EKF identification technique. (a) Measured and estimated ship state $\dot{\psi}$. (b) Measured and estimated ship state ψ. (c) Measured and estimated ship state δ.*

Table 1. *Performance evaluation of the different controllers.*

Controller	t_r	MAED	MAID	MAES	MAIS
Adaptive CGS controller	18.511	28.023	2.351	0.065	0.504
PD controller	70.648	26.766	1.079	0.015	0.017
CGS Controller	110.181	26.771	0.086	0.030	0.004

a similar coursekeeping performance. However, the violent maneuver of the proposed adaptive CGS steering controller is obvious, and further improvement is required.

5.5. Trajectory Tracking Analysis and Comparison

The performance of ship trajectory tracking is usually determined by the ship coursechanging capability of the ship steering control system. The trajectories shown in Figure 8 were registered during the ship steering trajectory tracking simulation, incorporating set course angles of 10° and 30°. The results show that the proposed adaptive CGS ship steering based on EKF on-line identification method

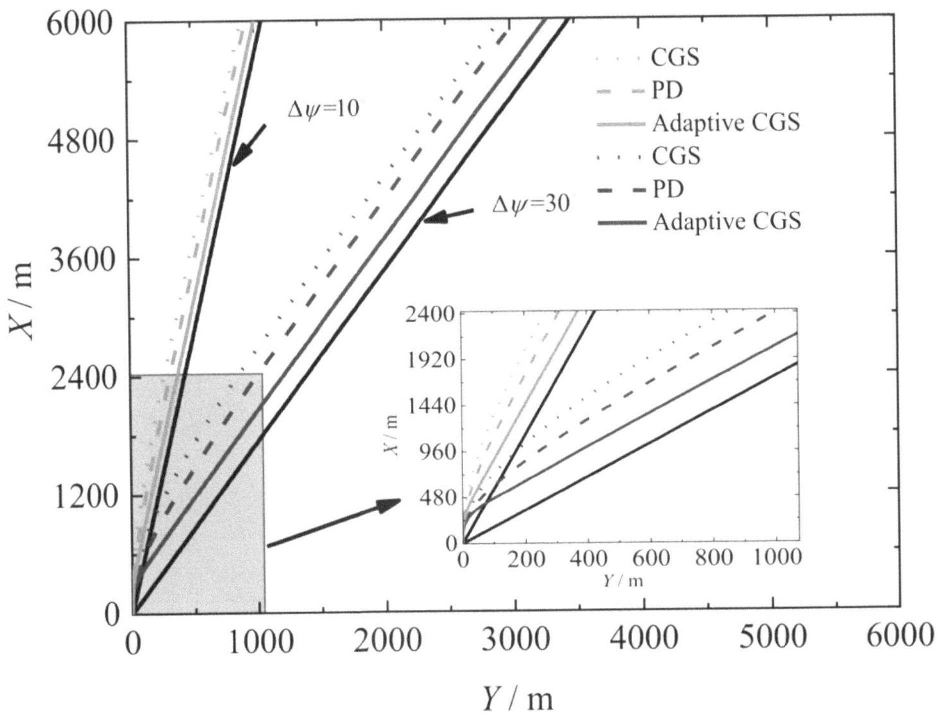

Figure 8. *Ship trajectory tracking performance for different ship steering controllers.*

had better trajectory tracking performance compared with the classical PD controller and traditional CGS controller.

6. CONCLUSIONS AND DISCUSSION

In this paper, the CGS controller design methodology combined with the online EKF identification technique was proposed to design a ship steering controller. Through the numeric simulations, it can be concluded that the dynamic response and trajectory tracking performance of the proposed ship steering adaptive CGS controller were significantly improved. This is the main contribution of this study. Furthermore, it can be inferred that the proposed steering controller design methodology might be a good solution for future practical MASS MC module design, especially as MASS call for a rapid course dynamic response maneuver under the MASS HDMC strategy.

However, due to the high sampling frequency of the EKF identification technique, the violent ship steering rudder operation and higher rudder control energy consumption are also prominent. This means that the effectiveness of the proposed ship steering adaptive CGS steering controller has great potential for improvement. Hence, the proposed adaptive CGS steering controller is the first step to applying advanced controllers to the practical MASS MC module design. The event trigger mechanics might be a potential scheme to provide a more reasonable steering operation frequency and remarkable rudder control energy savings.

Author Contributions

Conceptualization, W.G., H.P. and X.Z.; methodology, H.P.; software, H.P.; validation, W.G., H.S. and X.Z.; formal analysis, W.G., H.P., H.S.; investigation, W.G. and H.P.; writing-original draft preparation, H.P.; writing-review and editing, W.G., H.S. and X.Z.; supervision, X.Z.; project administration, W.G.; funding acquisition, W.G. All authors have read and agreed to the published version of the manuscript.

REFERENCES

1. Felski, A.; Zwolak, K. The Ocean-Going Autonomous Ship-Challenges and Threats. J. Mar. Sci. Eng. 2020, 8, 41.
2. Guan, W.; Zhou, H.; Su, Z.; Zhang, X.; Zhao, C. Ship Steering Control Based on Quantum Neural Network. Complexity 2019, 2019, 1-10.
3. Wang, L.; Wu, Q.; Liu, J.; Li, S.; Negenborn, R. State-of-the-Art Research on Motion Control of Maritime Autonomous Surface Ships. J. Mar. Sci. Eng. 2019, 7, 438.
4. Minorsky, N. Directional Stability of Automatic Steered Bodies. J. Am. Soc. Nav. Eng. 1922, 34, 280-309.
5. Skjetne, R.; Fossen, T.I.; Kokotović, P.V. Adaptive Maneuvering, with Experiments, for a Model Ship in a Marine Control Laboratory. Automatica 2005, 41, 289-298.
6. Lei, Z.; Guo, C. Disturbance Rejection Control Solution for Ship Steering System with Uncertain Time Delay. Ocean Eng. 2015, 95, 78-83.
7. Zhang, H.; Zhang, X.; Bu, R. Active Disturbance Rejection Control of Ship Course Keeping Based on Nonlinear Feedback and ZOH Component. Ocean Eng. 2021, 233, 109136.
8. Johansen, T.A.; Fuglseth, T.P.; Tøndel, P.; Fossen, T.I. Optimal Constrained Control Allocation in Marine Surface Vessels with Rudders. Control. Eng. Pract. 2008, 16, 457-464.
9. Dong, Y.; Wu, N.; Qi, J.; Chen, X.; Hua, C. Predictive Course Control and Guidance of Autonomous Unmanned Sailboat Based on Efficient Sampled Gaussian Process. J. Mar. Sci. Eng. 2021, 9, 1420.
10. Peng, X.Y.; Jia, S.L.; Hu, Z.H. Nonlinear H-infinity Inverse Optimal Output Feedback Control for Ship Course. Control Theory Appl. 2014, 31, 215-222.
11. Perera, L.P.; Guedes Soares, C. Pre-filtered Sliding Mode Control for Nonlinear Ship Steering Associated with Disturbances. Ocean Eng. 2012, 51, 49-62.
12. Tzeng, C.Y.; Goodwin, G.C.; Crisafulli, S. Feedback Linearization Design of a Ship Steering Autopilot with Saturating and Slew Rate Limiting Actuator. Int. J. Adapt. Control Signal Process 1999, 13, 23-30.
13. Fossen, T.I.; Strand, J.P. Tutorial on Nonlinear Backstepping: Applications to ship control. Model. Identif. Control A Nor. Res. Bull. 1999, 20, 83-135.
14. Zhang, X.K.; Yang, G.P.; Zhang, Q.; Zhang, G.Q.; Zhang, Y.Q. Improved Concise Backstepping Control of Course Keeping for Ships Using Nonlinear Feedback Technique. J. Navig. 2017, 70, 1401-1414.
15. Zhang, X.K.; Han, X.; Guan, W.; Zhang, G.Q. Improvement of Integrator Backstepping Control for Ships with Concise Robust Control and Nonlinear Decoration. Ocean Eng. 2019, 189, 106349.1-106349.7.

16. Le, T.T. Ship heading control system using neural network. J. Mar. Sci. Technol. 2021, 26, 963-972.
17. Borkowski, P. Inference Engine in an Intelligent Ship Course-Keeping System. Comput. Intell. Neurosci. 2017, 2017, 2561383.
18. Rigatos, G.; Tzafestas, S. Adaptive Fuzzy Control for the Ship Steering Problem. Mechatronics 2006, 16, 479-489.
19. Zhang, Z.; Zhang, X.; Zhang, G. ANFIS-based Course-Keeping Control for Ships Using Nonlinear Feedback Technique. J. Mar. Sci. Technol. 2018, 24, 1326-1333.
20. Qin, H.; Tan, P.; Chen, Z.; Sun, M.; Sun, Q. Deep Reinforcement Learning Based Active Disturbance Rejection Control for Ship Course Control. Neurocomputing 2021. Available online: https://www.sciencedirect.com/science/article/abs/pii/S0925231221015812 (accessed on 1 January 2022).
21. Guan, W.; Cao, W.; Sun, J.; Su, Z. Steering Controller Design for Smart Autonomous Surface Vessel Based on CSF L2 Gain Robust Strategy. IEEE Access 2019, 7, 109982-109989.
22. Perera, L.P.; Oliveira, P.; Guedes Soares, C. System Identification of Nonlinear Vessel Steering. J. Offshore Mech. Arct. Eng. 2015, 137, 031302.
23. Gelb, A. Applied Optimal Estimation; The MIT Press: Cambridge, MA, USA, 1974.
24. Raman-Nair, W.; Gash, R. Least Squares Identification of Linear Sway-Yaw Manoeuvring Coefficients and Drag-Area Parameters of Ships. Proc. Inst. Mech. Eng. Part M J. Eng. Marit. Environ. 2021, 235, 809-815.
25. Xie, S.; Chu, X.; Liu, C.; Liu, J.; Mou, J. Parameter Identification of Ship Motion Model Based on Multi-Innovation Methods. J. Mar. Sci. Technol. 2019, 25, 162-184.
26. Jiang, Y.; Wang, X.G.; Zou, Z.J.; Yang, Z.L. Identification of Coupled Response Models for Ship Steering and Roll Motion Using Support Vector Machines. Appl. Ocean Res. 2021, 110, 102607.

CHAPTER

4

Model-Free High Order Sliding Mode Control with Finite-Time Tracking for Unmanned Underwater Vehicles

Josué González-García[1], Néstor Alejandro Narcizo-Nuci[1], Luis Govinda García-Valdovinos[2], Tomás Salgado-Jiménez[2], Alfonso Gómez-Espinosa[1], Enrique Cuan-Urquizo[1] and Jesús Arturo Escobedo Cabello[1]

[1]Tecnologico de Monterrey, Escuela de Ingenieria y Ciencias, Av. Epigmenio González 500, Fracc. San Pablo, Querétaro 76130, Mexico
[2]Energy Division, Center for Engineering and Industrial Development-CIDESI, Santiago de Queretaro, Queretaro 76125, Mexico

ABSTRACT

Several strategies to deal with the trajectory tracking problem of Unmanned Underwater Vehicles are encountered, from traditional controllers such as Proportional Integral Derivative (PID) or Lyapunov-based, to backstepping, sliding mode, and neural network approaches. However, most of them are model-based controllers where it is imperative to have an accurate knowledge of the vehicle hydrodynamic parameters. Despite some sliding mode and neural network-based controllers are reported as model-free, just a few of them consider a solution with finite-time convergence, which brings strong robustness and fast convergence compared with asymptotic or exponential solutions and it can also help to reduce the power consumption of the vehicle thrusters. This work aims to implement a model-free high-order sliding-mode controller and synthesize it with a time-base generator to achieve finite-time convergence. The time-base was included by parametrizing the control gain at the sliding surface. Numerical simulations validated the finite-time convergence of the controller for different time-bases even in the presence of high ocean currents. The performance of the obtained solution was also evaluated by the Root Mean Square (RMS) value of the control coefficients computed for the thrusters, as a parameter to measure the power consumption of the vehicle when following a trajectory. Computational results showed a reduction of up to 50% in the power consumption from the thrusters when compared with other solutions.

Keywords: *unmanned underwater vehicles; trajectory tracking; sliding Mode Control; finitetime convergence*

1. INTRODUCTION

Unmanned Underwater Vehicles (UUVs) are used in multiple tasks such as ship's hull inspection [1], ecological surveys [2,3], underwater inspection and mapping [4], among others. The use of UUVs avoid risks for human divers and reduce operational costs. In addition, they can perform some tasks that cannot be done by humans in an underwater environment. The motion of these vehicles can be remotely operated (ROVs) or autonomous (AUVs). For ROVs, there is a human pilot that commands the vehicle's movements and actions, usually through a tether. Human presence makes complex exploration and intervention tasks possible since humans can react to changes in the mission plan caused by the unpredictable nature of the underwater environments. However, certain operations such as high precision navigation require some form of autonomy from the vehicle. That is a complex task to achieve and one of the main problems for AUVs [5] since their six Degree of Freedom (DOF) dynamics is highly non-linear and, therefore, difficult to control [6]. Motion control of a UUV has different goals such as waypoint tracking, path following, and trajectory tracking [7]. For waypoint tracking, the controller should guide the vehicle from a starting position to the desired destination throughout a set of pre-defined waypoints. For path following, the aim is that the vehicle accurately follows the desired path in terms of a geometric description while the trajectory tracking problem requires the vehicle to converge to a desired time-parametrized trajectory. For trajectory tracking, convergence can be either asymptotic, exponential, or in finite-time.

Several control strategies have been applied in the past to achieve high precision trajectory tracking. The Proportional Integral Derivative (PID) control is one of the most employed control strategies for UUVs due to their simple structure. They work well for systems where it is possible to control all outputs through the inputs-meaning they are not underactuated-and are widely used for heading, depth, and surge control [8]. However, one limitation for PID controllers is that they are tuned to deal with specific conditions and, if those conditions change, the performance will be affected. In addition, PID controllers do not consider nonlinearities, which will eventually deteriorate the controller performance. Some other strategies are used with the PID controller to compensate some of the uncertainties faced in the underwater environment. Dong et al. [9] designed a fuzzy rules-based strategy to tune a PID depth controller. Experiments were carried out to test the controller, showing a depth control accuracy of up to 3 cm. Yang et al. [10] adopted a cascade PID structure to deal with horizontal displacements in the heading control. Experimental results showed better performance when compared with a traditional PID controller since oscillations were reduced and even eliminated when controlling the heading of the ROV. Zhang et al. [11] used Particle Swarm Optimization (PSO) to optimize the parameters of a PID controller for horizontal displacements. Simulation results showed better performance than a PID controller, but weaker compared with a fuzzy controller. Hernández-Alvarado et al. [12] proposed a Neural

Network (NN) based self-tuning PID controller where the NN automatically estimates a suitable set of gains for the PID controller to achieve system stability. The controller was validated with numerical simulations and experimentation, showing better performance when subject to unknown disturbances and over 3% of energy-saving against a traditional PID controller.

Backstepping Control (BC) is a model-based nonlinear controller that is commonly used for trajectory tracking of UUVs. It is based on the Lyapunov theory [13] and works by sectioning the dynamic system of the vehicle into subsystems, so the controller begins with a known stable system and then iterates to obtain proper controllers for the rest of the subsystems. Yan et al. [14] applied backstepping control to the trajectory tracking of an AUV to a docking station. Simulations of their controller showed that all tracking errors converge to a small neighborhood of the origin. Zhou et al. [15] proposed a solution for underactuated AUVs tracking using a combination of a Neural Network (NN) algorithm and a backstepping sliding mode controller. Simulations showed a smooth tracking of the desired three-dimensional trajectory. Yan et al. [13] proposed a backstepping sliding mode controller with fuzzy switching gain. Simulation results showed that the controller designed can guarantee that the UUV accurately track the desired path in the presence of time-varying interference and effectively eliminate the chattering in the traditional Sliding Mode Control (SMC) method.

SMC is another typical approach for trajectory tracking of UUVs. It has been extensively applied due to its simple form, robustness, and tolerance to model uncertainty. However, SMC has the problem of chattering. For UUV control, the chattering causes high-frequency changes in the speed of the thrusters, which leads to high energy consumption, increase friction, and thruster damage [16]. Ramezani-al et al. [17] proposed an SMC with adaptative gain. In their controller, a continuous term is used instead of a discrete traditional sign function, a large gain for the sliding mode is applied to satisfy the reaching condition, as well as a small gain for the sliding mode to avoid the chattering phenomena. Simulation results showed the controller proposed can decrease the amplitude of chattering in the control signal and increase the convergence rate to the desired trajectory. Lv et al. [18] proposed a fault-tolerant control method integrated with thrust allocation based on the sliding mode theory. Experimental results showed a reduction of up to 94% of the steady error compared with a conventional SMC. García-Valdovinos et al. [19] proposed a 2nd order SMC combined with a backpropagation NN control scheme for UUVs to deal with external disturbances and parameter variations. Experimental results showed that the controller was able to reduce the position error to zero in a robust way while the online adjustments of the weights in the NN compensated for the variations in the vehicle hydrodynamics and perturbations.

Despite the numerous approaches for trajectory tracking of UUVs encountered, just a few of them have considered a finite-time convergence. Finite-time control is an effective control strategy with strong robustness and fast convergence. It has many advantages such as faster convergence time, higher accuracy, and better anti-disturbance capability [20]. Yan et al. [21] adopted the globally finite-time tracking control strategy, along with a PID-SMC model-based controller, to track a predefined trajectory for a UUV. Simulations showed that the control laws can achieve strongly robust and preferably control performance for the horizontal trajectory tracking control in the presence of parameter perturbation and unknown currents. Their work was extended to three-dimensional trajectory tracking by Yu et al. [22]. Simulation results show that the vehicle can converge to the trajectory and remain in a small neighborhood even in the presence of ocean currents, wherein a small fluctuation in the tracking errors is observed. Chu et al. [23] proposed a local recurrent NN for finite-time trajectory tracking of ROVs with an unknown dynamical model. The authors designed a state observer based on sliding-mode to estimate those variables that cannot be measured by the ROV and achieved finite-time convergence with the adaptative controller proposed. Qiao et al. [24] proposed two control schemes based on sliding mode to the tracking control of UUVs in presence of uncertainties and disturbances: an adaptive integral terminal sliding mode control and an adaptive fast integral terminal sliding mode control. Each controller is composed of a kinematic and a dynamic controller, where the kinematic can achieve local finite-time convergence of the tracking error to zero. Guerrero et al. [25] proposed an adaptative high order SMC with finite-time convergence for an AUV. The authors validated the proposed controller with real experiments for depth and yaw trajectories, considering parametric uncertainties and external disturbances.

Finite-time tracking controllers have been widely and successfully used to control fully actuated robot arms [26,27], but little research has been done for underactuated unmanned underwater vehicles (UUVs), whose particular nonlinear dynamics have other challenges, for instance, the hydrodynamics effects inherent to the vehicle's geometry and water density. Considering the advantages of a controller with finite-time convergence, and that most of the controllers designed to achieve that for trajectory tracking in UUVs navigation are model-based 1st order SMC, this paper presents the formulation and validation of a model-free 2 nd order SMC based on a Time Base Generator (TBG) that guarantees finite-time three-dimension al trajectory tracking of underactuated UUVs subject to ocean currents. The resulting control law does neither depend on the hydrodynamics nor other parameters of the vehicle; therefore, it is considered as a model-free controller. In addition, formulation of the control law of this 2nd order SMC eliminates the chattering effect which is common in conventional 1st order SMC [28,29]. To achieve finite-time convergence of the tracking errors of three-dimensional trajectories, the 2nd order SMC is parametrized by a TBG which allows the user to set an arbitrary base-time in which the vehicle will meet the desired trajectory. The performance of the proposed controller is evaluated by numerical simulations including the effects of external disturbances. All this is structured in this paper as: Section 2 describes the BlueROV2 UUV and its dynamic model, as well as the proposed control scheme including the TBG. Section 3 describes the method and parameters used to validate the proposed controller. Section 4 shows the results of the testing and validation, and Section 5 contains the concluding remarks.

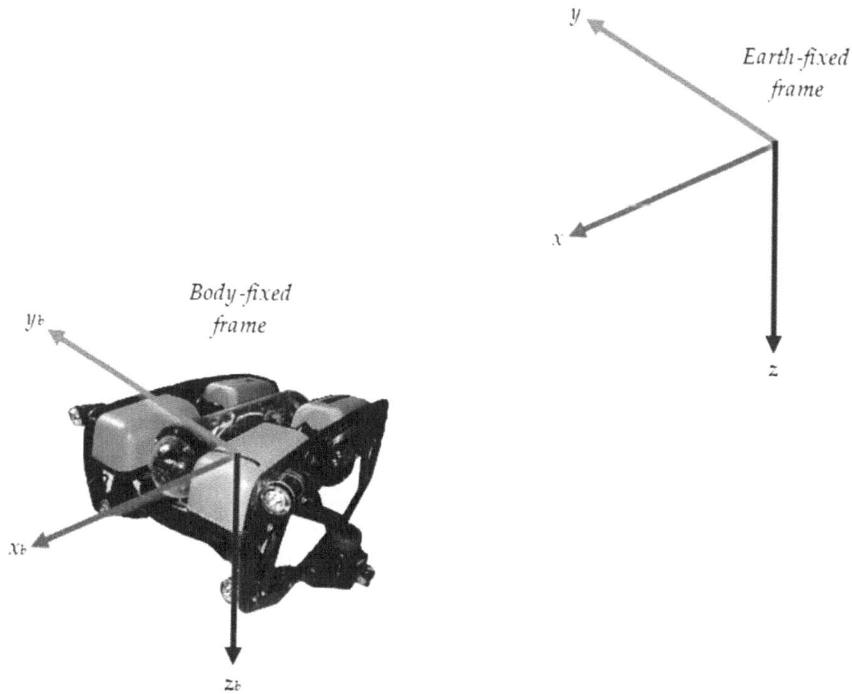

Figure 1. *Reference frames needed to describe the BlueROV2.*

2. MATERIALS AND METHODS

2.1. Unmanned Underwater Vehicle Model

2.1.1. Kinematic Model

As described by Fossen [30], two reference frames are needed to build the mathematical model that represents the underwater vehicle: one referenced to the earth (labeled as Earth-fixed frame) and the other referenced to the vehicle (labeled as body-fixed frame). The orthonormal axes are denoted by x, y, z for the Earth-fixed frame, and by x_b, y_b, z_b for the body-fixed frame. These reference frames are shown in Figure 1.

The Society of Naval Architects and Marine Engineers (SNAME) has a convention to express the position and orientation of vehicles, and the forces and moments applied to them. The SNAME nomenclature for position, velocity, and forces in underwater vehicles [30] is presented in Table 1.

The position of the vehicle with respect to the Earth-fixed frame η and its velocity with respect the body-fixed frame v can be represented as

$$\eta = (x, y, z, \varphi, \theta, \psi)^T, \tag{1}$$

$$v = (u, v, w, p, q, r)^T, \tag{2}$$

Table 1. *Notation for underwater vehicle's movements, positions, velocities, and forces.*

Movement	Name	Position	Velocity	Force/Moment
X translation	Surge	x	u	X
Y translation	Sway	y	v	Y
Z translation	Heave	z	w	Z
X rotation	Roll	φ	p	K
Y rotation	Pitch	θ	q	M
Z rotation	Yaw	ψ	r	N

The forces and moments of the vehicle with respect to the body-fixed frame are described as

$$\tau = (X, Y, Z, K, M, N)^T, \tag{3}$$

2.1.2. Hydrodynamic Model

The hydrodynamic model of a UUV can be described by Newton-Euler equations [30] as

$$Miv + C(v)v + D(v)v + g(\eta) = \tau + \omega, \tag{4}$$

$$\tau = B_t u_t, \tag{5}$$

where
$M \in \mathbb{R}^{6\times6}$ is the inertial and added mass matrix,
$C \in \mathbb{R}^{6\times6}$ is the rigid body and added mass centripetal and Coriolis matrix,
$D \in \mathbb{R}^{6\times6}$ is the hydrodynamic damping matrix,
$g \in \mathbb{R}^{6\times1}$ is the restitution forces vector,
$B_t \in \mathbb{R}^{6\times6}$ is the thruster allocation matrix,
$u_t \in \mathbb{R}^{6\times1}$ is a vector containing the force generated by the thrusters, and
$\omega \in \mathbb{R}^{6\times1}$ represents environmental disturbances.

2.1.3. BlueROV2 Model

The vehicle considered for the simulations in this work was the BlueROV2 from Blue Robotics ®. It has the six-thruster vectored configuration that can be seen in Figure 2.

Matrices and vectors involved in the vehicle's hydrodynamics contain numerous unknown parameters, making it infeasible to estimate. Nevertheless, there are some assumptions and considerations that can be made because of the BlueROV2 geometry and operation, resulting in a considerable reduction of unknown parameters. These conditions are listed below:

- Since BlueROV2 operates at relatively low speeds (i.e., less than 2 m/s), lift forces can be neglected.
- BlueROV2 is assumed to have port-starboard symmetry and fore-aft symmetry, and the center of gravity (CG) is in the symmetry planes.

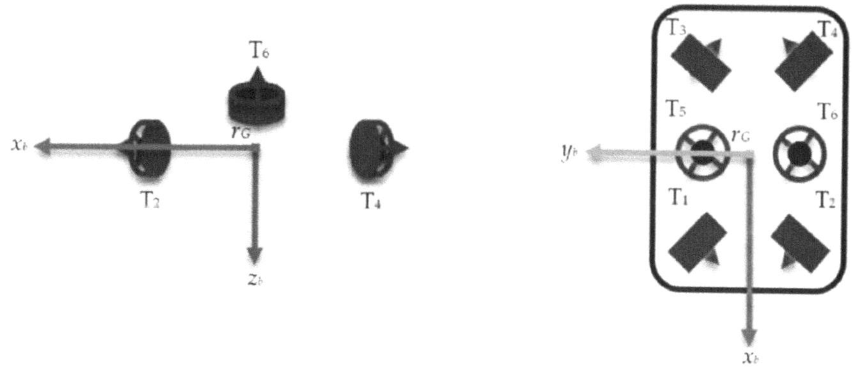

(a) (b)

Figure 2. *BlueROV2 thruster configuration. (a) Right view. (b) Top view.*

- BlueROV2 is assumed to operate below the wave-affected zone. As a result, disturbances of waves ω on the vehicle are negligible.
- The thruster allocation of BlueROV2 does not permit active control of the pitch orientation q. However, the motion around this axis is considered self-regulated due to the vehicle buoyant restoring moments, resulting in the following reduced system

$$v = (u, v, w, p, 0, r)^T \tag{6}$$

Considering the reduced system detailed before, the physical and hydrodynamic parameters of BlueROV2 are summarized in Table 2 as identified in [31-33].

The matrices that comprise the hydrodynamical model in Equation (4) are considered as defined in [30].

Considering the thruster distribution in the BlueROV2, the thruster allocation matrix is assembled as

$$B_t = \begin{bmatrix} 0.7071 & 0.7071 & -0.7071 & -0.7071 & 0 & 0 \\ -0.7071 & 0.7071 & -0.7071 & 0.7071 & 0 & 0 \\ 0 & 0 & 0 & 0 & -1 & -1 \\ 0 & 0 & 0 & 0 & 0.115 & -0.115 \\ 0 & 0 & 0 & 0 & 0 & 0 \\ -0.1773 & 0.1773 & -0.1773 & 0.1773 & 0 & 0 \end{bmatrix} \tag{7}$$

2.2. Model-Free High-Order SMC with Finite-Time Convergence

The dynamics of the UUV (Equation (4)) can be expressed in the Earth-fixed frame applying the following kinematic transformations [30]

$$\dot{\eta} = J(\eta_2)\, v \leftrightarrow v = J^{-1}(\eta_2)\, \dot{\eta} \tag{8}$$

$$\ddot{\eta} = J(\eta_2)\, \dot{v} + \dot{J}(\eta_2)\, v \leftrightarrow \dot{v} = J^{-1}(\eta_2)\left[\ddot{\eta} - J(\eta_2)\, v\right], \tag{9}$$

Table 2. *BlueROV2 parameters.*

Parameters	Symbol	Value	Unit
Mass	m	10	kg
Buoyancy	B	100.06	N
Weight	W	98.1	N
Center of gravity	$r_G = (x_g, y_g, z_g)$	$r_G = (0,0,0)$	m
Center of buoyancy	$r_B = (x_b, y_b, z_b)$	$r_B = (0,0,0.02)$	m
Inertia moment	I = diagonal (I_x, I_y, I_z)	I = diagonal (0.16, 0.16, 0.16)	kg·m^2
Added mass parameters	$X_{\dot{u}}$	−5.5	kg
	$Y_{\dot{v}}$	−12.7	kg
	$Z_{\dot{w}}$	−14.57	kg
	$K_{\dot{p}}$	−0.12	kg·$\frac{m^2}{rad}$
	$M_{\dot{q}}$	−0.12	kg·$\frac{m^2}{rad}$
	$N_{\dot{r}}$	−0.12	kg·$\frac{m^2}{rad}$
Linear Damping parameters	X_u	−4.03	N·$\frac{s}{m}$
	Y_v	−6.22	N·$\frac{s}{m}$
	Z_w	−5.18	N·$\frac{s}{m}$
	K_p	−0.07	N·$\frac{s}{rad}$
	M_q	−0.07	N·$\frac{s}{rad}$
	N_r	−0.07	N·$\frac{s}{rad}$
Quadratic Damping parameters	$X_{u\|u\|}$	−18.18	N·$\frac{s^2}{m^2}$
	$Y_{v\|v\|}$	−21.66	N·$\frac{s^2}{m^2}$
	$Z_{w\|w\|}$	−36.99	N·$\frac{s^2}{m^2}$
	$K_{p\|p\|}$	−1.55	N·$\frac{s^2}{rad^2}$
	$M_{q\|q\|}$	−1.55	N·$\frac{s^2}{rad^2}$
	$N_{r\|r\|}$	−1.55	N·$\frac{s^2}{rad^2}$

where the transformation matrix $J(\eta_2)$ is defined as

$$J(\eta_2) = \begin{bmatrix} J_1(\eta_2) & 0_{3\times3} \\ 0_{3\times3} & J_2(\eta_2) \end{bmatrix}, \tag{10}$$

with

$$J_1(\eta_2) = \begin{bmatrix} c_\psi c_\theta & -s_\psi c_\phi + s_\phi s_\theta c_\psi & s_\psi s_\phi + s_\theta c_\psi c_\phi \\ s_\psi c_\theta & c_\psi c_\phi + s_\phi s_\theta s_\psi & -c_\psi s_\phi + s_\theta s_\psi c_\phi \\ -s_\theta & s_\phi c_\theta & c_\phi c_\theta \end{bmatrix}, J_s(\eta_2) = \begin{bmatrix} 1 & s_\phi t_\theta & c_\phi t_\theta \\ 0 & c_\phi & -s_\phi \\ 0 & s_\phi / c_\theta & c_\phi / c_\theta \end{bmatrix}, \text{ and } \eta_2 = [\phi, \theta, \psi]^T,$$

where $c_{\text{angle}} = \cos(\text{ angle }), s_{\text{angle}} = \sin(\text{ angle })$, and $t_{\text{angle}} = \tan(\text{ angle })$.
After some mathematical manipulation, the resulting expression is

$$M_\eta(\eta)\ddot{\eta} + C_\eta(v,\eta)\dot{\eta} + D_\eta(v,\eta)\dot{\eta} + g_\eta(\eta) = \tau_\eta, \tag{11}$$

where

$$\begin{aligned} M_\eta(\eta) &= J^{-T}(\eta)MJ^{-1}(\eta), \\ C_\eta(v,\eta) &= J^{-T}(\eta)\left[C(v) - MJ^{-1}(\eta)\dot{J}(\eta)\right]J^{-1}(\eta), \\ D_\eta(v,\eta) &= J^{-T}(\eta)D(v)J^{-1}(\eta), \\ g_\eta(\eta) &= J^{-T}(\eta)g(\eta) \\ \tau_\eta(\eta) &= J^{-T}(\eta)\tau \end{aligned} \tag{12}$$

García-Valdovinos et al. [26,27] established that Equation (10) is linearly parametrizable by the product of a regressor $Y(\eta,\dot{\eta},\ddot{\eta}) \in \mathbb{R}^{n\times p}$ composed of known nonlinear functions and a vector $\theta \in \mathbb{R}^p$ containing constant parameters. Then, this parametrization can be rewritten in terms of a nominal reference $\dot{\eta}_r$ and its time derivative $\ddot{\eta}_r$ as follows

$$M_\eta(\eta)\ddot{\eta}_r + C_\eta(v,\eta)\dot{\eta}_r + D_\eta(v,\eta)\dot{\eta}_r + g_\eta(\eta) = Y\left(\eta,\dot{\eta},\dot{\eta}_r,\ddot{\eta}_r\right)\theta, \tag{13}$$

Subtracting Equation (12) from both sides of Equation (10) leads to the open-loop error dynamics expression

$$M_\eta(\eta)\dot{S}_r + C_\eta(v,\eta)S_r + D_\eta(v,\eta)S_r = \tau_\eta - Y\left(\eta,\dot{\eta},\dot{\eta}_r,\ddot{\eta}_r\right)\theta, \tag{14}$$

where $S_r = \dot{\eta} - \dot{\eta}_r$ is known as the extended error.
Now, consider the following nominal reference

$$\dot{\eta}_r = \dot{\eta}_d - \alpha\tilde{\eta} + S_d - K_i\int_0^t \text{sign}\left(S_\eta\right)d\sigma, \tag{15}$$

where $\tilde{\eta} = \eta - \eta_d$ represents the tracking error of the position, η_d is the desired trajectory, α and K_i are diagonal positive definite $n\times n$ gain matrices, sign(x) stands for the signum function of vector x and

$$S = \dot{\tilde{\eta}} + \alpha\tilde{\eta}, \tag{16}$$

$$S_d = S\left(t_0\right)e^{-\kappa t}, \tag{17}$$

$$S_\eta = S - S_d, \tag{18}$$

with $\kappa > 0$. So that the extended error S_r can be rewritten as follows

$$S_r = S_\eta + K_i\int_0^t \text{sign}\left(S_\eta\right)d\sigma \tag{19}$$

The result is the model-free high-order SMC with the following control law

$$\tau_\eta = -K_d S_r \tag{20}$$

Notice that the so-called model-free control law does not depend on any parameter, matrix, or vector describing the hydrodynamics of the vehicle, it is defined by a diagonal definite positive $n \times n$ gain matrix K_d, and the extended error S_r. Furthermore, since the signum function is not used directly but through its integral, the chattering effect common in conventional SMC is not present. A stability analysis of the controller just described can be found in [34] where the existence of the sliding mode was proven.

A TBG will be used to achieve finite-time convergence. A TBG is a scalar time function that provides a smooth transition from 0 to 1 , with a duration controllable by a time-base t_b. Its derivative provides a bell-shaped speed profile. The TBG implemented in this work was introduced by Parra-Vega [26] and García-Valdovinos [27] for finite-time convergence of a robotic arm. Consider the fifth-order polynomial

$$\xi(t) = 10\frac{(t-t_0)^3}{(t_b-t_0)^3} - 15\frac{(t-t_0)^4}{(t_b-t_0)^4} + 6\frac{(t-t_0)^5}{(t_b-t_0)^5} \tag{21}$$

and its time derivative

$$\dot{\xi}(t) = 30\frac{(t-t_0)^2}{(t_b-t_0)^3} - 60\frac{(t-t_0)^3}{(t_b-t_0)^4} + 30\frac{(t-t_0)^4}{(t_b-t_0)^5} \tag{22}$$

with t_0 representing the initial time and subjected to the following conditions: $\xi(t_0) = \dot{\xi}(t_0) = \dot{\xi}(t_b) = 0$. The time-base t_b can be chosen arbitrarily and does not depend on the initial condition.

Now, $\xi(t)$ will be used to parameterize a time-varying gain $\alpha(t)$ as follows

$$\alpha(t) = \begin{cases} \alpha_0 \frac{\dot{\xi}}{(1-\xi)+\delta}, & 0 \le t \le t_b \\ \alpha_c, & t > t_b \end{cases} \tag{23}$$

with $\alpha_0 = 1+\varepsilon, 0 < \varepsilon \ll 1, 0 < \delta \ll 1$, and $\alpha_c > 0$.

To achieve finite-time convergence in the trajectory tracking of the BlueROV2, the constant gain α in Equation (15) is replaced with the time-varying gain $\alpha(t)$ given by Equation (22). Then, the control law described in Equation (19) is applied. The control scheme is represented in the block diagram depicted in Figure 3.

To obtain the forces needed in the thrusters, τ_η needs to be converted from the Earthfixed frame to the body-fixed frame as follows

$$\tau = J^{-1}(\eta_2)\,\tau_\eta, \tag{24}$$

Then, the control coefficient vector u for the thrusters is computed as

$$u = B_t^{-1} K_T^{-1} \tau \tag{25}$$

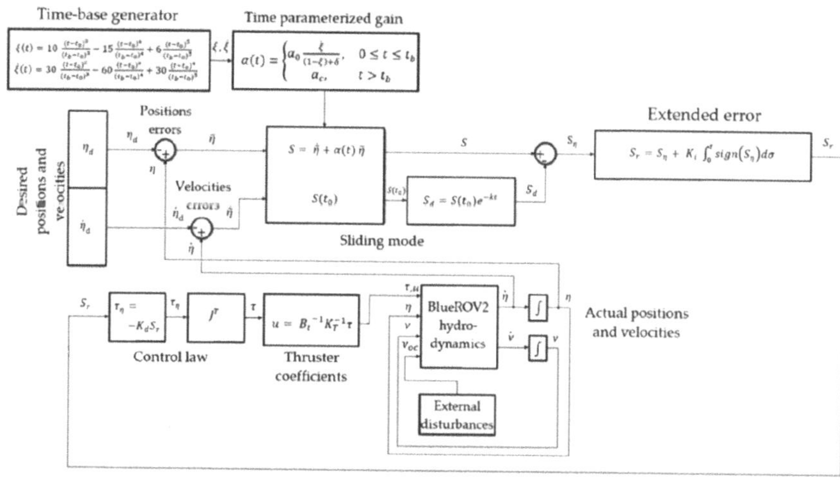

Figure 3. *Model-free high-order Sliding Mode Control (SMC) with Time Base Generator (TBG) scheme.*

where K_T is a diagonal matrix containing the maximum force (N) each thruster can produce. Then, the control coefficient vector u is limited to a range of $-1 \leq u \leq 1$ to avoid thruster saturation and to keep the control signal introduced to the model in a range that can be achievable by the thrusters.

3. VALIDATION METHOD FOR THE CONTROLLER

To test and validate the proposed control scheme the hydrodynamics of the BlueROV2as presented in Section 2.1-was programmed in a simulator developed in MATLAB / Simulink®. This simulator also considers some external disturbances such as ocean currents.

3.1. Parameterized Trajectory of the UUV

A spiral trajectory was selected to be able to control all DOFs of the BlueROV2 (x, y, z, φ, ψ) while moving simultaneously at surge, sway, heave, and yaw. The desired positions η_d and velocities $\dot{\eta}_d$ are given by the following equations

$$\eta_d = (x_d, y_d, z_d, \varphi_d, \psi_d) \tag{26}$$

where

$$x_d = x_i + R\cos(\omega t),$$
$$y_d = y_i + R\sin(\omega t),$$
$$z_d = z_i + \left(z_f - z_i\right)\frac{t}{t_s},$$
$$\varphi_d = 0, \text{ and}$$
$$\psi_d = \psi_i + \omega t,$$

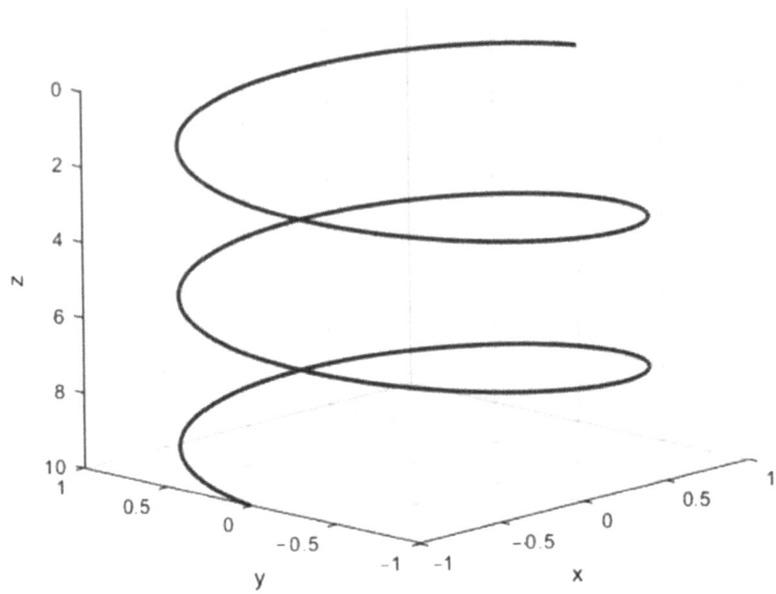

Figure 4. *Parameterized desired spiral trajectory of the Unmanned Underwater Vehicle (UUV) BlueROV2.*

$$\dot{\eta}_d = (\dot{x}_d, \dot{y}_d, \dot{z}_d, \dot{\varphi}_d, \dot{\psi}_d), \tag{27}$$

where

$$\begin{aligned} \dot{x}_d &= -\omega R \sin(\omega t), \\ \dot{y}_d &= \omega R \cos(\omega t), \\ \dot{z}_d &= \frac{(z_f - z_i)}{t_s}, \\ \dot{\varphi}_d &= 0, \text{ and} \\ \dot{\psi}_d &= \omega, \end{aligned}$$

where R is the radius for the spiral trajectory, ω is a constant, (x_i, y_i) is the center of the circumference of the spiral, z_i is the initial depth for the trajectory, z_f is the final depth of the trajectory, ψ_i is the initial yaw, and t_s is the simulation time.

The desired spiral trajectory with $R = 1$ meter, $\omega = \frac{\pi}{6}\frac{\text{rad}}{\text{s}}$, $(x_i, y_i) = (0,0)$, $z_i = 0$, $z_f = 10$ meters, and a simulation time $t_s = 30$ s is plotted in Figure 4.

3.2. Considering Ocean Currents as Disturbances

The forces of the ocean currents can be considered constant and irrotational according to Fossen [30]; and can be implemented using relative velocity, which is the difference between the real velocity and the ocean current velocity

$$v_{rel} = v - v_{oc}, \tag{28}$$

where v is the vehicle velocity and v_{oc} is the ocean current velocity.

The generalized vector for an irrotational ocean current velocity is given by

$$v_{oc} = \left[u_{oc}, v_{oc}, w_{oc}, 0, 0, 0\right]^T, \tag{29}$$

where u_{oc} is an ocean current velocity from the north, v_{oc} is the ocean current velocity from the east, and w_{oc} is the ocean current velocity from below.

Defining the angles α_{oc} as the angle of attack, and β_{oc} as the slide slip angle, every element of the velocity vector can be calculated as

$$\begin{aligned} u_{oc} &= v_{oc} \cos \alpha_{oc} \cos \beta_{oc}, \\ v_{oc} &= v_{oc} \sin \beta_{oc}, \text{ and} \\ w_{oc} &= v_{oc} \sin \alpha_{oc} \cos \beta_{oc}, \end{aligned} \tag{30}$$

Since the velocity vector is expressed in the Earth-Fixed frame, Equation (4) can be modified resulting in

$$M\dot{v}_{rel} + C\left(v_{rel}\right) v_{rel} + D\left(v_{rel}\right) v_{rel} + g(\eta) = \tau \tag{31}$$

Zhang et al. [35] performed some experiments where high ocean currents affecting a UUV were measured. Taking them into account and for purposes of the simulations presented in this work, the following ocean currents will be included as constants

$$\begin{aligned} u_{oc} &= 0.75\frac{\text{m}}{\text{s}}, \\ v_{oc} &= 0.25\frac{\text{m}}{\text{s}}, \text{ and} \\ w_{oc} &= 0.25\frac{\text{m}}{\text{s}}, \end{aligned} \tag{32}$$

3.3. Comparing with Classic Controllers

Other classic control approaches such as PID, Feed-Back Linearization (FBL), and Lyapunov-based controllers [30], [36] will be simulated for the trajectory tracking of the BlueROV2 to compare the performance of the proposed model-free 2nd order SMC.

3.3.1. PID Controller

The control signal used for the PID controller is given by

$$\tau_\eta = -K_p\tilde{\eta} - K_i \int \tilde{\eta} dt - K_d\dot{\tilde{\eta}} \tag{33}$$

where K_p, K_i, and K_d are the proportional, integral, and derivative gain matrices, respectively. An initial set of gain matrices was proposed and then several simulations were performed aiming to find the one that produces the best results for the trajectory tracking of all DOFs. As a result, the gain matrices for this controller are defined as

$$\begin{aligned} K_p &= \text{diag}[140, 140, 140, 140, 0, 140], \\ K_i &= \text{diag}[120, 120, 120, 120, 0, 120], \text{ and} \\ K_d &= \text{diag}[180, 180, 180, 180, 0, 180], \end{aligned} \tag{34}$$

3.3.2. Feed-Back Linearization Controller

The control signal used for the FBL controller is given by:

$$\tau = Ma^b + C(v)v + D(v)v + g(\eta) \tag{35}$$

where a^b is the commanded acceleration. Here, a PID controller will be used as an auxiliary controller to compute a^b as

$$a^b = \dot{v}_d - K_p\tilde{v} - K_i \int \tilde{v}dt - K_d\dot{\tilde{v}} \tag{36}$$

where

$$\tilde{v} = v - v_d \tag{37}$$

K_p, K_i, and K_d will take the same values as in Equation (33).

3.3.3. Lyapunov-Based Controller

The control signal used for the Lyapunov-based controller is given by:

$$\tau = M\dot{v}_r + Cv_r + Dv_r + g(\eta) - K_dS \tag{38}$$

where

$$S = v - v_r \tag{39}$$

$$v_r = J^{-1}\dot{\eta}_{r'} \tag{40}$$

$$\dot{v}_r = J^{-1}\left[\ddot{\eta}_r - \dot{J}v_r\right], \text{ and} \tag{41}$$

$$\dot{\eta}_r = \dot{\eta}_d - \alpha\tilde{\eta} \tag{42}$$

where K and α are gain matrices which, after several iterations to find a suitable performance of the controller, will take the following values

$$\begin{aligned} K_d &= \text{diag}[80, 80, 80, 80, 0, 80], \text{ and} \\ \alpha &= [0.80, 0.80, 0.80, 0.80, 0, 0.80], \end{aligned} \tag{43}$$

The values for the gain matrices used in the simulations of this work are summarized in Table 3.

Table 3. *Gain matrices for classic controllers and the proposed model-free 2nd order SMC with finite-time convergence.*

Controller	K_p	K_i	K_d	α	κ
PID	140	120	180	-	-
Feed-back linearization	140	120	180	-	-
Lyapunov-based	-	-	80	0.80	-
Model-free 2nd order SMC with finite-time convergence	-	5	800	$\alpha(t)$	5

Table 4. *Controller parameters for simulation of a desired spiral trajectory.*

Parameter	Value	Parameter	Value
t_0	0	δ	0.001
t_b	5	κ	5
α_0	1.01	K_i	$diag[0, 5, 5, 5, 5, 5]$
α_c	20	K_d	$diag[800, 800, 800, 800, 0, 800]$

4. RESULTS AND DISCUSSION

A simulation was performed so that a computational model of the BlueROV2 follow the spiral trajectory described before, the parameters and gains for the proposed model-free 2nd order SMC with finite-time convergence are shown in Table 4. The initial position for the vehicle was set to $x = 2$ meters, $y = -1$ meter, and $z = 2$ meters so it started slightly away from the desired trajectory. The effects of ocean currents were neglected for this simulation.

Results for trajectory tracking and errors of the proposed controller are presented in Figure 5. The controlled trajectory smoothly converged with the desired trajectory at the time-base proposed and without overshoots, after that, the vehicle followed the desired trajectory without deviations. The errors decreased over time to become zero at the designated t_b except for the pitch, which, as established in Section 2.1, was not actuated and the resulting increase in its error-smaller than $0.5°$ - was caused by the vehicle's dynamics while moving forward.

A three-dimensional trajectory of the BlueROV2 simulation along with the control trajectory can be seen in Figure 6.

The control coefficients for each of the BlueROV2 thrusters are shown in Figure 7. These control coefficients were in a range of $-1 \leq u \leq 1$, where $|u| = 1$ represents that the controller is demanding 100% of the thruster force. The maximum force was demanded at the start of the simulation and, after reaching the time-base, the control signals remained constant and no chattering was observed.

Since t_b can be tuned arbitrarily for the user, further simulations were performed for the proposed controller under the same gains and parameters but for different time-bases. Simulation results for trajectory tracking and errors are shown in Figures 8 and 9 for a time-base of 3 and 7 s, respectively. In both cases, the vehicle met the desired trajectory at the designated time-base t_b.

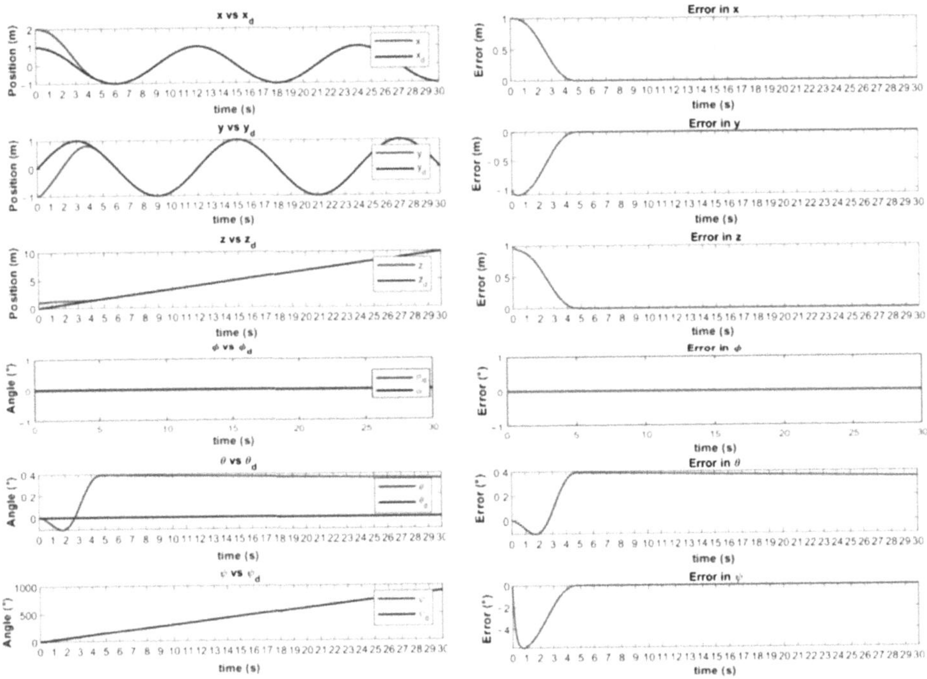

(a) (b)

Figure 5. *Results with* $t_b = 5$ s. *(a) Trajectory tracking. (b) Position tracking errors.*

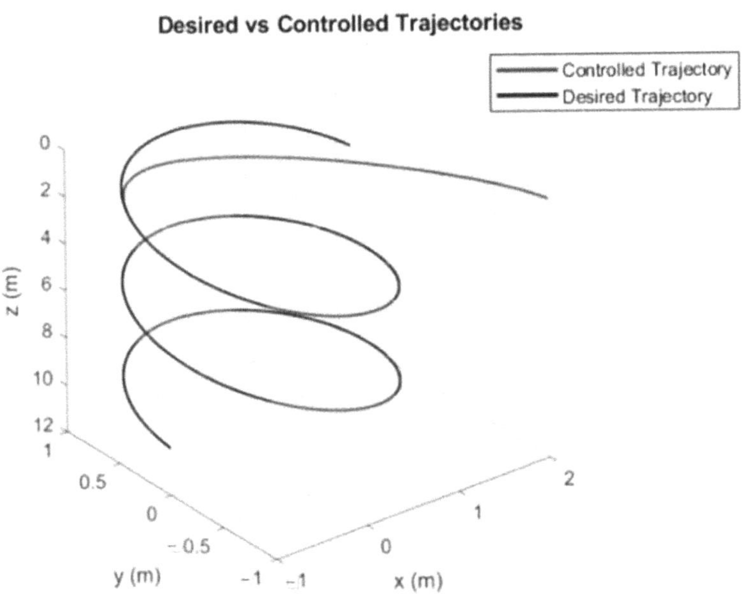

Figure 6. *Simulated three-dimensional trajectory of the BlueROV2 with* $t_b = 5$ s.

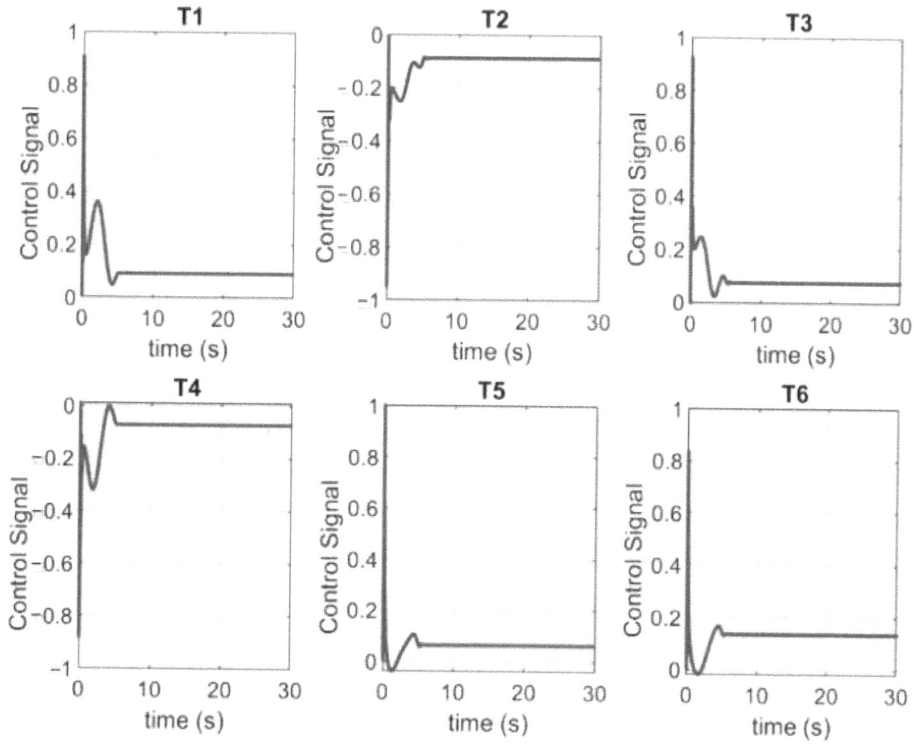

Figure 7. *Control signal for each thruster. Simulation with* $t_b = 5$ s.

The controllers presented and described in Section 3.3 were simulated for comparing their error convergence to zero and the control coefficient computed for the BlueROV2 thrusters. As can be seen in Figure 10, the surge and sway error of the proposed controller converges to zero at the desired time. Meanwhile, the rest of the controllers converged at different times and, for the PID and FBL, there was not a stationary error but an oscillation in its value was observed through the simulated time. For the heave error, despite the oscillations resulting in the PID and the FBL, all the controllers converged to zero. Finally, for the yaw orientation, the FBL exhibited some troubles converging to zero during the first 20 s to finally converge to it exponentially. In general, the performance of the proposed model-free 2nd order SMC with finite-time convergence surpassed the rest of the controllers, even when FBL and Lyapunov-based were model-based.

To evaluate the performance of the proposed controller in terms of energy consumption from the thrusters, the Root Mean Square (RMS) value of the control coefficients computed for the thrusters was obtained and results are shown in Table 5. The mean column of the table represents the mean value of the coefficients considering all six thrusters. As it can be seen in the table, the average thruster force demanded by the proposed modelfree 2nd order SMC was in the range of 0.1004 and 0.1296 − 1.0 being the maximum-this was up to 50% less than the forces

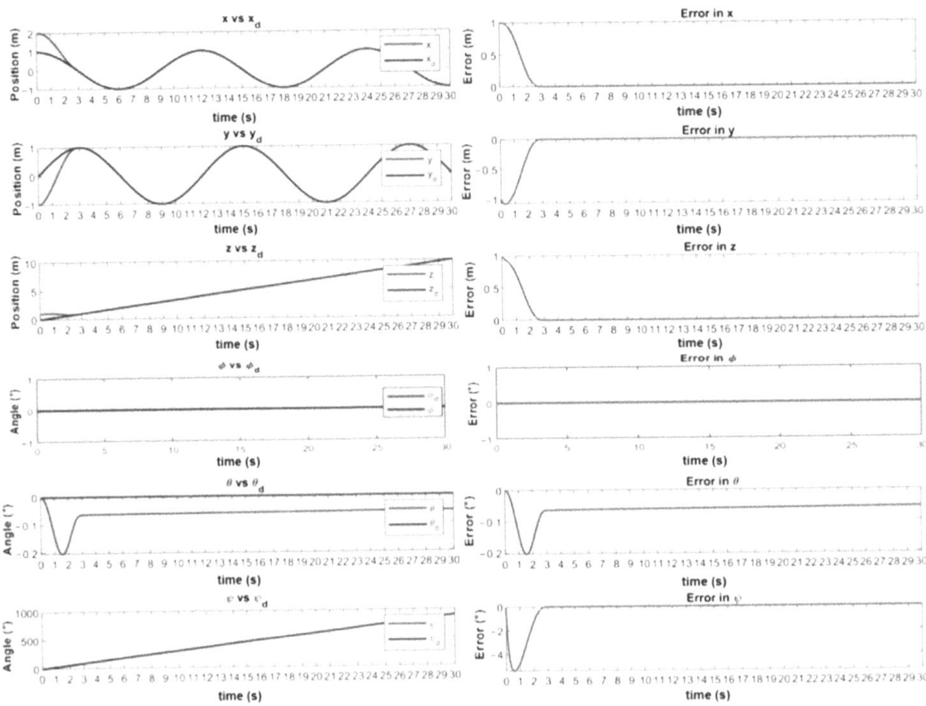

Figure 8. *Results with $t_b = 3$ s. (a) Trajectory tracking. (b) Position tracking errors.*

Table 5. *Root Mean Square (RMS) values for thrusters control coefficients.*

Controller	Thruster 1	Thruster 2	Thruster 3	Thruster 4	Thruster 5	Thruster 6	Mean
PID	0.2078	0.1946	0.2010	0.2105	0.1311	0.2041	0.1915
Feed-back linearization	0.1676	0.1876	0.1909	0.1693	0.2851	0.2760	0.2127
Lyapunov-based	0.1278	0.1500	0.1380	0.1192	0.0678	0.1358	0.1231
Model-free 2nd Order SMC with $t_b = 3$ s	0.1449	0.1400	0.1333	0.1378	0.0801	0.1418	0.1296
Model-free 2nd Order SMC with $t_b = 5$ s	0.1264	0.1175	0.0986	0.1059	0.0740	0.1324	0.1091
Model-free 2nd Order SMC with $t_b = 7$ s	0.1132	0.1067	0.0880	0.0921	0.0726	0.1295	0.1004

demanded by a traditional PID or FBL controller even when it converged faster. The Lyapunov-based controller converged to the trajectory approximately at 7 s, and its mean value for control coefficient of the thrusters was 23% bigger than the 2 nd Order SMC with $t_b = 7$ s.

To test the controller response to external disturbances, the effects of ocean currents were included as described in Section 3.2. Results for trajectory tracking and errors of a simulation with the proposed model-free 2 nd order SMC with $t_b = 5$ s are shown in Figure 11. Despite the high ocean currents introduced, the

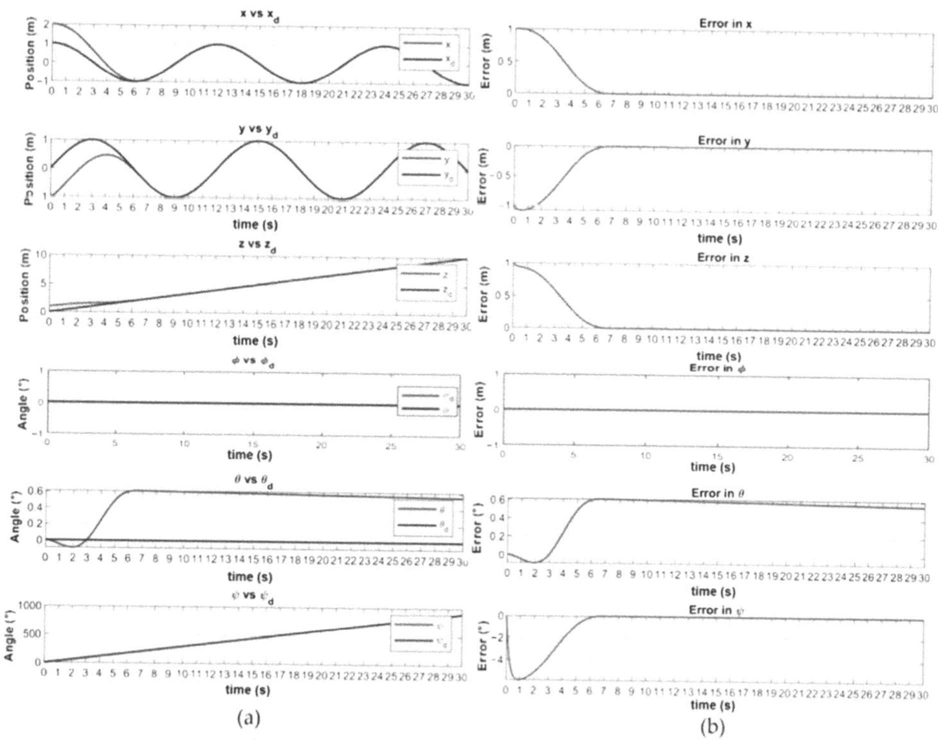

Figure 9. *Results with* $t_b = 7$ s. *(a) Trajectory tracking. (b) Position tracking errors.*

BlueROV2 still smoothly converged to the desired trajectory at the designated t_b, there were not overshoots at all, and the vehicle followed the designed trajectory with an error of zero after the time-base was reached.

A three-dimensional trajectory of the BlueROV2 simulation along with the control trajectory when subject to ocean currents can be seen in Figure 12.

Control signals for each of the BlueROV2 thrusters are shown in Figure 13. Contrary to the simulation results shown in Figure 7 , the control signal did not remain constant after the time-base was reached. Thruster control signals changed over time to overcome the perturbations introduced by the ocean currents, which allowed the BlueROV2 to follow the desired trajectory as designed.

Ocean currents were also included in simulations for the rest of the controllers considered in the past section. Results for tracking errors are presented in Figure 14. None of the classic controllers tested were able to overcome the perturbations introduced, while the proposed model-free 2 nd order SMC was able to compensate for them and keeping the finite-time convergence as designated by the user.

To evaluate the performance of the proposed controller in terms of energy consumption from the thrusters, the Root Mean Square (RMS) value of the control coefficients computed for the thrusters was obtained when the vehicle was subject to high ocean currents. Results are shown in Table 6.

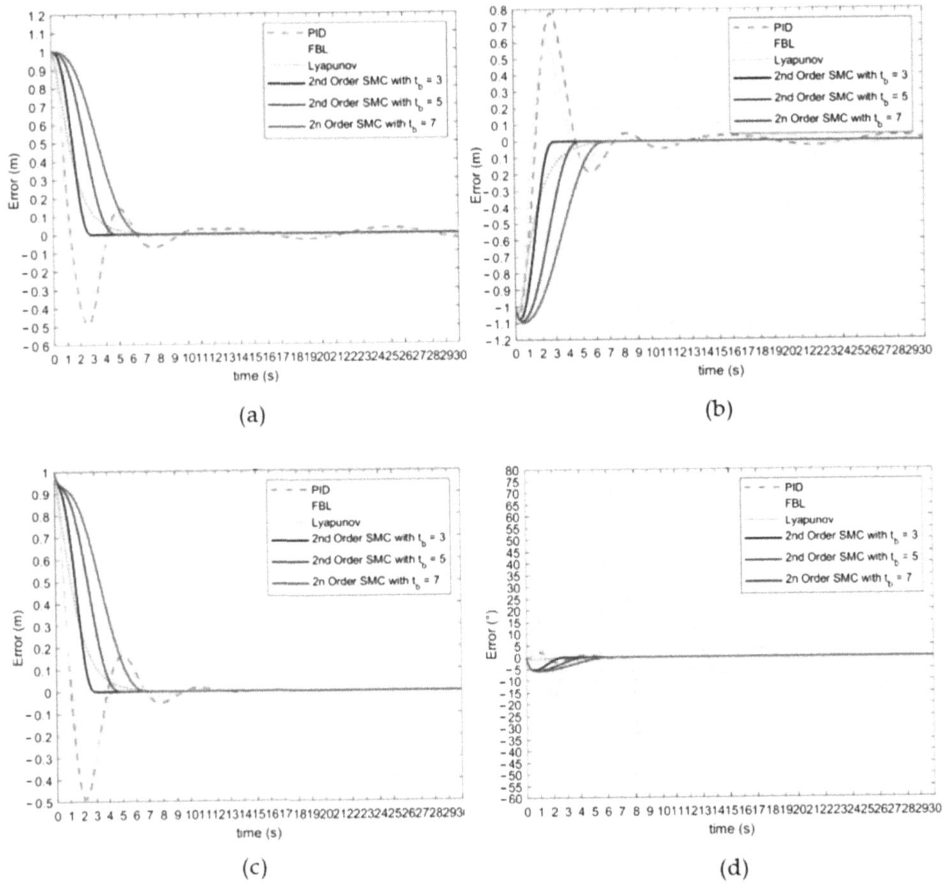

Figure 10. *Error convergence for different controllers. (a) Surge error. (b) Sway error. (c) Heave error. (d) Yaw error.*

Table 6. *RMS values for thrusters control coefficients when the vehicle is subject to high ocean currents.*

Controller	Thruster 1	Thruster 2	Thruster 3	Thruster 4	Thruster 5	Thruster 6	Mean
PID	0.3254	0.3131	0.3534	0.3163	0.1135	0.1761	0.2663
Feed-back linearization	0.3978	0.4008	0.4056	0.3892	0.7889	0.7749	0.5262
Lyapunov-based	0.27321	0.2479	0.2607	0.2306	0.0578	0.0843	0.1924
Model-free 2nd Order SMC with $tb = 3$ s	0.2790	0.2684	0.3093	0.2646	0.0578	0.0840	0.2105
Model-free 2nd Order SMC with $tb = 5$ s	0.2909	0.2670	0.2801	0.2535	0.0500	0.0556	0.1995
Model-free 2nd Order SMC with $tb = 7$ s	0.2897	0.2525	0.2625	0.2421	0.0479	0.0469	0.1903

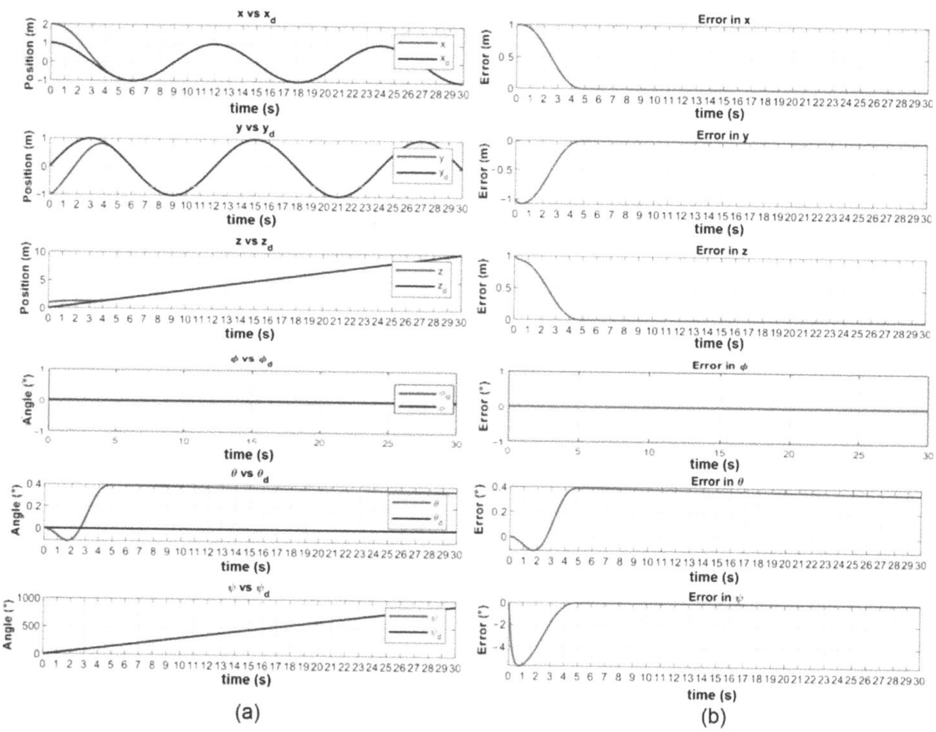

Figure 11. *Results for the finite-time convergence controller subject to ocean currents and* $t_b = 5$ s. *(a) Trajectory tracking. (b) Position tracking errors.*

The average thruster force demanded by the model-free 2nd order SMC with finitetime convergence was in a range of 0.1903 and 0.2105 , which resulted in up to 30% less than the average force demanded by the PID controller and up to 64% less than the demanded by the FBL controller, even when they were not able to meet the trajectory and maintain the vehicle there.

5. CONCLUSIONS

Finite-time trajectory tracking of underactuated UUVs is a considerable challenge. Their nonlinear dynamics and the challenging conditions of the underwater environment make it more complex for the controller to achieve a stable and robust performance as they do in other systems and environments. In this work, for the first time, a model-free 2nd order SMC and a TBG were synthesized and implemented to achieve finite-time convergence of an underactuated nonlinear UUV to a desired trajectory. Such a controller has some advantages compared with other finite-time model-based 1st order SMC reported in literature since it neither requires the dynamic model structure nor the knowledge of any hydrodynamical parameter from the vehicle. In addition, this controller does not present the chattering effect common in other conventional SMC. The TBG was included in the controller by the parametrization of the α gain at the sliding surface, making

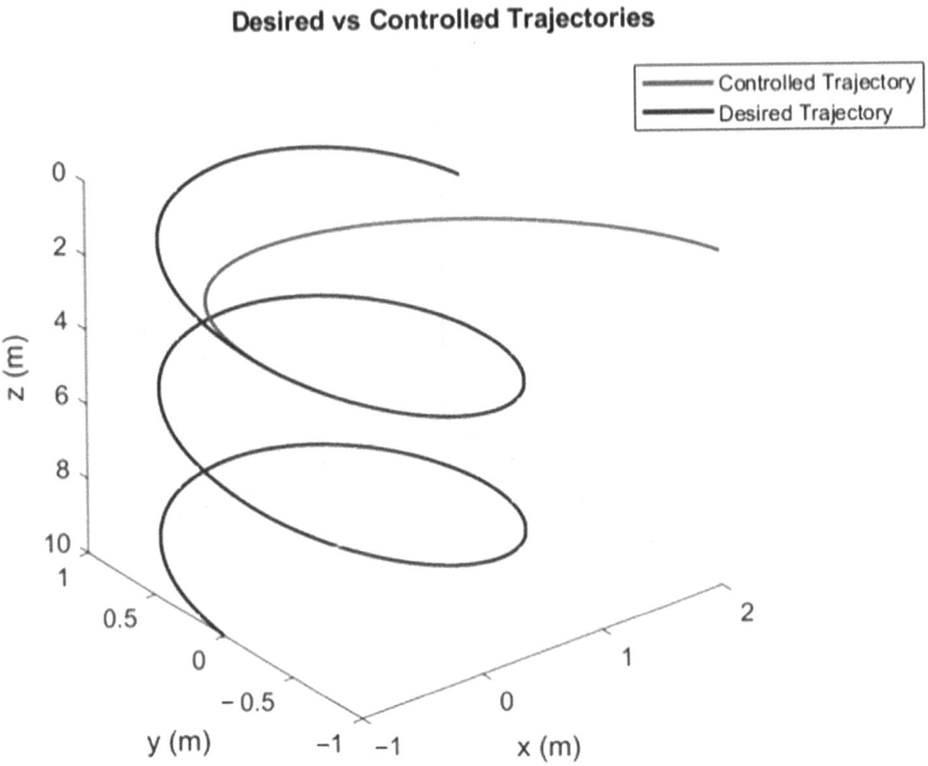

Figure 12. *Simulated three-dimensional trajectory of the BlueROV2 when subject to ocean currents and* $t_b = 5$ s.

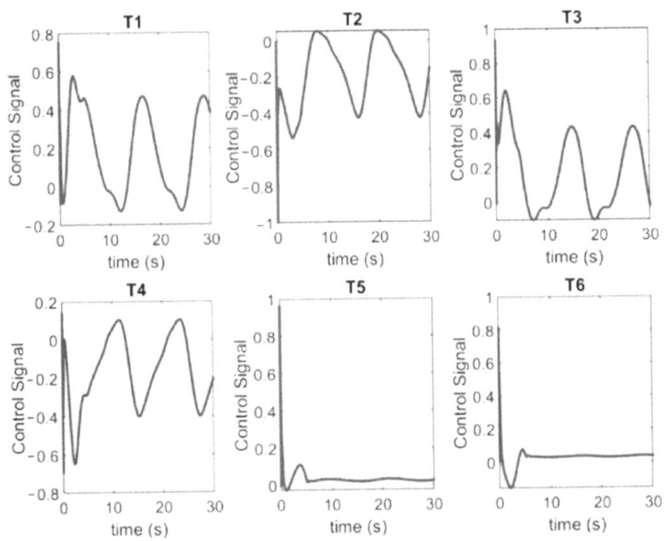

Figure 13. *Control signal for each thruster when subject to ocean currents and* $t_b = 5$ s.

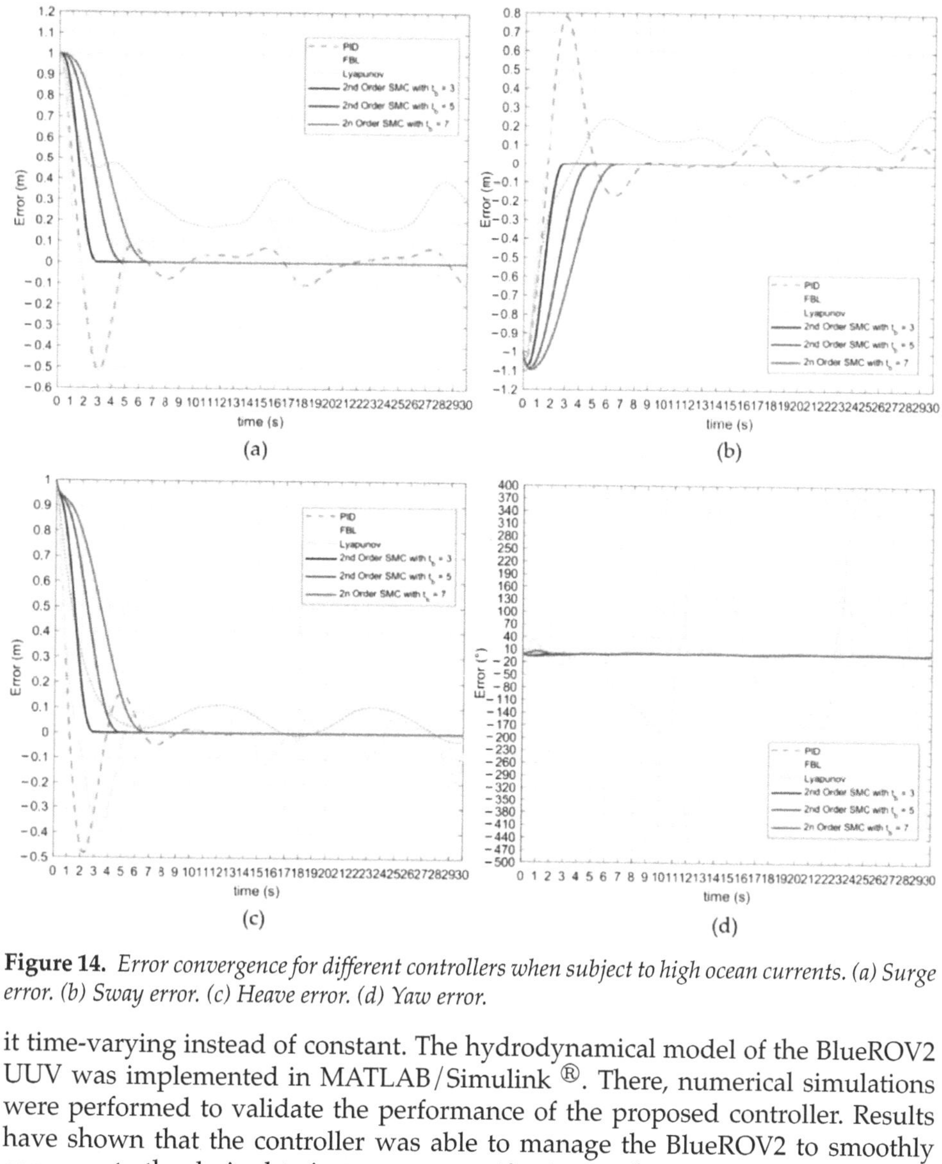

Figure 14. *Error convergence for different controllers when subject to high ocean currents. (a) Surge error. (b) Sway error. (c) Heave error. (d) Yaw error.*

it time-varying instead of constant. The hydrodynamical model of the BlueROV2 UUV was implemented in MATLAB/Simulink ®. There, numerical simulations were performed to validate the performance of the proposed controller. Results have shown that the controller was able to manage the BlueROV2 to smoothly converge to the desired trajectory at a specific time without overshoots, and then follow it with a tracking error of zero. Other classic controllers such as PID and FBL could not achieve a similar performance. Since the time-base can be tuned arbitrarily for the user, simulations were repeated with different time-bases and the results remained consistent and satisfactory. The performance of the solution proposed was also evaluated in terms of the energy consumption regarding the thruster forces demanded by the different controllers. It was shown that the proposed model-free 2nd order SMC demands up to 50% less power from the thrusters than other of the controllers simulated and it does not present the chattering effect common in some SMC controllers. Finally, the vehicle was subject to high ocean currents. Simulation results showed that the controller stills able to meet the desired trajectory at the designated time, smoothly and without overshoots, and then maintain the vehicle in the desired trajectory when other classic controllers could not. In future work, authors expect to verify the finite-time convergence of the model-free 2nd order SMC by performing experimentation under controlled conditions with the BlueROV2 vehicle whose hydrodynamical model was considered for the simulations presented in this work. Further research will also include the obstacle avoidance problem during the trajectory tracking.

Author Contributions

Conceptualization, L.G.G.-V., T.S.-J., and A.G.-E.; methodology, L.G.G.-V., and T.S.-J.; investigation, J.G.-G., N.A.N.-N., and L.G.G.-V.; software, J.G.-G., and N.A.N.-N.; writingoriginal draft preparation, J.G.-G.; visualization, J.G.-G., and N.A.N.-N.; writing-review and editing, L.G.G.-V., T.S.-J., A.G.-E., J.G.-G., N.A.N.-N., E.C.-U., and J.A.E.C.; supervision, L.G.G.-V., T.S.-J., and A.G.-E.; All authors have read and agreed to the published version of the manuscript.

Acknowledgments

The authors would like to acknowledge support from CONACyT for PhD studies of first author and MSc studies of second author.

REFERENCES

1. Hartono, M.T.; Budiyanto, M.A.; Dhelika, R. Micro class underwater ROV (remotely operated vehicle) as a ship hull inspector: Development of an initial prototype. AIP Conf. Proc. 2020, 2227, 020025.
2. Buscher, E.; Mathews, D.L.; Bryce, C.; Bryce, K.; Joseph, D.; Ban, N.C. Applying a Low Cost, Mini Remotely Operated Vehicle (ROV) to Assess an Ecological Baseline of an Indigenous Seascape in Canada. Front. Mar. Sci. 2020, 7, 1-12.
3. Sward, D.; Monk, J.; Barrett, N. A systematic review of remotely operated vehicle surveys for visually assessing fish assemblages. Front. Mar. Sci. 2019, 6, 1-19.
4. Sandøy, S.S.; Hegde, J.; Schjølberg, I.; Utne, I.B. Polar Map: A Digital Representation of Closed Structures for Underwater Robotic Inspection. Aquac. Eng. 2020, 89, 102039.
5. González-García, J.; Gómez-Espinosa, A.; Cuan-Urquizo, E.; García-Valdovinos, L.G.; Salgado-Jiménez, T.; Cabello, J.A.E. Autonomous Underwater Vehicles: Localization, Navigation, and Communication for Collaborative Missions. Appl. Sci. 2020, 10, 1256.
6. Kumar, N.; Rani, M. An efficient hybrid approach for trajectory tracking control of autonomous underwater vehicles. Appl. Ocean Res. 2020, 95, 102053.
7. Elmokadem, T.; Zribi, M.; Youcef-Toumi, K. Terminal sliding mode control for the trajectory tracking of underactuated Autonomous Underwater Vehicles. Ocean Eng. 2017, 129, 613-625.
8. Rojas, J.; Baatar, G.; Cuellar, F.; Eichhorn, M.; Glotzbach, T. Modelling and Essential Control of an Oceanographic Monitoring Remotely Operated Underwater Vehicle. IFAC-PapersOnLine 2018, 51, 213-219.

9. Dong, M.; Li, J.; Chou, W. Depth control of ROV in nuclear power plant based on fuzzy PID and dynamics compensation. Microsyst. Technol. 2020, 26, 811-821.
10. Yang, M.; Sheng, Z.; Che, Y.; Hu, J.; Hu, K.; Du, Y. Design of Small Monitoring ROV for Aquaculture. In Proceedings of the OCEANS 2019-Marseille, Marseille, France, 17-20 June 2019; pp. 1-9.
11. Zhang, L.; Zhang, L.; Liu, S.; Zhou, J.; Papavassiliou, C. Low-level control technology of micro autonomous underwater vehicle based on intelligent computing. Cluster Comput. 2019, 22, 8569-8580.
12. Hernández-Alvarado, R.; García-Valdovinos, L.G.; Salgado-Jiménez, T.; Gómez-Espinosa, A.; Fonseca-Navarro, F. Neural network-based self-tuning PID control for underwater vehicles. Sensors 2016, 16, 1429.
13. Yan, Z.; Yang, Z.; Zhang, J.; Zhou, J.; Jiang, A.; Du, X. Trajectory tracking control of uuv based on backstepping sliding mode with fuzzy switching gain in diving plane. IEEE Access 2019, 7, 166788-166795.
14. Yan, Z.; Wang, M.; Xu, J. Integrated guidance and control strategy for homing of unmanned underwater vehicles. J. Franklin Inst. 2019, 356, 3831-3848.
15. Zhou, J.; Zhao, X.; Chen, T.; Yan, Z.; Yang, Z. Trajectory Tracking Control of an Underactuated AUV Based on Backstepping Sliding Mode with State Prediction. IEEE Access 2019, 7, 181983-181993.
16. Huang, B.; Yang, Q. Double-loop sliding mode controller with a novel switching term for the trajectory tracking of work-class ROVs. Ocean Eng. 2019, 178, 80-94.
17. Ramezani-al, M.R.; Tavanaei Sereshki, Z. A novel adaptive sliding mode controller design for tracking problem of an AUV in the horizontal plane. Int. J. Dyn. Control 2019, 7, 679-689.
18. Lv, T.; Zhou, J.; Wang, Y.; Gong, W.; Zhang, M. Sliding mode based fault tolerant control for autonomous underwater vehicle. Ocean Eng. 2020, 216, 107855.
19. García-Valdovinos, L.G.; Fonseca-Navarro, F.; Aizpuru-Zinkunegi, J.; Salgado-Jiménez, T.; Gómez-Espinosa, A.; Cruz-Ledesma, J.A. Neuro-Sliding Control for Underwater ROV's Subject to Unknown Disturbances. Sensors 2019, 19, 2943.
20. Qin, H.; Li, C.; Sun, Y.; Li, X.; Du, Y.; Deng, Z. Finite-time trajectory tracking control of unmanned surface vessel with error constraints and input saturations. J. Frankl. Inst. 2020, 357, 11472-11495.
21. Yan, Z.; Yu, H.; Zhang, W.; Li, B.; Zhou, J. Globally finite-time stable tracking control of underactuated UUVs. Ocean Eng. 2015, 107, 132-146.
22. Yu, H.; Guo, C.; Yan, Z. Globally finite-time stable three-dimensional trajectory-tracking control of underactuated UUVs. Ocean Eng. 2019, 189, 106329.
23. Chu, Z.; Zhu, D.; Yang, S.X. Observer-based adaptive neural network trajectory tracking control for remotely operated vehicle. IEEE Trans. Neural Networks Learn. Syst. 2017, 28, 1633-1645.
24. Qiao, L.; Zhang, W. Double-Loop Integral Terminal Sliding Mode Tracking Control for UUVs with Adaptive Dynamic Compensation of Uncertainties and Disturbances. IEEE J. Ocean. Eng. 2019, 44, 29-53.
25. Guerrero, J.; Torres, J.; Creuze, V.; Chemori, A. Trajectory tracking for autonomous underwater vehicle: An adaptive approach. Ocean Eng. 2019, 172, 511-522.
26. Parra-Vega, V.; Garcia-Valdovinos, L.; Dominguéz-Ramiréz, O.A. Sliding PID Control for Tracking in Finite Time for Robot Arms. In Proceedings of the 11th International

Conference on Advanced Robotics, Coimbra, Portugal, 30 June-3 July 2003; pp. 1526-1531.

27. García-Valdovinos, L.G.; Parra-Vega, V.; Arteaga, M.A. Bilateral Cartesian sliding PID force/position control for tracking in finite time of master-slave systems. Proc. Am. Control Conf. 2006, 2006, 369-375.
28. Garcia-Valdovinos, L.G.; Parra-Vega, V.; Mendez-Iglesias, J.A.; Arteaga, M.A. Cartesian sliding PID force/position control for transparent bilateral teleoperation. IECON Proc. Ind. Electron. Conf. 2005, 2005, 1979-1985.
29. Perruquetti, W.; Barbot, J.-P. Sliding Mode Control in Engineering; Marcel Dekker Inc.: New York, NY, USA, 2002; ISBN 9780429222832.
30. Fossen, T.I. Handbook of Marine Craft Hydrodynamics and Motion Control; John Wiley & Sons, Ltd.: Chichester, UK, 2011; ISBN 9781119994138
31. BlueRobotics. BlueROV2_Datasheet. 2019. Available online: https://bluerobotics.com/wp-content/uploads/2020/02/br_bluerov2_datasheet_rev6.pdf (accessed on 11 December 2020).
32. Lack, S.; Rentzow, E.; Jeinsch, T. Experimental Parameter Identification for an open-frame ROV: Comparison of towing tank tests and open water self-propelled tests. IFAC-PapersOnLine 2019, 52, 271-276.
33. Wu, C. 6-DoF Modelling and Control of a Remotely Operated Vehicle. Master's Thesis, Flinders University, Adelaide, Australia, 2018.
34. García-Valdovinos, L.G.; Salgado-Jiménez, T.; Bandala-Sánchez, M.; Nava-Balanzar, L.; Hernández-Alvarado, R.; Cruz-Ledesma, J.A. Modelling, Design and Robust Control of a Remotely Operated Underwater Vehicle. Int. J. Adv. Robot. Syst. 2014, 11, 1.
35. Zhang, S.; Yu, J.; Zhang, A.; Zhang, F. Spiraling motion of underwater gliders: Modeling, analysis, and experimental results. Ocean Eng. 2013, 60, 1-13.
36. Do, K.D.; Pan, J. Control of Ships and Underwater Vehicles; Springer Science & Business Media: London, UK, 2009; ISBN 978-1-84882-729-5.

CHAPTER

5

Stability and Manoeuvrability Simulation of a Semi-Autonomous Submarine Free-Running Model SUBOFF with an Autopilot System

Yu-Hsien Lin, Yu-Ting Lin and Yen-Jun Chiu

Department of Systems & Naval Mechatronic Engineering, National Cheng-Kung University, Tainan City 70101, Taiwan

ABSTRACT

On the basis of a full-appendage DARPA SUBOFF model (DTRC model 5470), a scale ($\lambda = 0.535$) semi-autonomous submarine free-running model (SFRM) was designed for testing its manoeuvrability and stability in the constrained water. Prior to the experimental tests of the SFRM, a six-degree-offreedom (6-DOF) manoeuvre model with an autopilot system was developed by using logic operations in MATLAB. The SFRM's attitude and its trim polygon were presented by coping with the changes in mass and trimming moment. By adopting a series of manoeuvring tests in empty tanks, the performances of the SFRM were introduced in cases of three sailing speeds. In addition, the PD controller was established by considering the simulation results of these manoeuvring tests. The optimal control gains with respect to each manoeuvring test can be calculated by using the PID tuner in MATLAB. Two sets of control gains derived from the optimal characteristics parameters were compared in order to decide on the most appropriate PD controller with the line-of-sight (LOS) guidance algorithm for the SFRM in the autopilot simulation. Eventually, the simulated trajectories and course angles of the SFRM would be illustrated in the post-processor based on the Cinema 4D modelling.

Keywords: *submarine free-running model; trim polygon; manoeuvrability; SUBOFF; PD controller; autopilot system*

1. INTRODUCTION

The manoeuvring performance of the submarine is an important parameter in the initial design, especially for the geometry of the appendages, which in turn is important for the submarine resistance and hence total performance of the submarine. Conventionally, captive experiments, i.e., rotating arm (RA) and planar motion mechanism (PMM), and numerical simulations, i.e., computerised planar motion mechanism (CPMM), are adopted for in the early design stage until a reasonable maturity of the design is reached. For an alternative option of matured designs, free-running manoeuvring models have further become available for validation of the manoeuvring performance.

The so-called submarine free-running model (SFRM) is a remotely controlled scale model replicating submarine manoeuvres. These models should be equipped with instrumentation to measure all the state variables required to describe its 6-DOF motion responses, including trajectories, attitudes, linear and angular velocities, and accelerations [1]. Over the last few decades, there have been numerous studies [2-5] investigating SFRMs in controlled environments. Although the SFRM can evaluate the manoeuvring performances of the full-scale model accurately, there are still some differences between the scale model and the full-scale model due to scale effect.

In order to keep the desired trajectories dynamically during the test, a robust control system is required for adjusting the control planes of the SFRM. Definitely, the input and output signals demand for the control planes are greatly affected by the manoeuvring characteristics. The CFD-based and coefficient-based methods are the two major mathematical models in exploring the hydrodynamic forces and moments exerted on the SFRM. As the name implies, the coefficient-based method uses the equation of motion (EOM) combined with the hydrodynamic coefficient as a mathematical model [6,7]. Taking advantage of the predefined coefficients in the mathematical model, it usually has less computation time than the CFD-based method. Furthermore, the coefficient-based method can approximate the hydrodynamic interaction effect through semi-empirical correction when the positions of the appendages are rearranged at a preliminary design stage [8]. However, the limitation of the coefficient-based method is that the unsteady viscous effect cannot be captured in the linear model, such as cross flow and vorticity [1]. In order to overcome the limitations of the coefficient-based method and improve the accuracy of manoeuvring characteristics, the CFD-based method provides another option, including the resistance test [9], self-propulsion test, and free-running test simulations [10].

Since the estimation of hydrodynamic coefficients has uncertainty and variation in different environmental conditions [11], hydrodynamic forces and moments exerted on the SFRM are changed accordingly. Therefore, an applicable controller needs to be a self-tuning and robust counter to the variation of SFRM parameters and also unpredictable environmental disturbances. In recent years, numerous control models have been proposed, e.g., linear controllers [12,13], sliding-mode controller (SMC) [14], adaptive control [15,16], FLC (Fuzzy Logic Control) [17], predictive control [18], static feedback control [19], neural-network-based control [20], and PID (proportional-integral-derivative) control [21]. The advantage of using a PID controller is that it is simple to implement and maintain, but it is primarily applicable for linear time-invariant systems.

The well-known classic equations of motion (EOMs) are those of Gertler and Hagen [6], Feldman [7], and Fossen and Fjellstad [22]. They were derived by adopting Newton-Euler equations or Lagrangian methods, which can be calculated in matrix forms. For example, Perrault, et al. [23] applied a fully nonlinear mathematical model based on Newton-Euler EOMs to investigate the sensitivity of an AUV response to changes in hydrodynamic coefficients, including those during simulated turning and zigzag tests. Coe [24] established a preliminary model for manoeuvrability computation of a submarine by using a database of hydrodynamic coefficients and empirical equations, thus enabling the prediction of submarine's general manoeuvrability in deep water. Although it is difficult to realise unsteady motions for extreme submarine manoeuvres [25], the coefficient-based model is considered to provide realistic estimations for moderate manoeuvres [1].

The main purpose of this study is to develop a 6-DOF manoeuvre model of the SFRM based on the generic model SUBOFF by means of a conventional PD controller. First of all, a six-degree-of-freedom (6-DOF) manoeuvre model was constructed for a series of fundamental manoeuvring tests. Secondly, the integrated control system, including rudder planes and stern planes, were modelled as 2nd order mass-spring systems that can be simulated as control inputs. Finally, the PD controller combined with the line-of-sight (LOS) [26] guidance algorithm was applied to track waypoints at different sailing speeds.

2. THE CONFIGURATION OF SFRM

2.1. Geometry

Our SFRM was developed based on the scale ($\lambda = 0.535$) DARPA SUBOFF (DTRC model 5470) [27]. The principal model architecture comprises of the bow, midsection, stern, sail, rudder, and stern planes. The model geometry and principal particulars were presented in Figure 1 and Table 1, respectively. In Table 1, the effective power is equal to the power output of the engine minus losses due to the gearbox, shafting, propeller, as well as interaction between the propeller and the hull.

2.2. Ballast Tank and Attitude Adjusting System

Our SFRM contains forward, aft ballast tanks, and an attitude adjusting system that allows the variation of mass, and movement of mass inside the model as shown in Figure 2. The designs of ballast tanks were maximised within the constraints of the hull geometry. The forward and aft ballast tanks are of the same design, approximately equal distance from midships, but each one can be independently controlled. The dimension of each ballast tank is 0.084 m $\times$ 0.084 m $\times$ 0.084 m, and the weight of the full tank is 600 g. In addition, both the inflow rate and the outflow rate are set to be 0.0002 m^3/s. For the attitude adjusting system, the trim angle and the heel angle of the SFRM can be controlled suitably by fine-tuning the locations of the lead weights. Overall speaking, the total mass, center of gravity (CG), trim angle and heel angle can be regulated suitably by using ballast tanks and the attitude adjusting system.

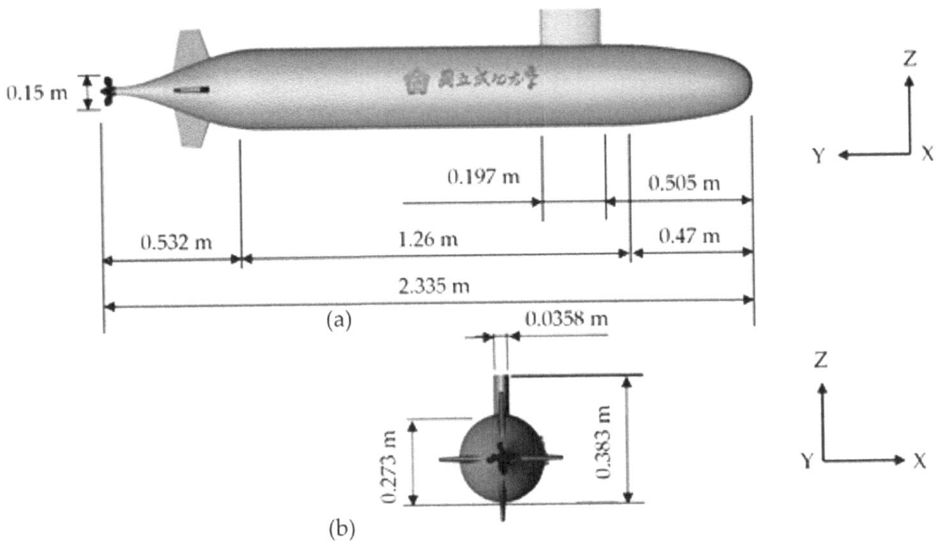

Figure 1. *(a) Side, and (b) rear view of the submarine free-running model (SFRM).*

Table 1. *Principal particulars of the SFRM and the propeller.*

SFRM	
Full length [m]	2.335
Diameter [m]	0.273
Bow length [m]	0.47
Midsection length [m]	1.26
Stern length [m]	0.532
Total height [m]	0.383
Sail length [m]	0.197
Sail width [m]	0.0358
Propeller	
Diameter [m]	0.15
Number of blades	4
Hub to diameter ratio	0.292
Pitch to diameter ratio	1.32
Revolution speed [RPM]	370 at $U = 0.5$ [m/s] 785 at $U = 1.0$ [m/s] 1210 at $U = 1.5$ [m/s]
Effective Power [W]	0.17 at $U = 0.5$ [m/s] 1.51 at $U = 1.0$ [m/s] 4.44 at $U = 1.5$ [m/s]

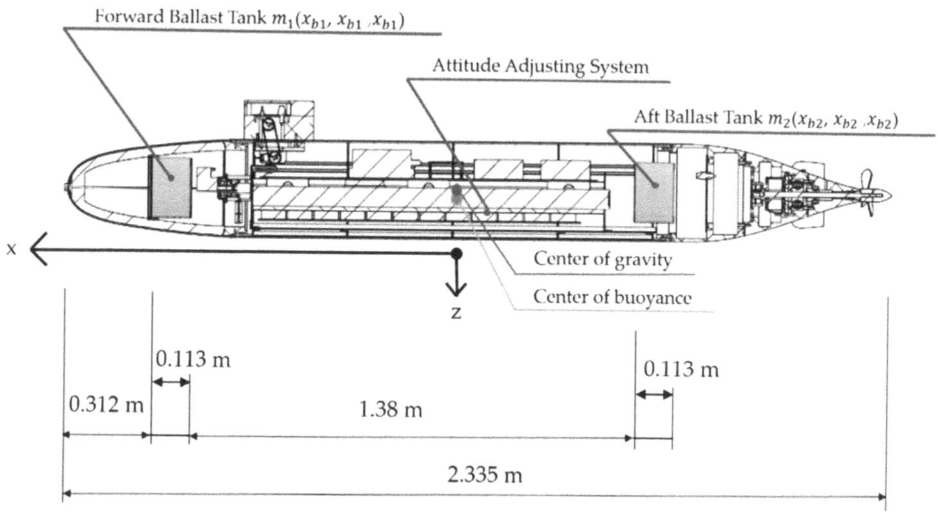

Figure 2. *The configuration of ballast tanks and the attitude adjusting system.*

2.3. Time-Varying Mass Model

The total mass and weight of the SFRM can be time-varied based on the following equations:

$$W(t) = m(t)g \tag{1}$$

$$m(t) = m_0 + \sum_{i=1}^{2} m_i(t) \tag{2}$$

$$m_i(t) = \int \dot{m}_i(t)dt = \rho Q_i(t), i = 1,2 \tag{3}$$

where t is defined as time; $W(t)$ and $m(t)$ are the total weight and mass of the SFRM; m_0 is the initial mass of the SFRM; $m_1(t)$ and $m_2(t)$ represent the masses of forward ballast tank and aft ballast tank; $\dot{m}_1(t)$ and $\dot{m}_2(t)$ denote the rates of masses of forward ballast tank and aft ballast tank; $Q_1(t)$ and $Q_2(t)$ denote the flow rates of forward ballast tank and aft ballast tank; ρ is the water density; t is defined as time.

Since the attitude of the SFRM is determined by the mass variation of the ballast tank, the coordinate of CG can be determined as below:

$$x_g(t) = \frac{m_0 x_{go} + x_{b1}(t)m_1(t) + x_{b2}(t)m_2(t)}{m_0 + m_1(t) + m_2(t)} \tag{4}$$

$$y_g(t) = \frac{m_0 y_{go} + y_{b1}(t)m_1(t) + y_{b2}(t)m_2(t)}{m_0 + m_1(t) + m_2(t)} \tag{5}$$

$$z_g(t) = \frac{m_0 z_{go} + z_{b1}(t)m_1(t) + z_{b2}(t)m_2(t)}{m_0 + m_1(t) + m_2(t)} \tag{6}$$

where (x_{go}, y_{go}, z_{go}) is the initial coordinate of CG; (x_{b1}, y_{b1}, z_{b1}) is the coordinate of CG for the forward ballast tank, and (x_{b2}, y_{b2}, z_{b2}) is the coordinate of CG for the aft ballast tank.

Subsequently, the inertia moments of the SFRM are defined by:

$$I_{xx}(t) = I_{x0} + [I_{xb1}(t) + (y_{b1}(t)^2 + z_{b1}(t)^2) \times m_1(t)] + [I_{xb2}(t) + (y_{b2}(t)^2 + z_{b2}(t)^2) \times m_2(t)] \tag{7}$$

$$I_{yy}(t) = I_{y0} + [I_{yb1}(t) + (x_{b1}(t)^2 + z_{b1}(t)^2) \times m_1(t)] + [I_{yb2}(t) + (x_{b2}(t)^2 + z_{b2}(t)^2) \times m_2(t)] \tag{8}$$

$$\begin{aligned} I_{zz}(t) =& I_{z0} + \left[I_{zb1}(t) + \left(x_{b1}(t)^2 + y_{b1}(t)^2\right) \times m_1(t)\right] \\ &+ \left[I_{zb2}(t) + \left(x_{b2}(t)^2 + y_{b2}(t)^2\right) \times m_2(t)\right] \end{aligned} \tag{9}$$

$$I_{xy}(t) = I_{xy0} + \left[I_{xyb1}(t) + x_{b1}(t)y_{b1}(t)m_1(t)\right] + \left[I_{xyb2}(t) + x_{b2}(t)y_{b2}(t)m_2(t)\right] \tag{10}$$

$$I_{xz}(t) = I_{xz0} + [I_{xzb1}(t) + x_{b1}(t)z_{b1}(t)m_1(t)] + [I_{xzb2}(t) + x_{b2}(t)z_{b2}(t)m_2(t)] \tag{11}$$

$$I_{yz}(t) = I_{yz0} + \left[I_{yzb1}(t) + y_{b1}(t)z_{b1}(t)m_1(t)\right] + \left[I_{yzb2}(t) + y_{b2}(t)z_{b2}(t)m_2(t)\right] \tag{12}$$

where $I_{x0}, I_{y0}, I_{z0}, I_{xy0}, I_{xz0}$, and I_{yz0} are the inertia moments of the SFRM corresponding to the origin of the initial coordinate; $I_{xb1}, I_{yb1}, I_{zb1}, I_{xyb1}, I_{xzb1}$, and I_{yzb1} are the inertia moments due to mass variation corresponding to the CG of the forward ballast tank; I_{xb2}, $I_{yb2}, I_{zb2}, I_{xyb2}, I_{xzb2}$, and I_{yzb2} are the inertia moments due to mass variation corresponding to the CG of the aft ballast tank.

In conditions of full ballast tanks, the weight of either the forward or aft water tank is about 600 g. Since the mass variation of ballast tanks causes the change of the pitch angle, the time-varying mass model would be adopted to understand the time series of the SFRM's attitude in the powerless state. Due to asymmetrical positions of ballast tanks on both sides of the SFRM, the time-varying model of each individual ballast tank was used to discuss the effect of ballast tanks on the inertia moments in this study.

3. MATHEMATICAL MODEL

3.1. Coordinate Systems

The earth-fixed coordinate system O-XYZ as shown in Figure 3 a and the body-fixed coordinate system o-xyz as shown in Figure $3b - d$ were considered in this study. The velocity component of the body-fixed coordinate system was (u, v, w, p, q, r), and the position in body-fixed coordinate system was (X, Y, Z). The three Euler angles of the body-fixed coordinate system relative to the earth-fixed coordinate system were expressed as ϕ (roll), θ (pitch), and ψ (yaw). The parameters of the six degree-of-freedom (6-DOF) motion are presented in Table 2.

In order to define the relationship between the body-fixed linear and angular velocities and the time derivative of the earth-fixed vehicle position coordinates, the transformation [28] can be related by

Table 2. *Parameters of the six-degree-of-freedom (6-DOF) motion.*

Degree of Freedom	Velocity/Angular Velocity	Position/Angle
Surge	u	X
Sway	v	Y
Heave	w	Z
Roll	p	ϕ
Pitch	q	θ
Yaw	r	ψ

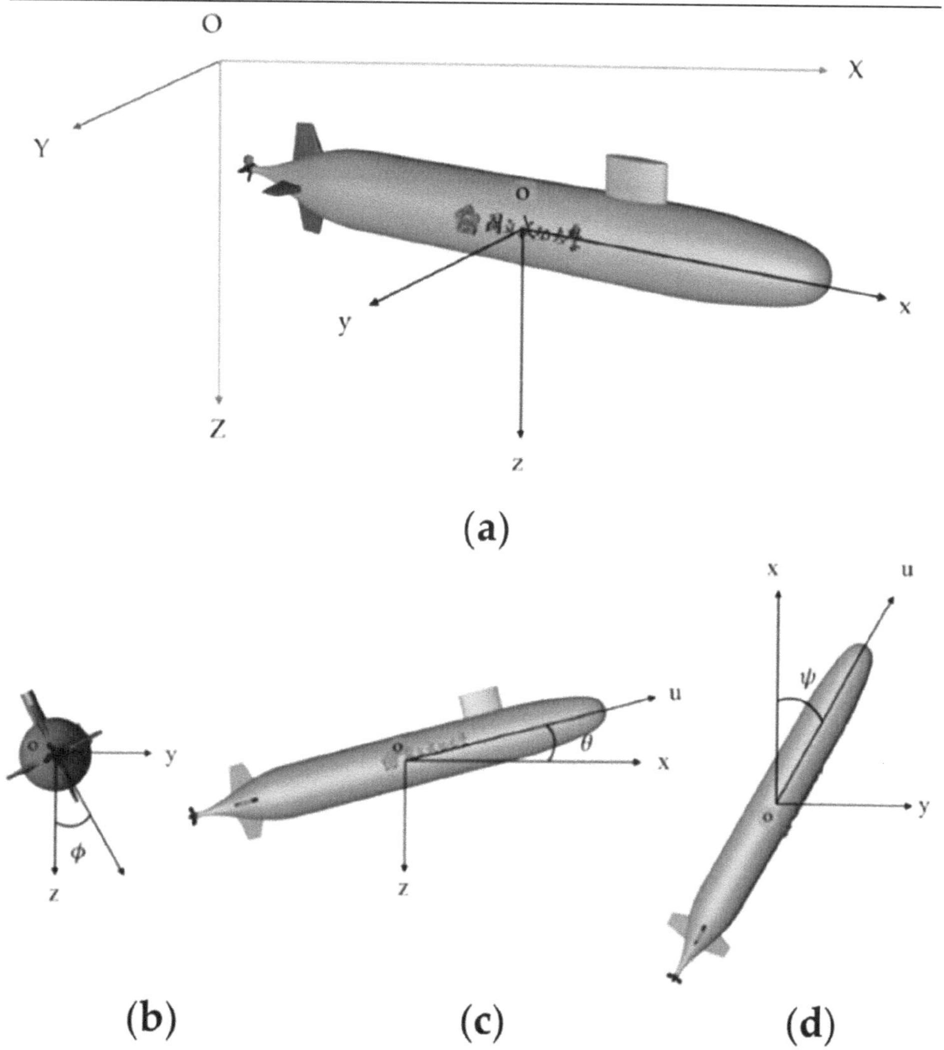

Figure 3. *(a) The earth-fixed coordinate system; (b) rear, (c) side, and (d) top views of the body-fixed coordinate system.*

$$\begin{bmatrix} \dot{X} \\ \dot{Y} \\ \dot{Z} \end{bmatrix} = \begin{bmatrix} \cos\psi\cos\theta & \cos\psi\sin\phi\sin\theta & \sin\phi\sin\psi + \\ & -\sin\psi\cos\phi & \cos\psi\cos\phi\sin\theta \\ \sin\psi\cos\theta & \sin\psi\sin\phi\sin\theta & \sin\psi\cos\phi\sin\theta \\ -\sin\theta & +\cos\psi\cos\phi & -\cos\psi\sin\phi \\ & \sin\phi\cos\theta & \cos\phi\cos\theta \end{bmatrix} \begin{bmatrix} u \\ v \\ w \end{bmatrix} \quad (13)$$

$$\begin{bmatrix} \dot{\phi} \\ \dot{\theta} \\ \dot{\psi} \end{bmatrix} = \begin{bmatrix} 1 & \sin\phi\tan\theta & \cos\phi\tan\theta \\ 0 & \cos\phi & -\sin\phi \\ 0 & \sin\phi/\cos\theta & \cos\phi/\cos\theta \end{bmatrix} \begin{bmatrix} p \\ q \\ r \end{bmatrix}. \quad (14)$$

3.2. Six-Degree-of-Freedom Motion Equations

According to Newton's Second Law, the equations of motion [7] to represent hydrodynamic forces and moments as functions of motion variables were introduced in Equations (15)-(20). The left-hand side of the equations is the rigid body dynamics, whereas the right-hand side represents the hydrodynamic force acting on the SFRM.

1 Surge

$$\begin{aligned} & m\left[\dot{u} - vr + wq - x_G\left(q^2 + r^2\right) + y_G(pq - \dot{r}) + z_G(pr + \dot{q})\right] \\ & = \frac{\rho}{2}L^4\left[X'_{qq}q^2 + X'_{rr}r^2 + X'_{rp}rp\right] + \frac{\rho}{2}L^3\left[X'_{\dot{u}}\dot{u} + X'_{vr}vr + X'_{wq}wq\right] \\ & + \frac{\rho}{2}L^2\left[X'_{uu}u^2 + X'_{vv}v^2 + X'_{ww}w^2\right] + \frac{\rho}{2}L^2u^2\left[X'_{\delta_r\delta_r}\delta_r{}^2 + X'_{\delta_s\delta_s}\delta_s^2 + X'_{\delta_b\delta_b}\delta_b{}^2\right] \\ & - (W - B)\sin\theta + X_P - X_R \end{aligned} \quad (15)$$

2 Sway

$$\begin{aligned} & m\left[\dot{v} + ur - wp + x_G(pq + \dot{r}) - y_G\left(p^2 + r^2\right) + z_G(qr - \dot{p})\right] \\ & = \frac{\rho}{2}L^4\left[Y'_{\dot{p}}\dot{p} + Y'_{\dot{r}}\dot{r} + Y'_{pq}pq + Y'_{p|p|}p|p|\right] \\ & + \frac{\rho}{2}L^3\left[Y'_{\dot{v}}\dot{v} + Y'_{wp}wp + Y'_{v|r|}\frac{v}{|v|}\left|\left(v^2 + w^2\right)^{\frac{1}{2}}\right||r| + Y_pup + Y_rur\right] \\ & + \frac{\rho}{2}L^2\left[Y'_{uu}u^2 + Y_vuv + Y'_{v|v|}v\left|\left(v^2 + w^2\right)^{\frac{1}{2}}\right| + Y_{wv}wv\right] + \frac{\rho}{2}L^2u^2Y'_{\delta_r}\delta_r \\ & + (W - B)\cos\theta\sin\phi + \frac{\rho}{2}L^2\left(F_y\right)_{vs}\frac{v^2 + w^2}{U^2}(-w)\sin\omega t + WF_s \end{aligned} \quad (16)$$

3 Heave

$$
\begin{aligned}
& m\left[\dot{w} - uq + vp + x_G(pr - \dot{q}) + y_G(qr + \dot{p}) - z_G\left(p^2 + q^2\right)\right] \\
& = \frac{\rho}{2}L^4\left[Z'_{\dot{q}}\dot{q} + Z'_{rr}r^2 + Z'_{rp}rp\right] + \frac{\rho}{2}L^3\left[Z'_{\dot{w}}\dot{w} + Z'_{vr}vr + Z'_{vp}vp\right] \\
& + \frac{\rho}{2}L^3\left[Z'_q uq + Z'_{w|q|}\frac{w}{|w|}\left|\left(v^2 + w^2\right)^{\frac{1}{2}}\right||q|\right] \\
& + \frac{\rho}{2}L^2\left[Z'_{uu}u^2 + Z'_w uw + Z'_{w|w|}w\left|\left(v^2 + w^2\right)^{\frac{1}{2}}\right|\right]'^{\,2}\Big) \\
& + \frac{\rho}{2}L^2\left[Z'_{|w|}u|w| + Z'_{ww}\left|w\left(v^2 + w^2\right)^{\frac{1}{2}}\right| + Z'_{vv}v^2\right] + \frac{\rho}{2}L^2u^2\left[Z'_{\delta_s}\delta_s + Z'_{\delta_b}\delta_b\right] \\
& + (W - B)\cos\theta\cos\phi + \frac{\rho}{2}L^2\left(F_z\right)_{vs}\frac{v^2 + w^2}{u^2}v\sin\omega t + WF_H
\end{aligned}
\tag{17}
$$

4 Roll

$$
\begin{aligned}
& I_x\dot{p} + \left(I_z - I_y\right)qr - (\dot{r} + pq)I_{xz} + m\left[y_G(\dot{w} - uq + vp) - z_G(\dot{v} + ur - wp)\right] \\
& = \frac{\rho}{2}L^5\left[K'_{\dot{p}}\dot{p} + K'_{\dot{r}}\dot{r} + K'_{qr}qr + K'_{p|p|}p|p|\right] \\
& + \frac{\rho}{2}L^4\left[K'_p up + K'_r ur + K'_{\dot{v}}\dot{v} + K'_{wp}wp\right] \\
& + \frac{\rho}{2}L^3\left[K'_{uu}u^2 + K'_v uv + K'_{vw}vw + K'_{v|v|}v\left|\left(v^2 + w^2\right)^{\frac{1}{2}}\right|\right] + \frac{\rho}{2}L^3u^2K'_{\delta_r}\delta_r \\
& + \left(Wy_G - By_B\right)\cos\theta\cos\phi - \left(Wz_G - Bz_B\right)\cos\theta\sin\phi + WF_R
\end{aligned}
\tag{18}
$$

5 Pitch

$$
\begin{aligned}
& I_y\dot{q} + \left(I_x - I_z\right)pr + \left(p^2 - r^2\right)I_{xz} - m\left[x_G(\dot{w} - uq + vp) - z_G(\dot{u} - vr + wq)\right] \\
& = \frac{\rho}{2}L^5\left[M'_{\dot{q}}\dot{q} + M'_{rr}r^2 + M'_{rp}rp\right] \\
& + \frac{\rho}{2}L^4\left[M'_q uq + M'_{|w|q}\left|\left(v^2 + w^2\right)^{\frac{1}{2}}\right|q + M'_{\dot{w}}\dot{w} + M'_{vr}vr + M'_{vp}vp\right] \\
& + \frac{\rho}{2}L^3\left[M'_{uu}u^2 + M'_w uw + M'_{w|w|}w\left|\left(v^2 + w^2\right)^{\frac{1}{2}}\right|\right] \\
& + \frac{\rho}{2}L^3\left[M'_{|w|}u|w| + M'_{ww}\left|w\left(v^2 + w^2\right)^{\frac{1}{2}}\right| + M'_{vv}v^2\right] + \frac{\rho}{2}L^3u^2\left[M'_{\delta_s}\delta_s + M'_{\delta_b}\delta_b\right] \\
& - \left(Wx_G - Bx_B\right)\cos\theta\cos\phi - \left(Wz_G - Bz_B\right)\sin\theta + WF_P
\end{aligned}
\tag{19}
$$

6 Yaw

$$
\begin{aligned}
&I_z\dot{r} + (I_y - I_x)\,pq + (rq - \dot{p})I_{xz} + m\left[x_G(\dot{v} + ur - wp) - y_G(\dot{u} - vr + wq)\right] \\
&= \frac{\rho}{2}L^5\left[N'_{\dot{r}}\dot{r} + N'_{pq}pq + N'_{\dot{p}}\dot{p}\right] \\
&+ \frac{\rho}{2}L^4\left[N'_r ur + N'_{|v|r}\left|\left(v^2 + w^2\right)^{\frac{1}{2}}\right| r + N'_p up + N'_{\dot{v}}\dot{v} + N'_{wp}wp\right] \qquad (20) \\
&+ \frac{\rho}{2}L^3\left[N'_{uu}u^2 + N'_v uv + N'_{v|r|}\left|\left(v^2 + w^2\right)^{\frac{1}{2}}\right| + N'_{wv}wv\right] + \frac{\rho}{2}L^3u^2N'_{\delta_r}\delta_r \\
&+ (Wx_G - Bx_B)\cos\theta\cos\phi + (Wy_G - By_B)\sin\theta + WF_Y
\end{aligned}
$$

where I_x, I_y, and I_z are the inertia moments about the axes of the body-fixed coordinate system, and I_{xz} is the cross-sectional area on plane $xz.X_P$ and X_R are defined as the propeller thrust and the resistance, respectively. WF_S, WF_H, WF_R, WF_P, and WF_Y are the exciting forces of waves that act on each DOF motion. Since the manoeuvre of the SFRM was given in the condition of neutral buoyancy, the effects of wave exciting forces have been disregarded.

3.3. Parameter Settings for Hydrodynamic Coefficients

The SFRM was designed and fabricated based on the DARPA SUBOFF model with full appendages (configuration 8). Due to the limitation of the towing tank dimension, the SFRM was scaled down. In addition, it is hypothesised that the scale effect has a limited effect on the hydrodynamic forces of the SFRM. Therefore, the dimensionless hydrodynamic coefficients for the vertical and horizontal planes were selected according to the published data and can be summarised in Table 3. Otherwise, the remaining terms in Equations (15)-(20) were set to be zero for simplicity.

4. AUTOPILOT SYSTEM

4.1. Line-of-Sight Guidance Algorithm

It is demonstrated that the line-of-sight (LOS) guidance algorithm is suitably applied to the surface ships [29]. In addition, the LOS guidance algorithm plays an important role in accurate path tracking control of the AUV [30,31]. Since the path tracking cannot be achieved due to lack of lateral control for the under-actuated AUV, using the LOS guidance algorithm converts the lateral error to the heading movement, which means that the desired path tracking is indirectly achieved by heading control [32].

In this study, the LOS guidance algorithm as shown in Figure 4 was adopted to estimate the desired displacements of the SFRM in the yaw and the pitch modes from the current position to the next waypoint in the earth-fixed coordinate system. The desired yaw and pitch angles can be represented as below:

Table 3. *Hydrodynamic coefficients of the DARPA SUBOFF model.*

Z'_w	−0.013910	Y'_v	−0.027834
M'_w	0.010324	N'_v	−0.013648
Z'_q	−0.007545	K'_v	−0.000584
M'_q	−0.003702	Y'_r	0.005251
$Z'_{\dot{w}}$	−0.014529	N'_r	−0.004444
$M'_{\dot{w}}$	−0.000561	$Y'_{\dot{v}}$	−0.016186
$Z'_{\dot{q}}$	−0.000633	$N'_{\dot{v}}$	0.000396
$M'_{\dot{q}}$	−0.000860	$Y'_{\dot{r}}$	0.000398
Z'_{δ_s}	−0.005603	$N'_{\dot{r}}$	−0.000897
M'_{δ_s}	−0.002409	Y'_{δ_r}	0.005929
		N'_{δ_r}	−0.002217
		K'_{δ_r}	−0.000005

$$\Psi_d = \tan^{-1}\left(\frac{Y_k - Y_C}{X_k - X_C}\right), k = 1, 2, \ldots, n \tag{21}$$

$$\theta_d = \tan^{-1}\left(\frac{Z_k - Z_C}{X_k - X_C}\right), k = 1, 2, \ldots, n \tag{22}$$

where both the SFRM's current position (X_C, Y_C, Z_C) and the next waypoint position (X_k, Y_k, Z_k) refer to the earth-fixed coordinate system; Ψ_d and θ_d denote the desired yaw angle and pitch angle, respectively; and n is the number of waypoints. Subsequently, a sphere of acceptance of the LOS guidance algorithm [26] was required for selecting the next waypoint as follows:

$$(X_k - X_C(t))^2 + (Y_k - Y_C(t))^2 + (Z_k - Z_C(t))^2 \leq R_0^2 \tag{23}$$

where the allowable radius R_0 was set to be two times the SFRM's total length L [26].

4.2. PD Controller

In Figure 5, the conventional PD control system is generally introduced in the following equations including two control gains, i.e., proportional gain (K_P) and derivative gain (K_D).

$$\delta_V(t) = K_{P1}\left(\Psi(t-1) - \Psi_d(t-1)\right) + K_{D1}\dot{\Psi}(t-1) \tag{24}$$

$$\delta_H(t) = K_{P2}\left(\theta(t-1) - \theta_d(t-1)\right) + K_{D2}\dot{\theta}(t-1) \tag{25}$$

where $\delta_V(t)$ and $\delta_H(t)$ denote control outputs of the rudder plane angle and the stern plane angle, and they can also be recognised as δ_r and δ_s, respectively. K_{P1} and K_{D1} are the yaw and yaw rate gains by the rudder plane, whereas K_{P2} and K_{D2}

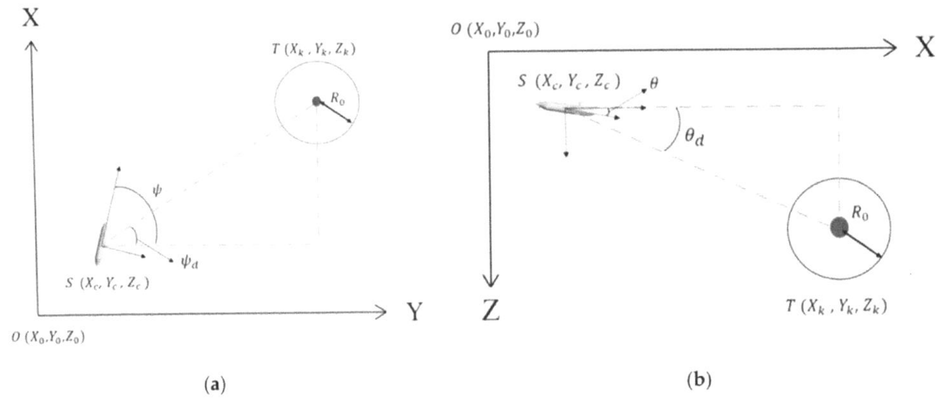

Figure 4. *The schematic diagrams of the line-of-sight (LOS) guidance algorithm in (a) the horizontal plane and (b) the vertical plane, respectively.*

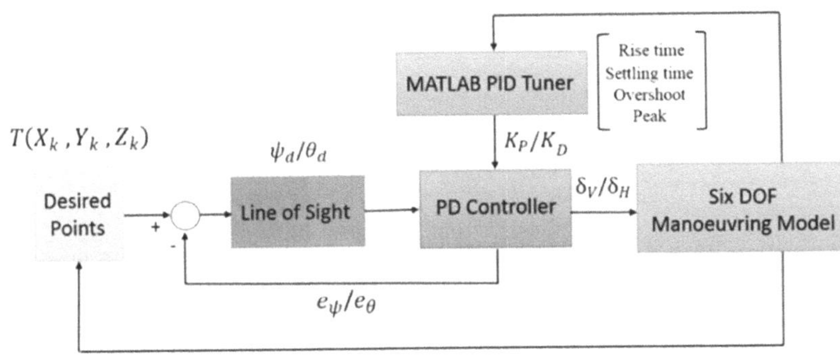

Figure 5. *The flow chart of the autopilot system for the submarine free-running model (SFRM) including the LOS guidance algorithm, the PD controller, the PID tuner, and the 6-DOF manoeuvring mathematical model.*

are the pitch and pitch rate gains by the stern plane. $\dot{\Psi}(t-1)$ and $\dot{\theta}(t-1)$ imply the yaw rate and pitch rate, respectively.

In the control system, the trial-and-error method has been adopted to determine an initial set of control tuning parameters. For an extensive variety of applied methods, it might be the most familiar one. Sometimes, the trial-and-error method does not offer fair tuning and tends to provide an evident overshoot. Hence, it usually needs returning before application. In order to tune the control constraints in demand for an optimal performance, the PID tuner in MATLAB/Simulink [33] has been suggested to achieve the design objectives by means of a balance between robustness and performance. Some important characteristics parameters that need to be considered in the PID tuner are the rise time, settling time, overshoot, and peak, as shown in Figure 6.

4.3. Post-Processor

Figure 7 introduces the post-processor integrated with the autopilot system for the SFRM. In the post-processing process, time histories of the SFRM's trajectories,

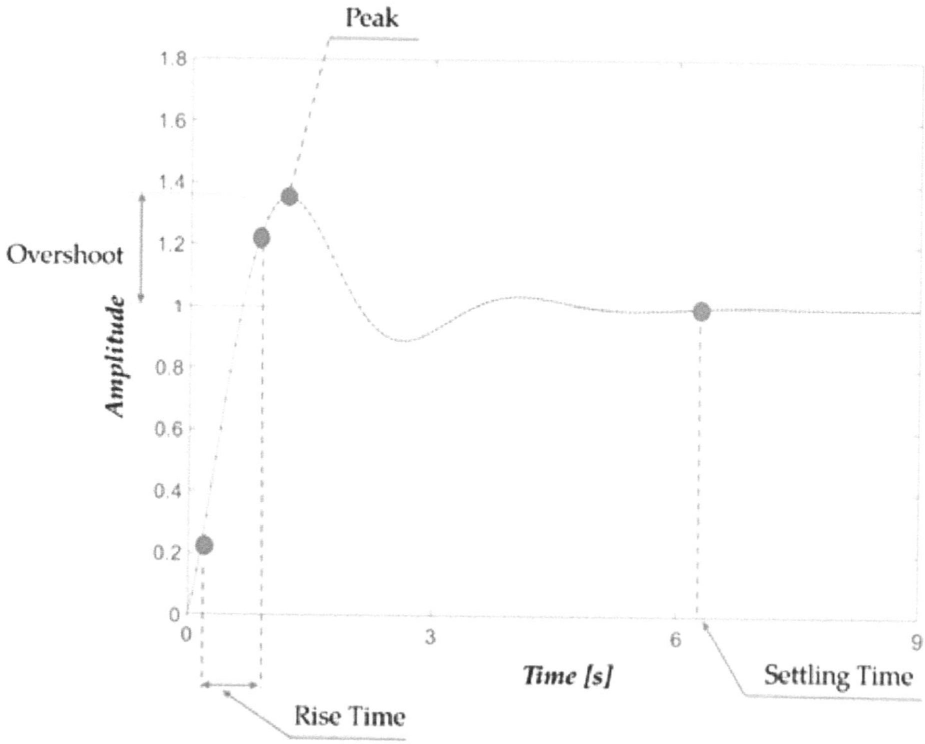

Figure 6. *Step response of the PD controller with the PID tuner.*

6-DOF motion responses, and rudder displacements were presented by conducting the Cinema 4D modelling [34]. The post-processor, i.e., Cinema 4D modelling, was mainly used to allow the animated model to match the simulated trajectories and courses graphically. At the beginning, point coordinates were sampled on the motion trajectories and then projected on a three-dimensional grid. Subsequently, the path-alignment function was used to simplify the problem of curve alignment, and parallel lines of equal proportions were drawn to fix the whole hull.

Figure 8a-f shows each step for the Cinema 4D modelling after the output results of the SFRM's trajectories and kinematics were computed. Layer-grid setting as shown in Figure 8a begins the initial modelling by adjusting the environment and defining the 3D coordinates. Figure 8 b presents the modelling-layer setting by which different viewing and ordering methods are applied to layers and labels. Furthermore, the material modules such as color, transparency, and roughness can be selected as illustrated in Figure 8c. Rendering and animation-recording setting as shown in Figure 8d determines the pixel aspect ratio in detail when recording animation. Figure 8e implies the setting of animation frame rate and coordinate record point, in which the inputs such as trajectories were presented by adjusting animation frame rate and time. Finally, the animation interface exhibits the real-time motion scenario in the simulated environment, as shown in Figure 8f.

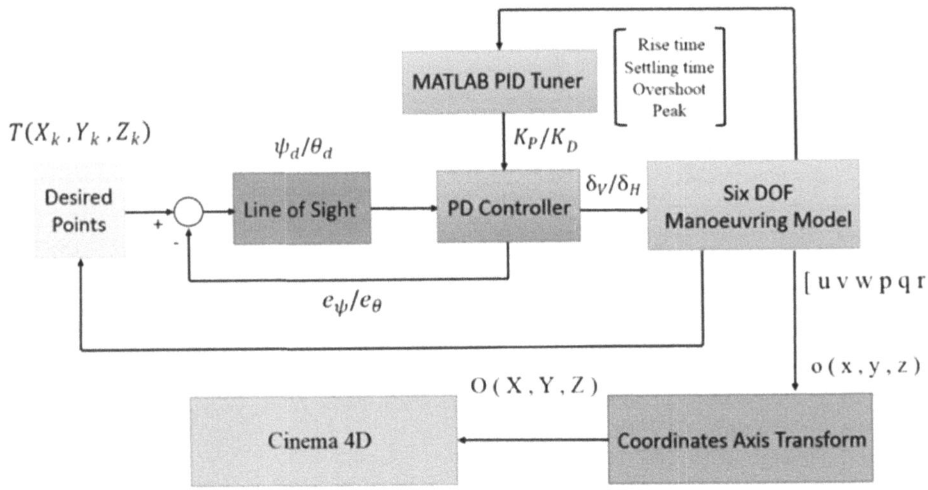

Figure 7. *The flow chart of the autopilot system combining with the Cinema 4D modelling.*

5. RESULTS AND DISCUSSION

For understanding the stability of the SFRM, the trim polygon would be illustrated by combining the main ballast tanks and the attitude adjusting system. There were several manoeuvring tests, including the turning circle test, horizontal zigzag test, vertical zigzag test, meander test, and spiral test, which were used for investigating the rudder performance of the SFRM with the empty ballast tanks. Meanwhile, the propeller revolution speed corresponding to each sailing speed is assumed constant. The descriptions of simulation conditions are indicated in Table 4. Subsequently, the optimal control gains on the rudder operations corresponding to the manoeuvring tests in different sailing speeds would be presented by the PID tuner. Lastly, autopilot simulations in a 3D map with numerous waypoints were conducted by adopting different settings of the PD controller.

5.1. Trim Polygon

Similar to the arrangement of a submarine, the SFRM ought to have several types of ballast tanks. The main ballast tanks are designed for diving and surfacing, whereas the trimming tanks are used for adjusting the submarine's attitude on the surface or underwater. Due to the limitation of the compartment space in the scale model, the trimming tanks were replaced by the attitude adjusting system without water compensation. Specifically, the design of the attitude adjusting system is a slider mechanism device installed in the midsection for adjusting the trim angle.

Figure 9 presents the time series of the mass changes in the forward and aft ballast tanks, respectively. Both the inflow and outflow rates were given as 20 cm^3/s during the periods of 20 to 50 s and 70 to 100 s. Figure 10 exhibits the time series of the corresponding pitch angles in the forward and aft ballast tanks, respectively.

Apparently, the maximum pitch angle of the SFRM reached $-8°$ when the forward ballast tank was full. There were some oscillations accompanied by the variation of the pitch angle. For the aft ballast tank, by using the same mass-variation

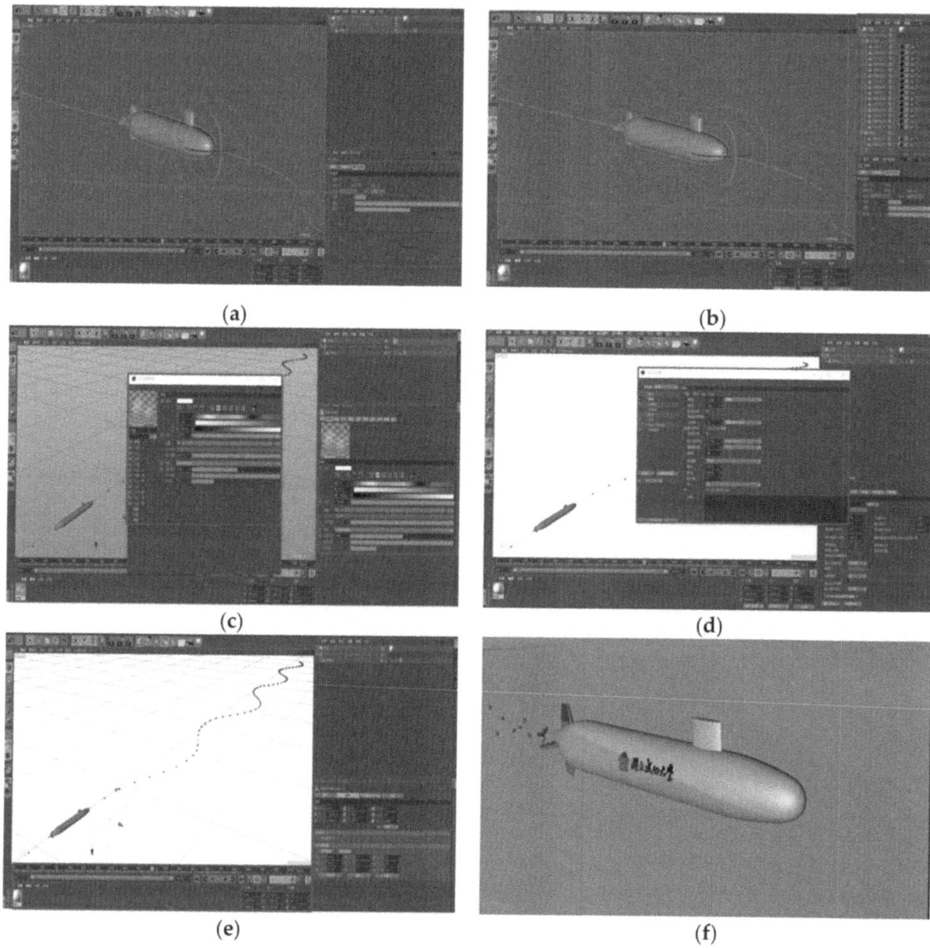

Figure 8. *The operation steps of the Cinema 4D modelling: (a) layer-grid setting, (b) modelling-layer setting, (c) material and texture setting, (d) rendering and animation-recording setting, (e) animation frame rate and coordinate record point setting, and (f) real-time animation interface.*

history of the forward ballast tank, the maximum pitch angle of the SFRM reached 5° with slight oscillations.

Figure 11 shows a typical trim polygon (equilibrium polygon), which is useful to assess the arrangement of the SFRM. Definitely, it is crucial to determine whether the ballast tanks and the attitude adjusting system are available to estimate the variation of SFRM's mass and the longitudinal CG. According to the procedure [35], the effects of the ballast tanks and the attitude adjusting system can be plotted as a function of mass and trimming moment as follows:

0 In the initial stage, the location with zero mass and zero trimming moment indicated all empty tanks and a zero pitch angle.

1 Move forward the weight of the slider mechanism device. In this stage, there is no mass variation, but there is an increase of trimming moment due to the movement of the longitudinal CG.

Table 4. *Standard manoeuvring tests for the SFRM.*

Test	Purpose	Measured Parameters
Trim Polygon	Determine all the possible changes in SFRM's mass and longitudinal CG	Mass and trimming moment
Turning Circle	Determine steady turning performance	Advance, transfer, and tactical diameter
Horizontal Zigzag	Demonstrate course-varying performance in the horizontal plane	Yaw overshoot, width of path overshoot, time at which each change of the rudder plane angle is initiated, and time to the maximum width excursion
Spiral	Determine control authority and straight-line stability	Steady yaw rate for each rudder angle
Vertical Zigzag	Demonstrate depth-varying performance in the vertical plane	Pitch overshoot, width of depth overshoot, time at which each change of the stern plane angle is initiated, and time to the maximum depth excursion
Meander	Determine motion stability and depth keeping performance	Path of the SFRM

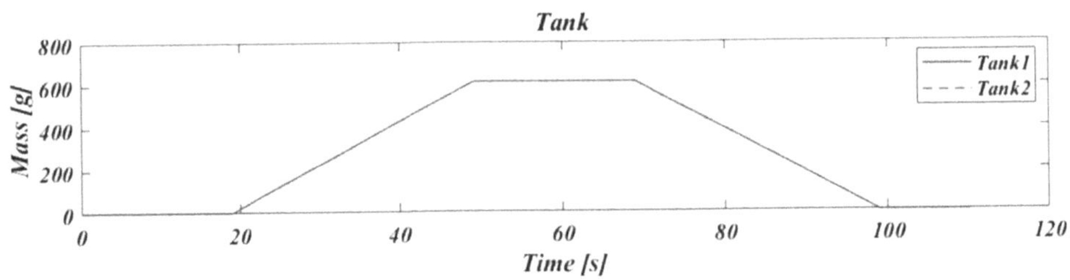

Figure 9. *Time series of the mass changes in the forward and aft ballast tanks, respectively.*

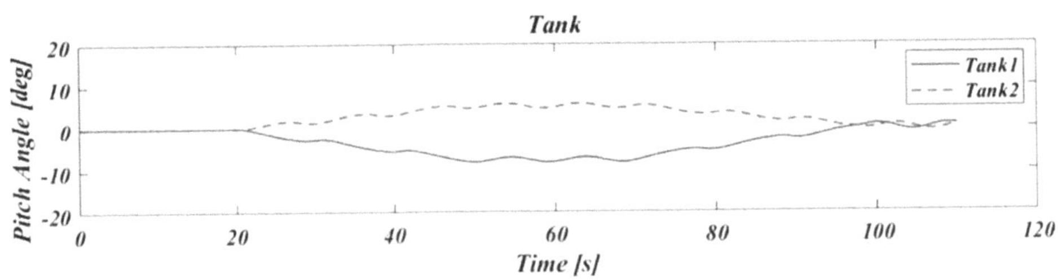

Figure 10. *Time series of the corresponding pitch angles in the forward and aft ballast tanks, respectively.*

2 The forward ballast tank is filled completely. There are increases in mass and trimming moment due to moving forward of the longitudinal CG.

3 The aft ballast tank is filled completely. This is continued until the previous procedure has been completed. There is an increase in mass but a slight decrease in trimming moment due to moving backward of the longitudinal CG.

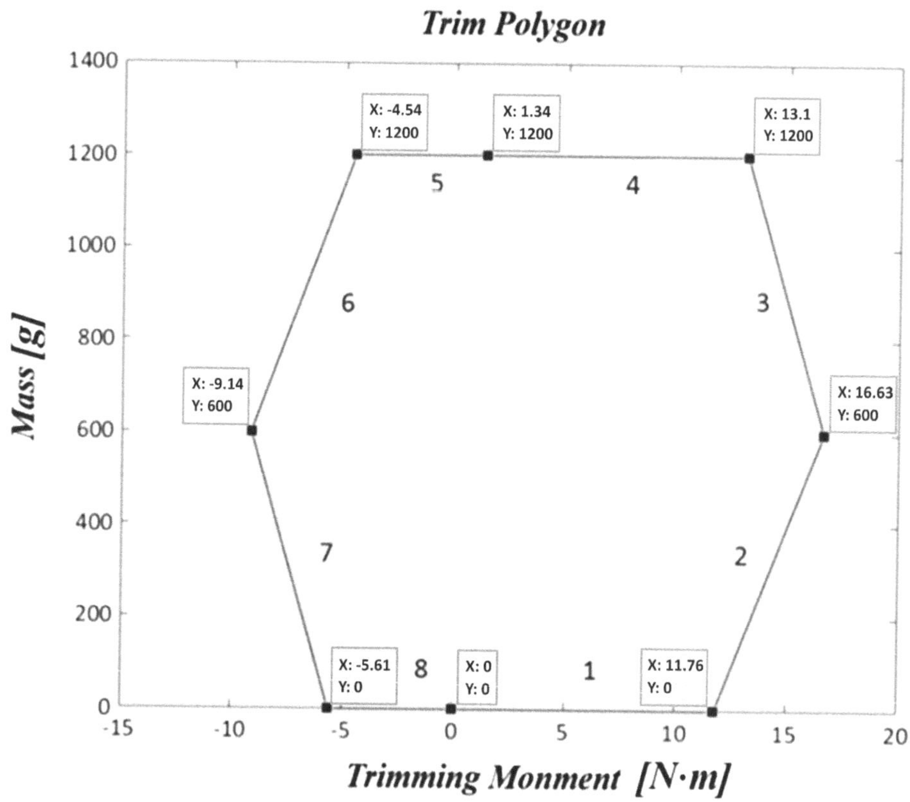

Figure 11. *The trim polygon of the SFRM.*

4 Move the weight of the slider mechanism device to the initial location. There is no change in mass, but there is a slight difference in trimming moment compared with the line in step 1. When both of the ballast tanks are full, the slight forward of the longitudinal CG causes a small forward trimming moment.

5 Move backward the weight of the slider mechanism device. The mass is still kept the same as the one in step 4, but the variation of the trimming moment is even less than the one in step 4.

6 The forward ballast tank is completely empty. There are decreases in mass and trimming moment due to moving backward of the longitudinal CG. The variation of mass is the same as the one in step 3, but the variation of the trimming moment is even larger than the one in step 3.

7 The aft ballast tank is completely empty. There is a decrease in mass, but an increase in trimming moment due to the moving forward of the longitudinal CG. The variation of mass is the same as the one in step 2, but the variation of the trimming moment is even smaller than the one in step 2. 8 Move the weight of the slider mechanism device to the initial location. When all the ballast tanks are empty, the line has returned to the initial location with zero mass and zero trimming moment.

Table 5. *Variations of trimming moment and mass in different steps.*

Step	Trimming Moment [N·m]	Mass [g]
0	0	0
1	11.76	0
2	16.63	600
3	13.1	1200
4	1.34	1200
5	−4.54	1200
6	−9.14	600
7	−5.61	0
8	0	0

Finally, the results of the trim polygon by adopting the above-mentioned procedure were introduced in Table 5.

5.2. Manoeuvring Tests

5.2.1. Turning Circle Test

When the rudder angle was given a constant value, the SFRM would run in a circle with the direction of its course changing 360°. In the turning circle test, the tactical diameter, advance, transfer, drift angle, and speed loss can be measured. At the beginning, the SFRM sailed straight for 100 s. Subsequently, the rudder was then pulled to 20° immediately for another 100 s for a total simulation time of 200 s. In the turning circle test, the exact turning time of the rudder was set to be 1 s.

The main purpose of the turning circle test was to evaluate the turning performance of the SFRM. Figure 12 compares the trajectories of the simulated SFRM at three sailing speeds: $U = 0.5, 1$, and 1.5 m/s. It is clearly found that the straight parts of these three trajectories are not coincident due to the variation of the critical point [35], which will affect the pitch angle and the straight-line stability. When executing the turning circle tests, the advances of the SFRM are 44.8, 95, and 145.6 m, corresponding to three sailing speeds, i.e., $U = 0.5, 1$, and 1.5 m/s. On the other hand, the transfers are identical to 4.61, 4.52 and 4.47 m, whereas the tactical diameters are 9.79, 9.77, and 0.68 m, respectively. In contrast to advance, both transfer and tactical diameter become small with the increase of the sailing speed.

When the speed was near 1.5 m/s, the SFRM started to exhibit slight vertical vibrations. It is obvious that the distance and trajectory of the SFRM were evidently affected by the sailing speed. With the increase of the sailing speed, the trajectory's vertical vibration becomes evident due to the increase of hydrodynamic trimming moment.

Figure 13 presents the time series of velocity components at different sailing speeds in each degree of freedom, including surge (u), sway (v), heave (w), roll (p), pitch (q), and yaw rates (r). The variations of the 6-DOF velocity components were found to be dominated by the sailing speeds. This was particularly evident

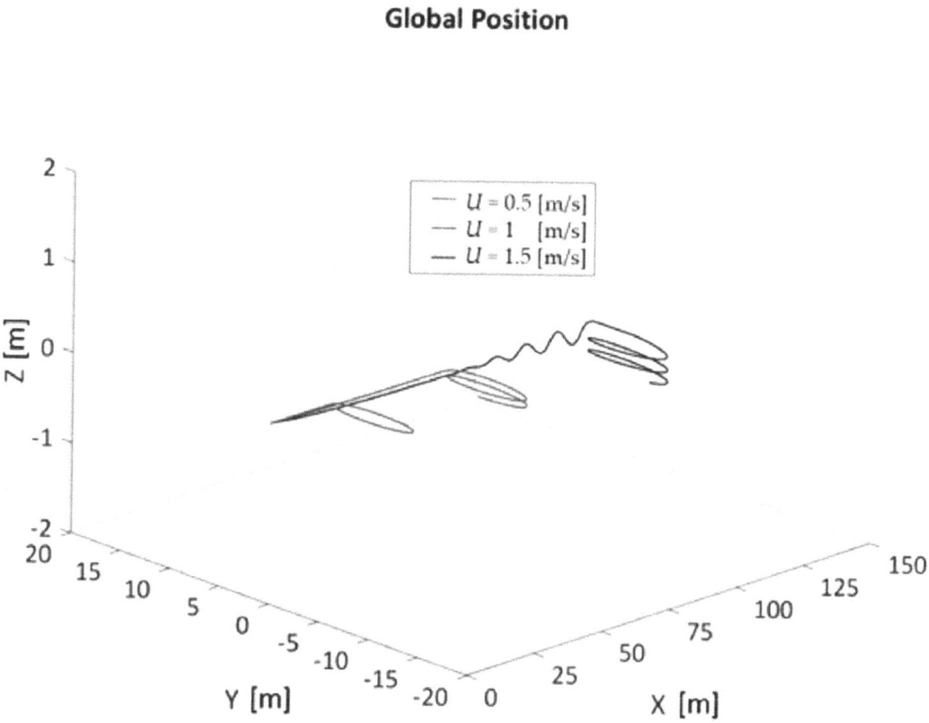

Figure 12. *The SFRM's trajectories at different sailing speeds in the turning circle test.*

after the rudder was steered, as lateral resistance caused a considerable decrease in surge velocity (Figure 13a). Additionally, different speed settings caused the heave velocity and the pitch velocity in Figure 13c,e, respectively, to undulate slightly when the sailing speed grew high but remained rather stable when the sailing speed was low.

5.2.2. Horizontal Zigzag Test

In the horizontal zigzag test, the rudder planes were immediately pulled to a fixed angle (δ_V) and remained in that position until the desired course angle (Ψ_d) of the SFRM is reached; then, the rudder was pulled to the counter rudder angle ($-\delta_V$) immediately and remained in that position until the desired course angle turned to $-\Psi_d$. Specifically, different rudder angles (δ) and course (yaw) angles (Ψ) were experimented on in the test. The course overshooting angle and the time required to reach maximum overshooting within the duration of each rudder angle change were measured to determine the SFRM's manoeuvrability in course change. In the simulation, the SFRM sailed straight for 100 s Then, the rudder plane was immediately turned to 20°, and the position of the rudder plane was maintained for a certain period. When the yaw angle reached 20°, the rudder plane was immediately turned to $-20°$, and the position of the rudder plane was maintained for a certain period. After the yaw angle changed to $-20°$, the previous

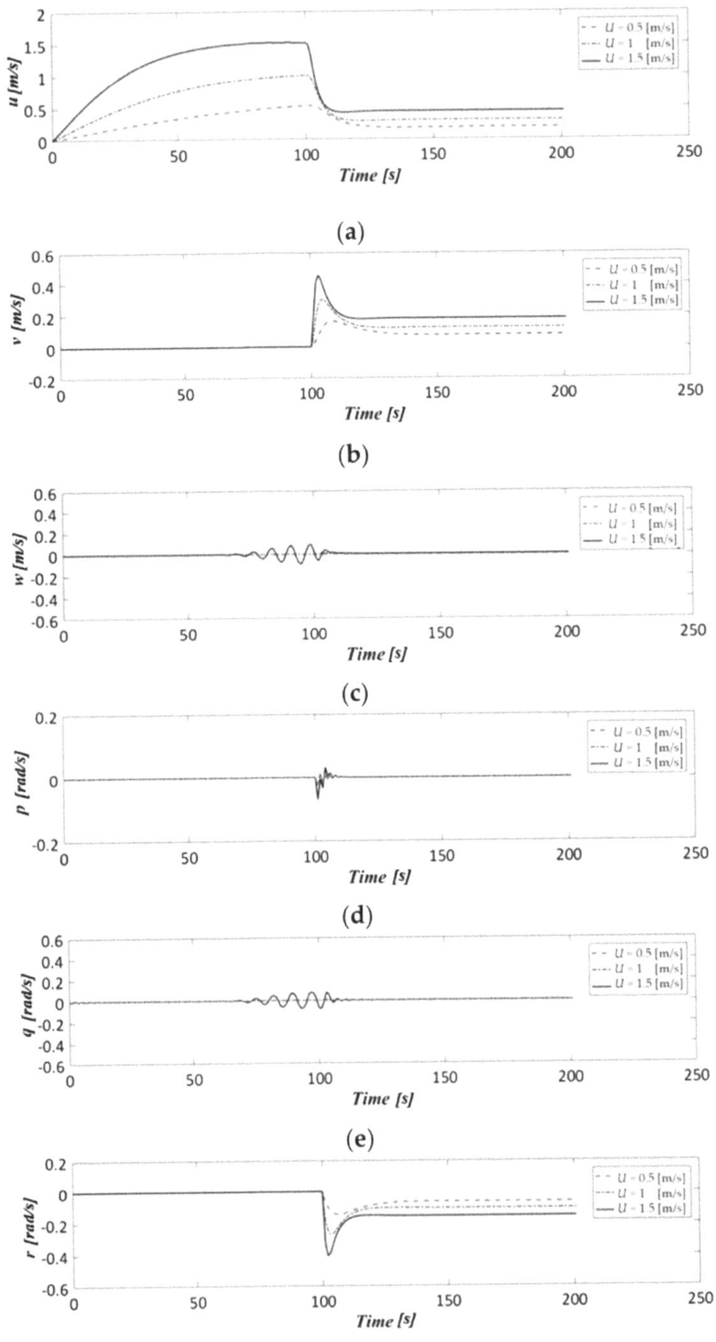

Figure 13. *Time series of 6-DOF velocity components during the turning circle test at different sailing speeds. (a) Surge rate:* $u[\mathrm{m/s}]$*, (b) sway rate:* $v[\mathrm{m/s}]$*, (c) heave rate:* $w[\mathrm{m/s}]$*, (d) roll rate:* $p[\mathrm{rad/s}]$*, (e) pitch rate:* $q[\mathrm{rad/s}]$*, and (f) yaw rate:* $r[\mathrm{rad/s}]$*.*

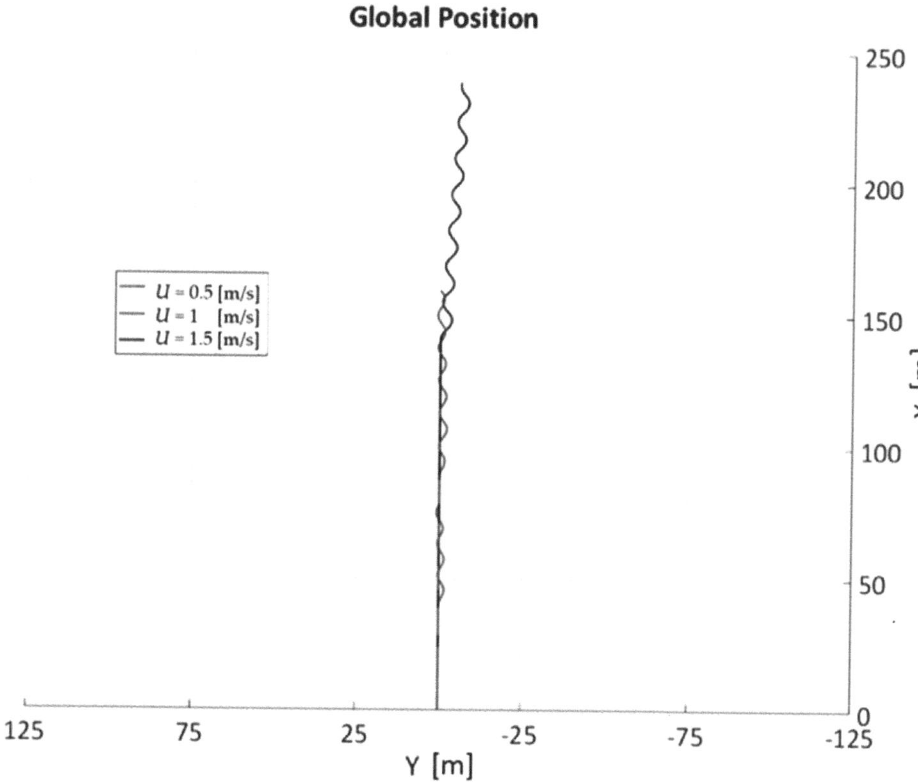

Figure 14. *The SFRM's trajectories in the horizontal zigzag test at different sailing speeds.*

procedure was repeated. In the horizontal zigzag test, the exact turning time of the rudder was set to be 1 s. Meanwhile, the maintaining time of the rudder angle at $\pm 20^\circ$ was given as 15,8 , and 6 s for $U = 0.5, 1$, and 1.5 m/s, respectively.

Figure 14 compares the SFRM's trajectories (top view) at three different sailing speeds: $U = 0.5, 1$, and 1.5 m/s, respectively. When $U = 1.5$ m/s, the trajectory was significantly longer than those of $U = 1$ m/s and $U = 0.5$ m/s. Additionally, the yawing amplitude (width of the path overshoot) was found to be larger for the case of the faster sailing speed and varied according to different periods of the rudder operations. For $U = 0.5, 1$, and 1.5 m/s, the yawing amplitudes are 0.89, 2.07, and 3.36 m. The trajectory of the SFRM at the highest sailing speed is not straight due to the slight variation of yaw overshoot in the subsequent horizontal zigzag maneouvres.

Figure 15a-c presents the time series of rudder and yaw angles at different sailing speeds. Figure 15a indicates that when $U = 0.5$ m/s, a full cycle of turning the yaw angle from 0° to -20° and then 20° required 21 s. Figure 15 b illustrates that when $U = 1$ m/s, a full turning cycle was 11 s long. Finally, Figure 15c indicates that when $U = 1.5$ m/s, a full turning cycle was 9 s long. Specifically, the time to the maximum width excursion of the first overshoot from the initial operation

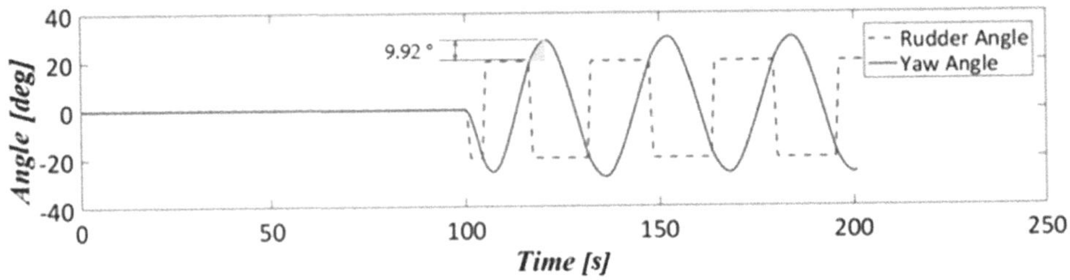

(a)

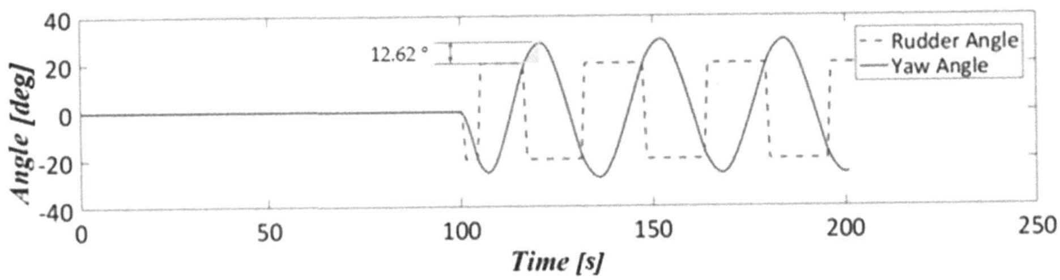

(b)

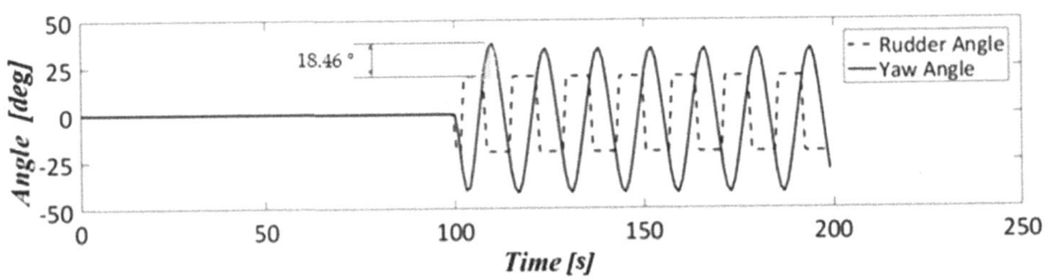

(c)

Figure 15. *Time series of rudder and yaw angles at* (**a**)$U = 0.5$ m/s, (**b**)$U = 1$ m/s, *and* (*c*) $U = 1.5$ m/s. *The maximum overshooting angle is highlighted by the gray area.*

of the rudder planes required 7, 4, and 3 s for $U = 0.5, 1.0$, and 1.5 m/s. On the other hand, time to the maximum width excursion of the second overshoot from the initial operation of the rudder planes required 21, 11, and 9 s for $U = 0.5, 1.0$, and 1.5 m/s.

The SFRM's reaction speed increased considerably at higher sailing speeds. In terms of overshooting angles, Figure 15 a indicates that when $U = 0.5$ m/s, the maximum overshooting angle was 9.92°. In Figure 15b, when $U = 1$ m/s, the maximum overshooting angle was 12.62°. Finally, Figure 15c indicates that when $U = 1.5$ m/s, the maximum overshooting angle was 18.46°. The overshooting angle increased considerably in accordance with higher speeds. Given that the time at which the deflection of rudder angle is initiated is the same, i.e., $t = 100$ s, the time to the maximum width excursion of the yaw angle is shorter for the higher speed than that for the lower speed. Additionally, more turning cycles were completed at higher speeds than at lower speeds. Therefore, sailing speed was found to be a critical parameter influencing the relationship between yaw angle, rudder angle, and overshooting angle.

Figure 16a-f display the time series of surge (u), sway (v), heave (w), roll (p), pitch (q), and yaw rates (r) at different sailing speeds and corresponding amplitude changes. In the horizontal zigzag test, the relationship between rudder angle and course angle was analysed. Accordingly, the velocities and angular velocities in the six degrees of freedom exhibited more significant changes when $U = 1.5\,\mathrm{m/s}$ than when $U = 1$ and $0.5\,\mathrm{m/s}$. In Figure 16c, e, when the sailing speed reaches close to $1.5\,\mathrm{m/s}$, some vertical vibrations will begin to appear in the heave and pitch rates.

5.2.3. Spiral Test

The main purpose of the spiral test was to evaluate the SFRM's ability to control the rudder plane and maintain linear stability. In the simulation, the rudder plane was immediately pulled to a certain angle, e.g., 25°, and the angle was gradually reduced until the stable yaw rate was reached. This was repeated to identify the relationship between the rudder plane angle and the yaw rate. Figure 17 illustrates the distributions of the spiral tests for different rudder plane angles at three sailing speeds: $U = 0.5, 1$, and $1.5\,\mathrm{m/s}$. Specifically, Figure 17 displays changes in yaw rate and rudder plane angles when the rudder plane was gradually reduced from 25° to −25° by 5° each time. From the simulation results, the yaw rate exhibited a relatively linear trend with different rudder plane angles and sailing speeds. For example, a larger rudder plane angle indicated a faster yaw rate. In other words, it explained that the yaw rate at each increment is dependent on the initial sailing speed. The trend also revealed that for small rudder angles, the rate of turn would depend upon whether the rudder angle is increasing or decreasing [36], especially in case of high sailing speed.

5.2.4. Vertical Zigzag Test

In the vertical zigzag test, the stern planes were pulled to a fixed angle (δ_H), and its position was maintained until the desired pitch angle (θ_d) was reached before the stern planes were pulled to the counter angle $(-\delta_H)$ and maintained until the desired pitch angle of the SFRM reached $-\theta_d$. Different stern plane angles (δ) and pitch angles (θ) were experimented on in the test. The overshooting angle of each pitch movement, the width of depth overshoot, and time to the maximum depth excursion of each stern plane angle change were measured to determine the SFRM's manoeuvrability in depth change. Vertical zigzag manoeuvre is widely used to evaluate the manoeuvrability of submarines, so this can be definitely used to evaluate the manoeuvrability of the SFRM. In the simulation, the SFRM first sailed straight for 100 s; then, the stern planes were turned to 20°, and its position was maintained for a certain period. When the pitch angle reached 20°, the

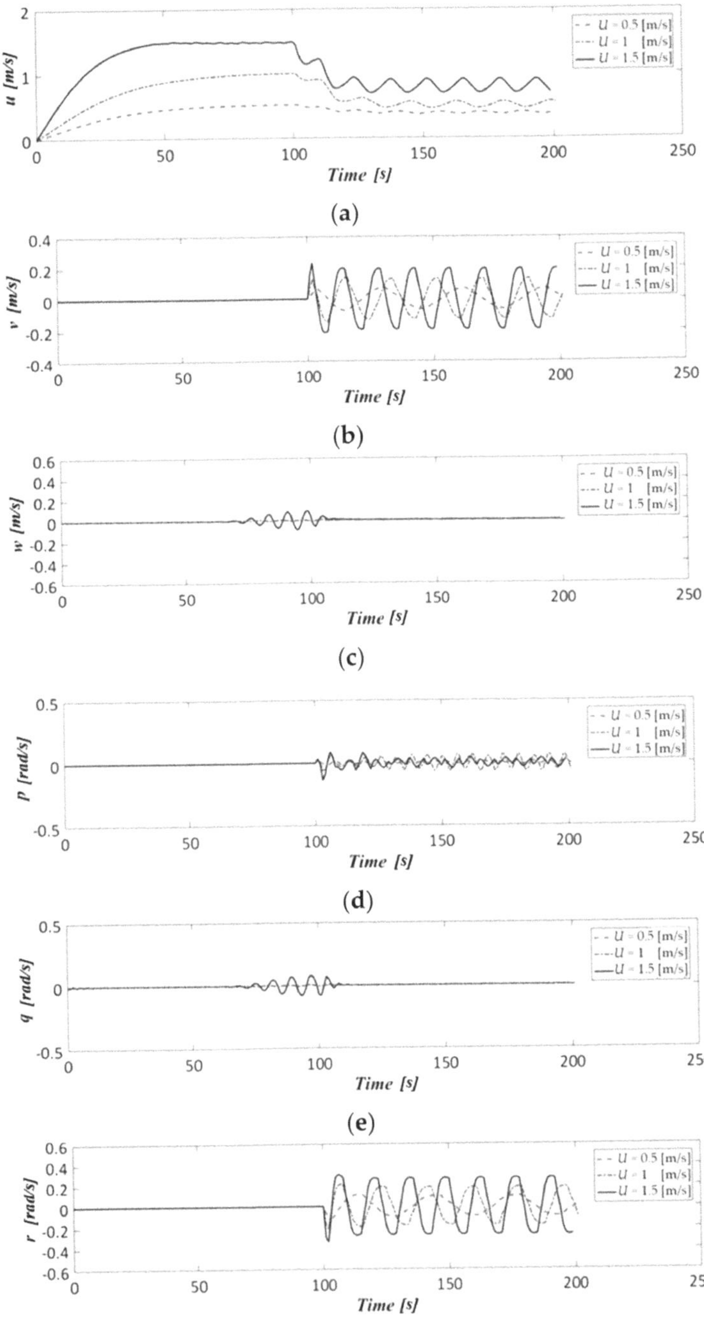

Figure 16. *6-DOF velocity components in the horizontal zigzag test at different sailing speeds. (a) Surge rate: u[m/s], (b) sway rate: v[m/s], (c) heave rate: w[m/s], (d) roll rate: p[rad/s], (e) pitch rate: q[rad/s], and (f) yaw rate: r[rad/s].*

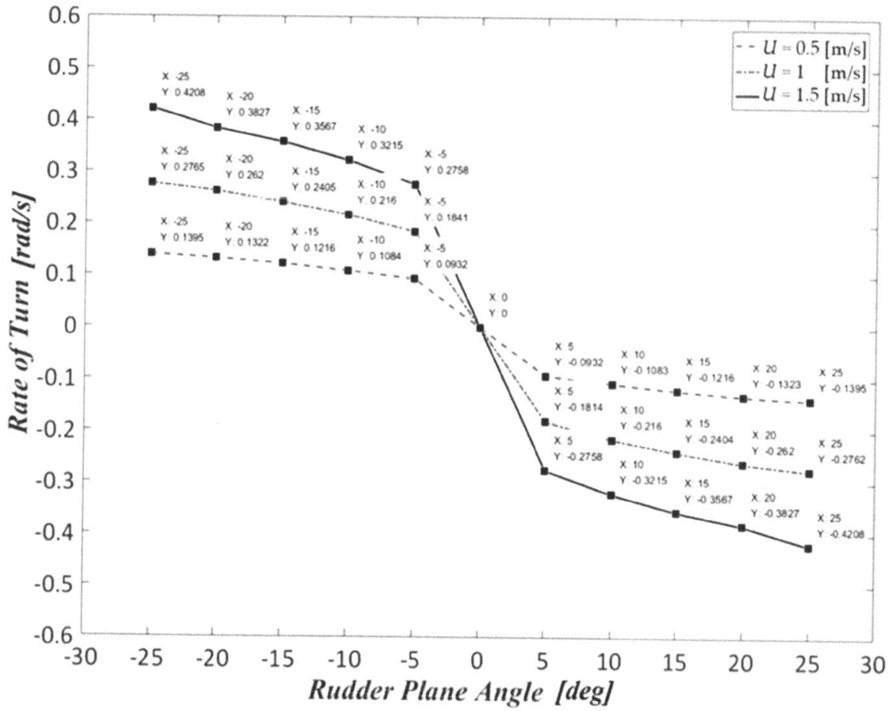

Figure 17. *Relationships between the yaw rates and the rudder plane angles at different sailing speeds for the spiral test.*

stern plane was immediately turned to $-20°$, and its position was maintained for a certain period. When the pitch angle reached $-20°$, the procedure was repeated. However, the pitch overshoot was reached only in case of $U = 1.5$ m/s. For comparison, the stern plane operations of $U = 0.5$ and 1.0 m/s were set to be the same as the ones of $U = 1.5$ m/s. In the vertical zigzag test, the exact turning time of the stern plane was set to be 1 s. In addition, the maintaining time of the stern plane angle at $\pm 20°$ was all given as 6 s for $U = 0.5, 1$, and 1.5 m/s.

Figure 18 compares the SFRM's trajectories in the vertical zigzag test at sailing speeds of $U = 0.5, 1$, and 1.5 m/s. When $U = 1.5$ m/s, the trajectory was significantly longer than those of $U = 1$ m/s and $U = 0.5$ m/s. When $U = 1.5$ m/s, the trajectory was significantly longer than those of $U = 1$ m/s and $U = 0.5$ m/s. The pitching amplitude (width of the depth overshoot) was found to be larger for the case of the faster sailing speed and varied according to different periods of the stern plane operations. For $U = 0.5, 1$, and 1.5 m/s, the pitching amplitudes are $0.44, 0.48$, and 1.14 m. The SFRM moved significantly farther at higher speeds, and a slight pitch motion occurred due to the movement of the critical point [35]. In addition, the simulated trajectories and time between stern plane position changes varied at different sailing speeds.

Figure 19a-c illustrates the variations of time series in stern plane and pitch angles at three sailing speeds. A full cycle of turning the pitch angle from $0°$ to $-20°$

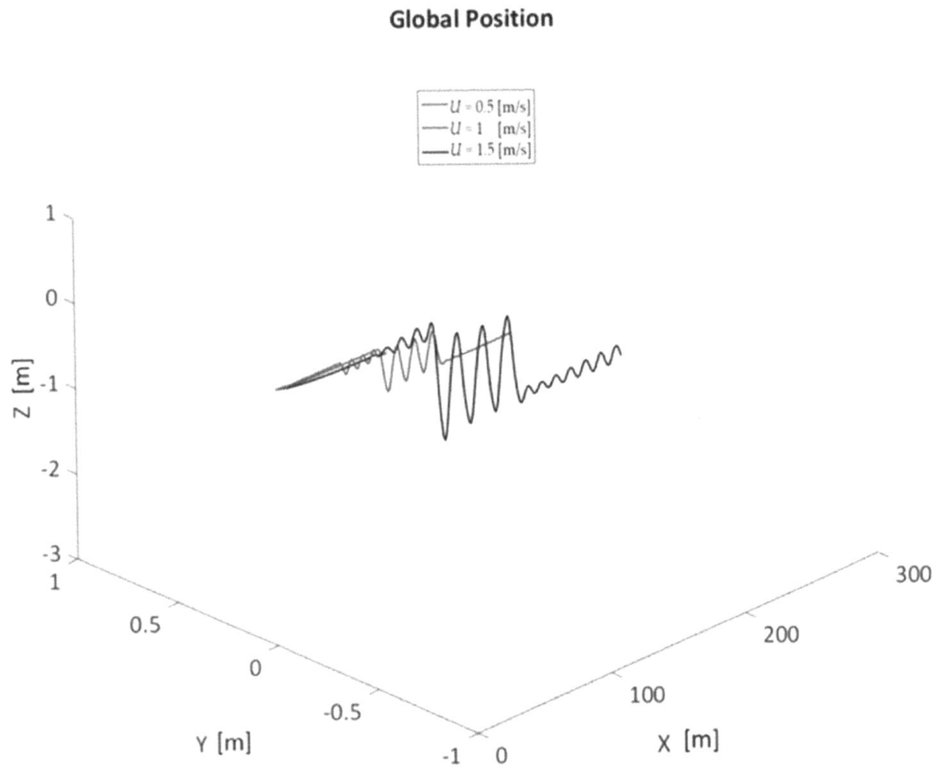

Figure 18. *The SFRM's trajectories in the vertical zigzag test at different sailing speeds.*

and then 20° required 10 s. Specifically, time to the maximum depth excursion of the first overshoot from the initial operation of the stern planes required 4 s. On the other hand, time to the maximum depth excursion of the second overshoot from the initial operation of the stern planes required 10 s.

In terms of overshooting angles, Figure 19 a indicates that when $U = 0.5\ \mathrm{m/s}$, the maximum overshooting angle was $-14.15°$. In Figure 19b, the maximum overshooting angle was $-6.94°$ when $U = 1\ \mathrm{m/s}$. Finally, Figure 19c demonstrates that when $U = 1.5\ \mathrm{m/s}$, the maximum overshooting angle was $2.1°$. Therefore, at high sailing speeds, the maximum overshooting angle of the SFRM increased significantly compared with that at lower sailing speeds. Given that the stern-plane operation period is the same, the vertical zigzag movement is verified to be highly correlated with sailing speed. Since the variation of ballast tanks has an influence on the vertical zig-zag test, the changes in the vehicle's weight, CG, buoyancy center, and moment of inertia must be considered in future work, in addition to the two aforementioned factors.

Figure 20a-f illustrate the variations of time series for surge (u), sway (v), heave (w), roll (p), pitch (q), and yaw rates (r) at different sailing speeds. In the vertical zigzag test, both the heave rate in Figure 20c and the pitch rate in Figure 20e had the most pronounced changes, and a slight undulation was observed

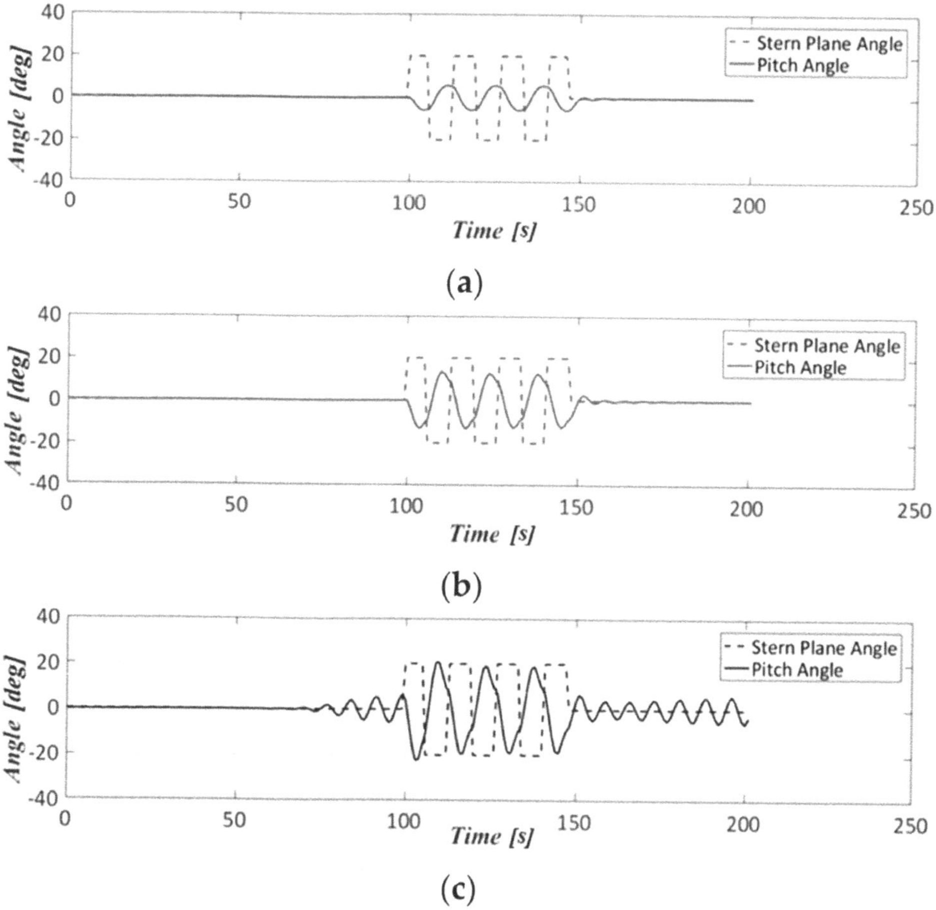

Figure 19. *Comparisons of stern plane and pitch angles at sailing speeds of (a) $U = 0.5$, (b) $U = 1$, and (c) $U = 1.5$ m/s, respectively.*

at high speeds. In Figure 20b,f, the vertical zigzag has less influence on the sway and the yaw rates of the horizontal plane motion. Figure 20d indicates that when $U = 1.5$ m/s between 100 and 150 s, there is a slight swing in the roll rate caused by the variation of the stern plane.

5.2.5. Meander Test

The meander test is similar to the vertical zigzag test, but the only difference is that the angle of the stern plane returns to zero when the desired pitch angle is reached. The main purpose of the meander test is to measure the SFRM's trajectory, motion stability, and ability to remain at a fixed depth. In the simulation, the model sailed straight for 100 s; then, the stern plane was pulled to 20° immediately, and its position was maintained for 10 s before the stern plane angle was reset to zero. In the meander test, the exact turning time of the stern plane was set to be 1 s.

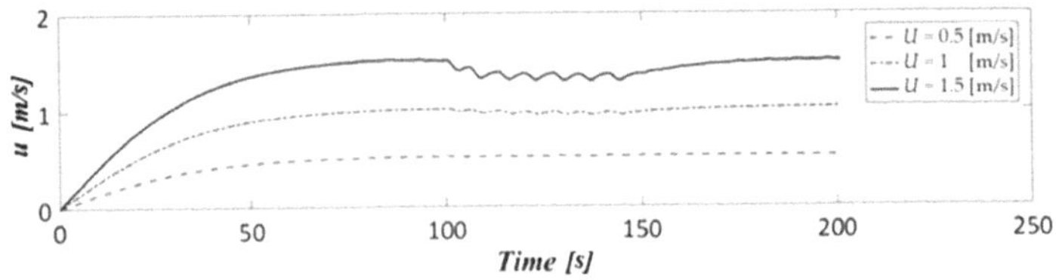

(a)

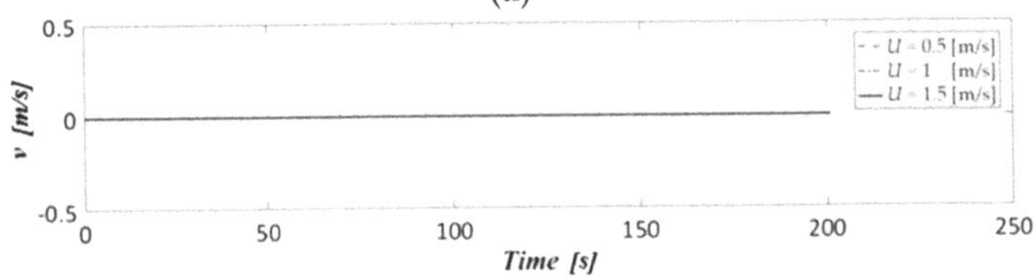

(b)

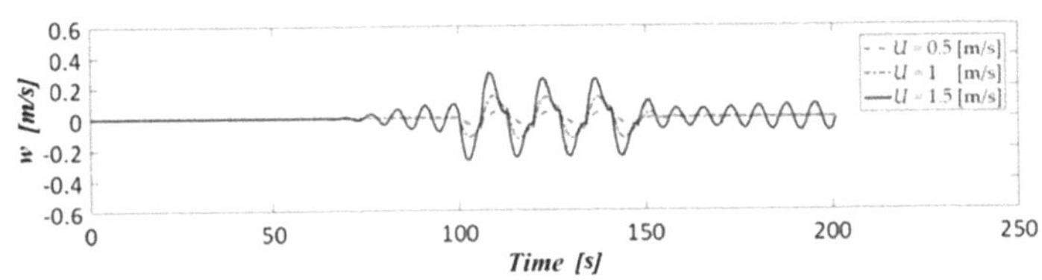

(c)

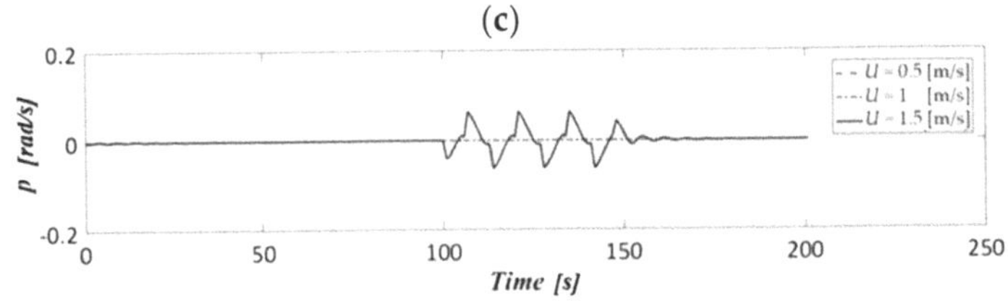

(d)

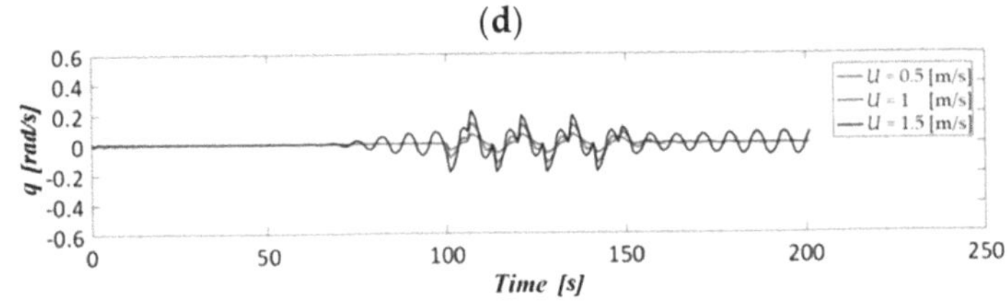

(e)

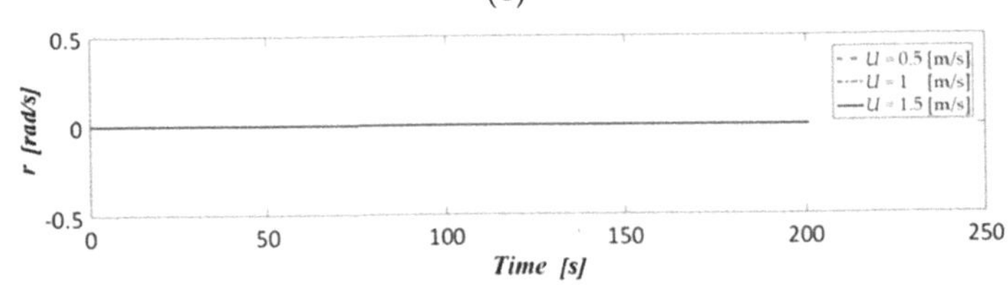

(f)

Figure 20. *6-DOF velocity components in the vertical zigzag test at different sailing speeds. (a) Surge rate:* $u[\mathrm{m/s}]$, *(b) sway rate:* $v[\mathrm{m/s}]$, *(c) heave rate:* $w[\mathrm{m/s}]$, **(d)** *roll rate:* $p[\mathrm{rad/s}]$, *(e) pitch rate:* $q[\mathrm{rad/s}]$, *and (f) yaw rate:* $r[\mathrm{rad/s}]$.

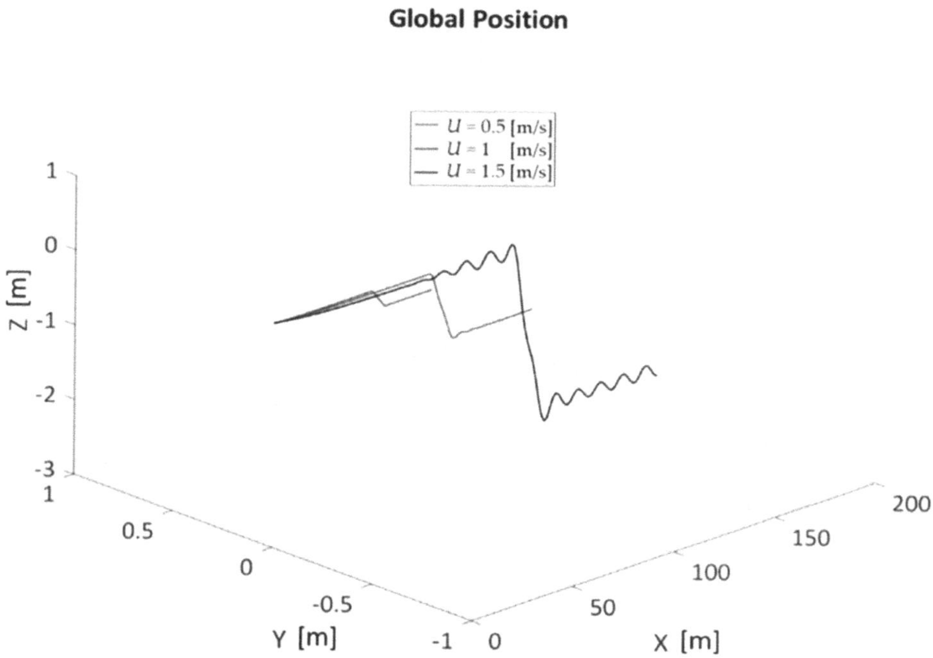

Figure 21. *The SFRM's trajectories in the meander test at different sailing speeds.*

Figure 21 illustrates the trajectories in the meander test when the simulated SFRM moved at three different sailing speeds: $U = 0.5, 1$, and 1.5 m/s. The SFRM moved farther at high sailing speeds, and a slight vertical pitching motion occurred when the stern plane was reset to zero in the fixed depth test. The depth excursions are 0.051, 0.992, and 2.365 m, corresponding to $U = 0.5, 1$, and 1.5 m/s. Moreover, the results of the meander tests indicated that motion stability and depth keeping ability would be unstable for the higher sailing speed by observing the variation of their trajectories.

Figure 22a-c illustrates the time series of the stern plane and pitch angles at different sailing speeds. The simulation results were similar to those of the vertical zigzag test and indicated a high correlation with sailing speed. Figure 22a,b indicates that when $U = 0.5$ and 1 m/s, the desired pitch angle could not be reached because of insufficient thrust force. Figure 22c indicates that when $U = 1.5$ m/s, the desired pitch angle was reached. These results were similar to those of the vertical zigzag test, suggesting that changes in pitch angle and ability to remain at a fixed depth have a crucial relationship with sailing speed adjustment. When conducting this test, changes in the vehicle's weight, CG, buoyancy center, and amount of inertia must be considered, and a state of neutral buoyancy must be ensured. Finally, at high speeds, relatively large undulation is likely to be observed in the pitch mode.

Figure 23a-f displays the time series of surge (u), sway (v), heave (w), roll (p), pitch (q), and yaw rates (r) under different sailing speed conditions. Since the

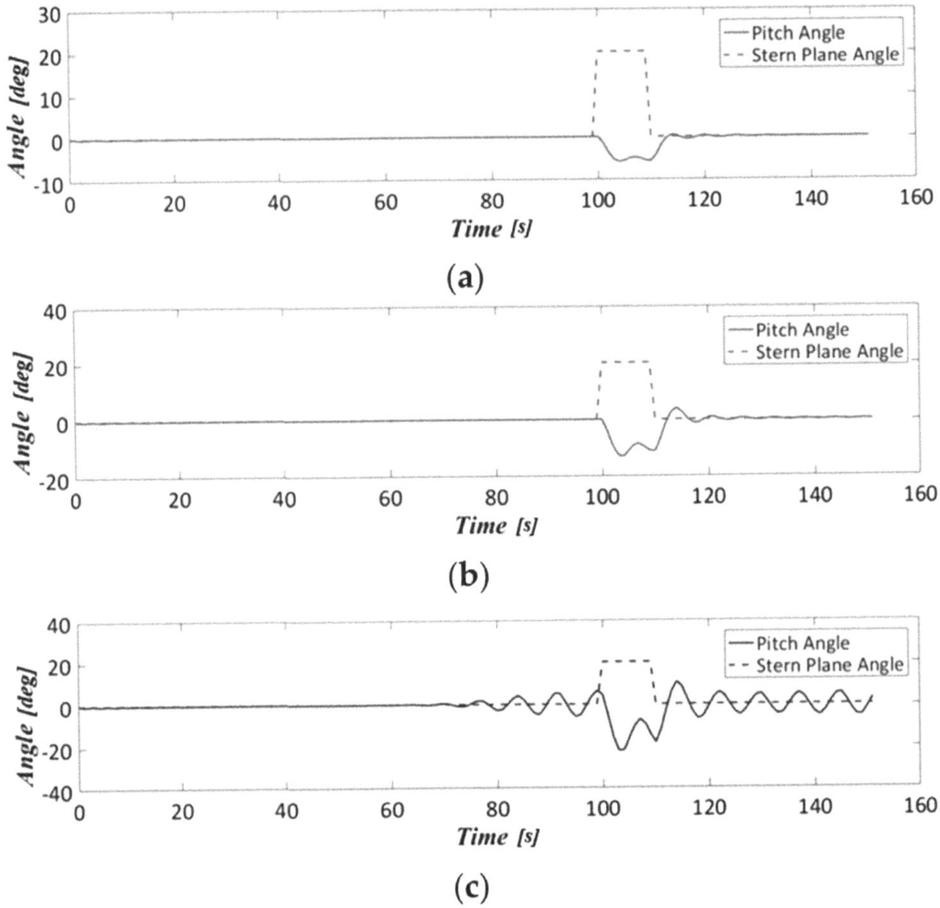

Figure 22. *Comparisons of rudder and pitch angles at different sailing speeds. (a)* $U = 0.5$, *(b)* $U = 1$, *and (c)* $U = 1.5$ m/s.

meander test is highly similar to the vertical zigzag test, changes of the heave rate in Figure 23c and the pitch rate in Figure 23e were evident. It is obvious in Figure 23b,d,f that the sway, roll, and yaw rates have little variations because the meander test is principally a test of vertical depth-determination ability.

5.3. Optimal PD Control Gains

After each manoeuvring test was conducted to determine the operation limit for rudder performance, the optimal PD control gains for each test would be calculated by using the PID tuner. Figure 24a-e show the step responses, i.e., the output of the control system for a step input, of different manoeuvring tests with three sailing speeds. Since the characteristics parameters, including rise time, settling time, and overshoot, change with the variation of the control gains (K_P, K_D), a set of optimal control gains can be obtained by examining the values of the characteristics parameters. To determine the efficacy of the proposed 6-DOF manoeuvring

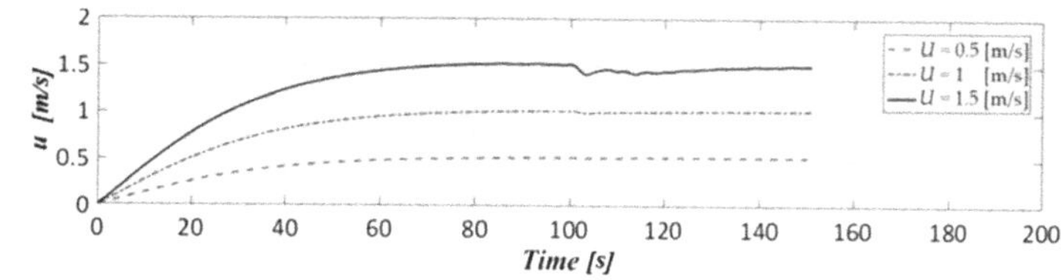

(a)

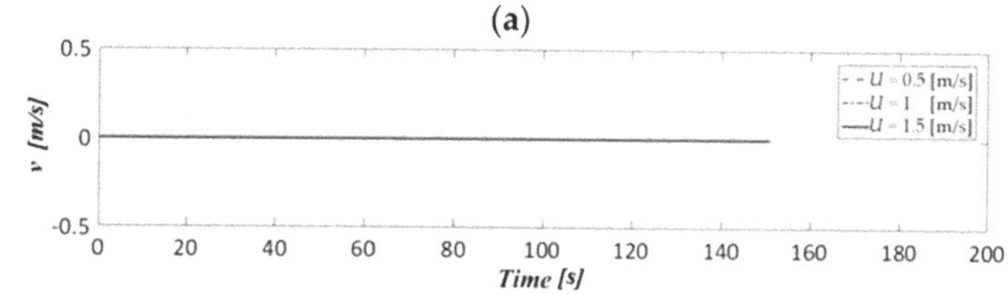

(b)

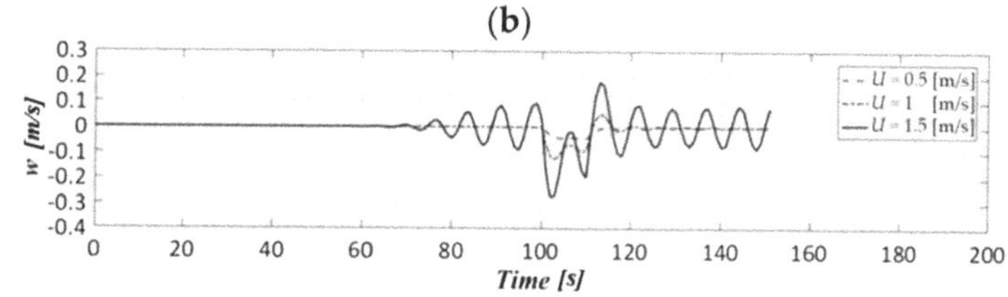

(c)

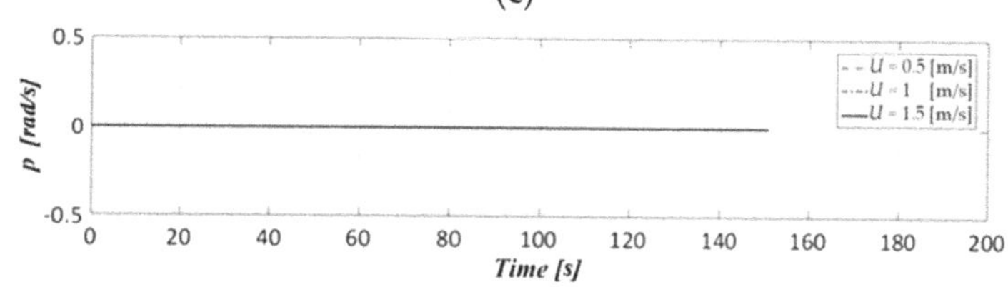

(d)

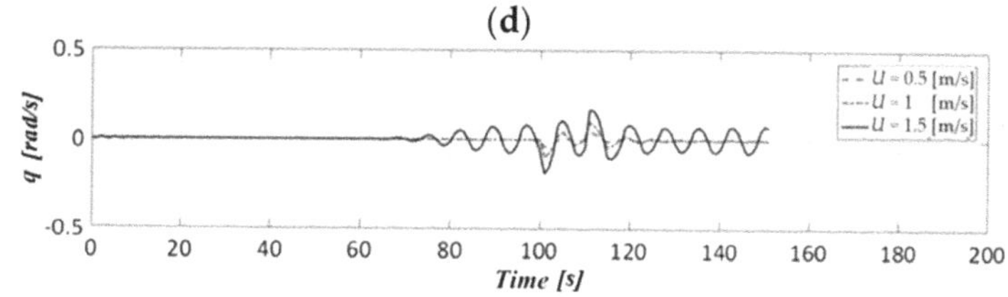

(e)

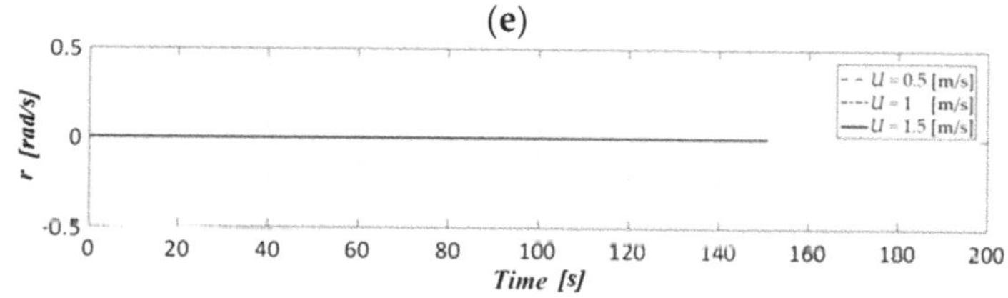

(f)

Figure 23. *Time series of 6-DOF velocity components in the meander test. (a) Surge rate:* $u[\mathrm{m/s}]$, *(b) sway rate:* $v[\mathrm{m/s}]$, *(c) heave rate:* $w[\mathrm{m/s}]$, *(d) roll rate:* $p[\mathrm{rad/s}]$, *(e) pitch rate:* $q[\mathrm{rad/s}]$, *and (f) yaw rate:* $r[\mathrm{rad/s}]$.

Table 6. *The characteristics parameters and the gains of the PD controller with the PID tuner for different manoeuvring tests.*

Test	U [m/s]	K_P	K_D	Rise Time [s]	Settling Time [s]	Overshoot [%]	Peak
Turning Circle	0.5	0.63	0.32	1.11	7.19	29.1	1.29
	1	0.82	0.46	0.918	8.06	34.4	1.32
	1.5	1.03	0.43	0.74	6.79	37.8	1.38
Horizontal ZigZag	0.5	0.98	0.32	1.09	6.66	17.8	1.19
	1	1.23	0.52	0.817	6.48	24.9	1.25
	1.5	1.47	0.61	0.764	6.42	31	1.31
Spiral	0.5	1.12	0.62	1.08	6.19	28.1	1.28
	1	1.33	0.67	0.93	6.83	32.4	1.31
	1.5	1.48	0.73	0.84	6.79	36.8	1.33
Vertical ZigZag	0.5	0.49	0.29	0.856	5.42	12.8	1.15
	1	0.78	0.75	0.981	6.71	22.8	1.25
	1.5	1.21	0.35	0.869	6.04	29.7	1.3
Meander	0.5	0.56	0.52	1.28	6.69	17.5	1.18
	1	0.92	0.82	0.866	6.03	19.5	1.22
	1.5	1.13	0.85	0.809	6.68	28.4	1.28

mathematical model combined with the PD controller, the PD control gains and the characteristics parameters of manoeuvring tests with the PID tuner are introduced in Table 6.

For the manoeuvring tests in the horizontal plane, i.e., turning circle, horizontal zigzag and spiral tests, the turning circle test has the largest percentage of overshoot on the basis of different sailing speeds. By contrast, the horizontal zigzag test and the spiral test have less amounts of overshoot than the turning circle test. Furthermore, the horizontal zigzag test has the minimum amount of overshoot, rise time, and settling time among all of the horizontal plane tests. In terms of the optimum action of the PD controller, the horizontal zigzag test has a better performance than the other two horizontal-plane tests.

For the manoeuvring tests in the vertical plane, i.e., vertical zigzag and meander tests, the amounts of overshoot for the meander test appear to be less than those for the vertical zigzag test mostly. On the other hand, the vertical zigzag test has a better performance of settling time than the meander test. In general, the values of rise time for the meander test are smaller than those for the vertical zigzag test. Consequently, it is difficult to judge the efficacy of the PID tuner by comparing the vertical zigzag test with the meander test.

5.4. Autopilot Simulation

In order to test turning and diving abilities of the SFRM with the PD controller, a 3D map with numerous waypoints was established. The coordinates of waypoints in the 3D map are described in Table 7. An autopilot system consists of a 6-DOF manoeuvring mathematical model and a course-keeping PD controller for an LOS-based guidance system. Simulation results of tuning PD control on the rudder planes and the stern planes would be presented by comparing with two types of controlling modes, i.e., A-type and B-type. It is worth mentioning that the PD

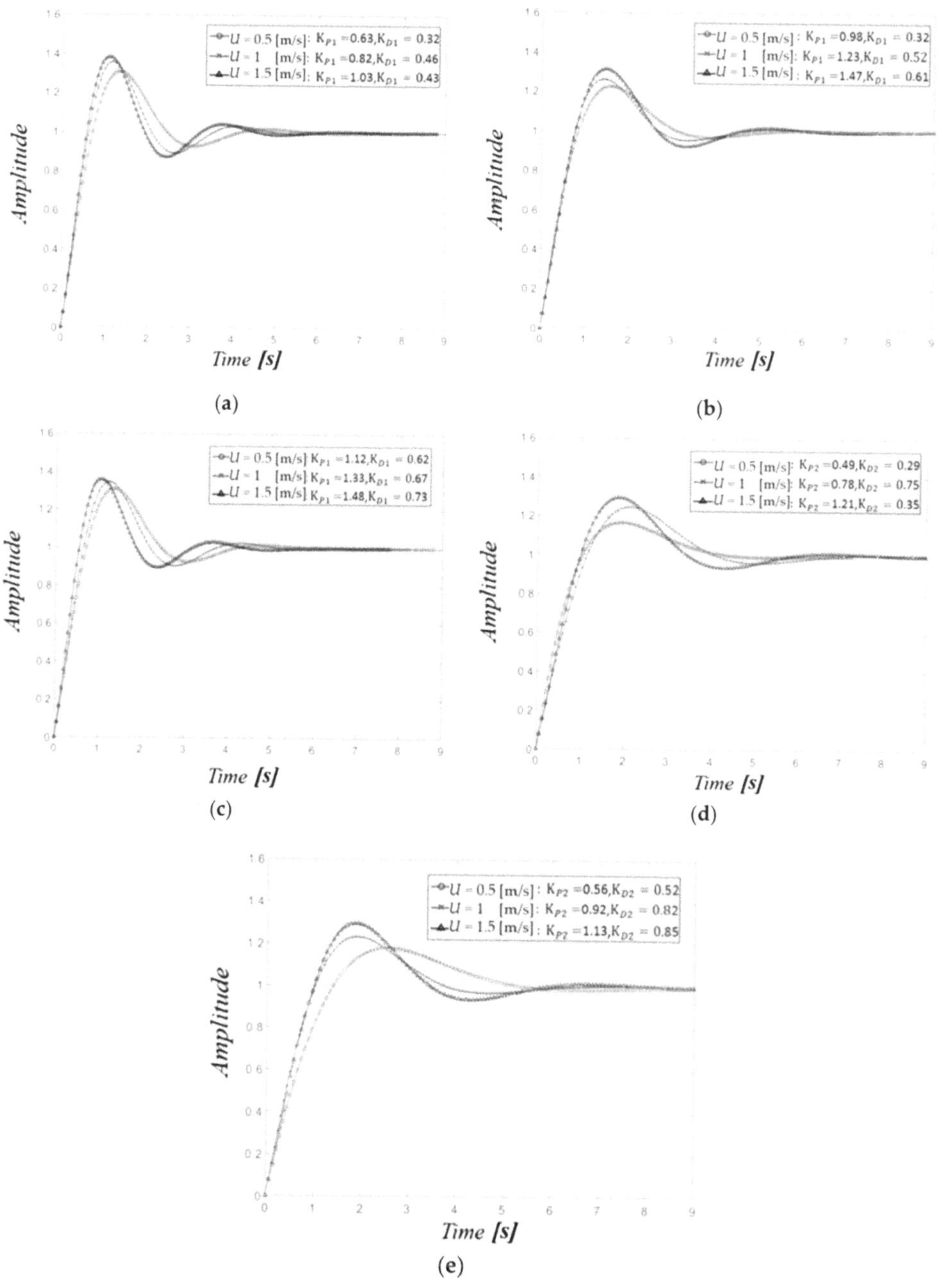

Figure 24. *Step responses of the PD controller with respect to three sailing speeds,* $U = 0.5, 1$, *and* 1.5 m/s, *for (a) the turning circle test, (b) the horizontal zigzag test, (c) the spiral test, (d) the vertical zigzag test, and (e) the meander test.*

Table 7. *The coordinates of waypoints in the 3D map. Unit: [m].*

Waypoint Number	Coordinate (X, Y, Z)
1	(20, 0, 0)
2	(20, −35, −5)
3	(55, −35, −10)
4	(55, 0, −15)
5	(90, 0, −20)
6	(90, −35, −25)
7	(125, −35, −30)
8	(125, 0, −35)
9	(145, 0, −40)

control gains of the A-type controller are tuned by the horizontal zigzag test and the vertical zigzag test, whereas the ones of the B-type controller are tuned by the horizontal zigzag test and the meander test, respectively.

Figure 25a-c shows the trajectories of the SFRM with three-view angles based on $U = 1.5\,\mathrm{m/s}$ in the 3D map by comparing the A-type with the B-type controllers. Comprehensively, the total sailing distance of the A-type controller is 382 m with the sailing time of 297 s, whereas the total sailing distance of the B-type controller is 387 m with the sailing time of 301 s. Compared with the B-type controller, the A-type controller is capable of saving 1.3% sailing distance and 0.7% sailing time. Although both simulated trajectories are very similar to each other, the A-type controller appears to have better performances of turning and diving abilities than those of the B-type.

Figure 26a,b presents variations of the rudder angle and the stern plane angle in the time domain. It is clearly found that both the rudder angle and the stern plane angle of the A-type controller are more sensitive to changes in the course angles during the passage.

On the other hand, time records of 6-DOF velocity components of the SFRM with the A-type and B-type controllers are illustrated in Figure 27a-f, respectively. There are large differences of heave rates (Figure 27c) and pitch rates (Figure 27e) between the A-type and the B-type controllers. Generally speaking, the A-type controller appears to have faster motion responses, especially for the heave rate and the pitch rate, than the B-type. The average power spent by the A-type controller is 5.02 W, which is better than 5.01 W spent by the B-type controller. In addition, the entire work done by using the A-type controller is 1490.94 J, whereas the result by using the B-type controller is 1508.01 J. In the case of efficiency, it would be more appropriate to adopt the A-type controller than the B-type in the autopilot simulation of the SFRM.

6. CONCLUSIONS

In this study, the 6-DOF manoeuvring mathematical model of the SFRM with the autopilot system was developed to conduct a preliminary evaluation of motion performance for optimal control gains. In order to realise the turning and diving

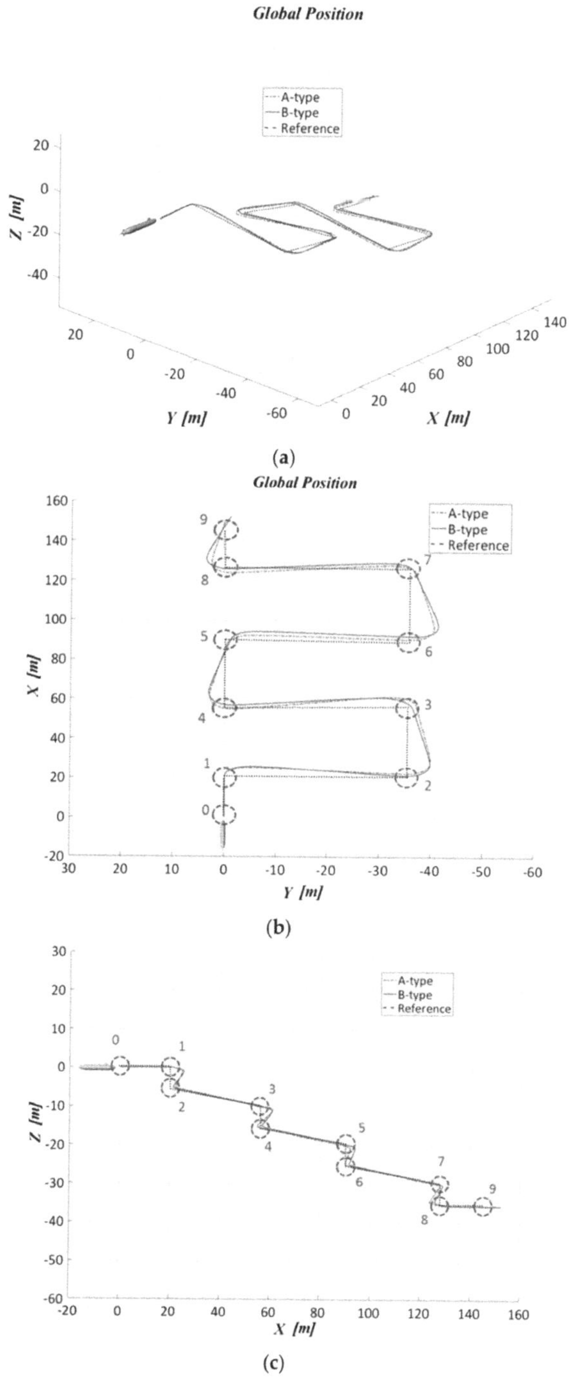

Figure 25. *Sailing trajectories of the SFRM by comparing the A-type controller with the B-type controller in (a) the stereo view; (b) the top view; and (c) the side view, respectively.*

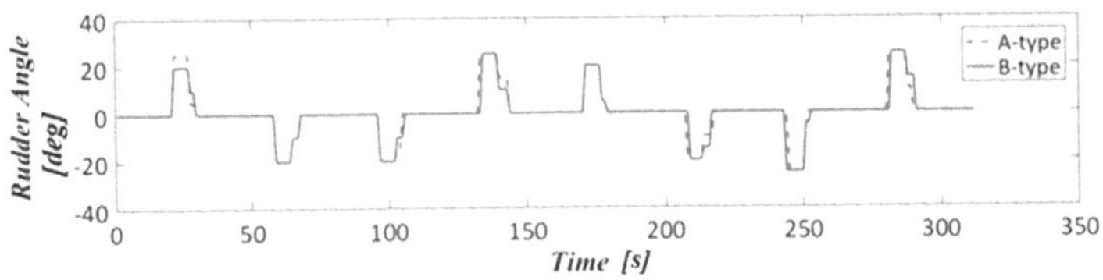

(a)

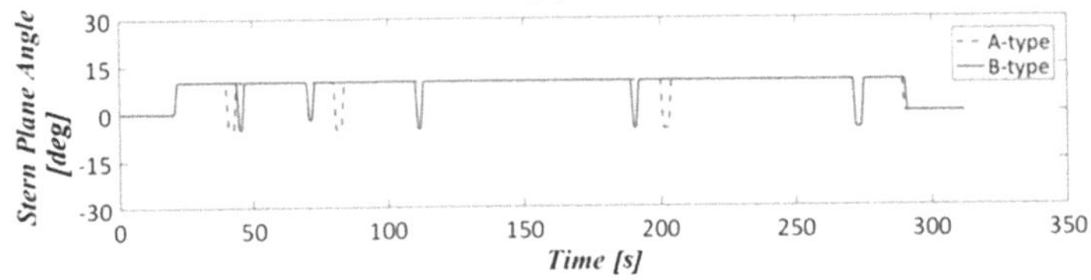

(b)

Figure 26. *Time series of (a) rudder angle, and (b) stern plane angles for the A-type and B-type controllers, respectively.*

performances of the SFRM, a series of manoeuvring tests, including turning circle, horizontal zigzag, vertical zigzag, meander, and spiral tests, would be implemented based on three sailing speeds with empty ballast tanks. Furthermore, the PID tuner was used to acquire the optimal control gains for each manoeuvring test by considering characteristics parameters. Therefore, two types of control gains by combining different horizontal plane motion tests with vertical plane motion tests were adopted for subsequent autopilot simulation. The post-processor based on the Cinema 4D modelling is capable of allowing the animated model to match the simulated trajectories and course angles graphically. The concluding remarks can be summarised as below:

1 There were only linear terms of hydrodynamic coefficients considered in the present simulation regardless of the scale effect. The sailing speed and the operation of control surface were considered as the influential parameters in the manoeuvring tests with empty ballast tanks.
2 Through a series of manoeuvring tests, the operation performances by adopting different combinations of rudder commands and sailing speeds were well analysed.
3 After evaluating the characteristics parameters of the PD controller for manoeuvring tests in the horizontal and vertical plane, both the A-type (horizontal zigzag plus vertical zigzag) and the B-type (horizontal zigzag plus meander) controllers were compared in the autopilot simulation.
4 When conducting the autopilot simulation in the 3D map, the A-type PD controller with the LOS guidance algorithm was proved to be more efficient than the B-type in the average power, total sailing distance, and consuming time.
5 In case of low-speed conditions, the SFRM can not reach the desired pitch angle with empty ballast tanks. In the future study, the PD control on the ballast tanks and the attitude adjusting system will be considered in the autopilot system of the SFRM for improving its stability and manoeuvrability.

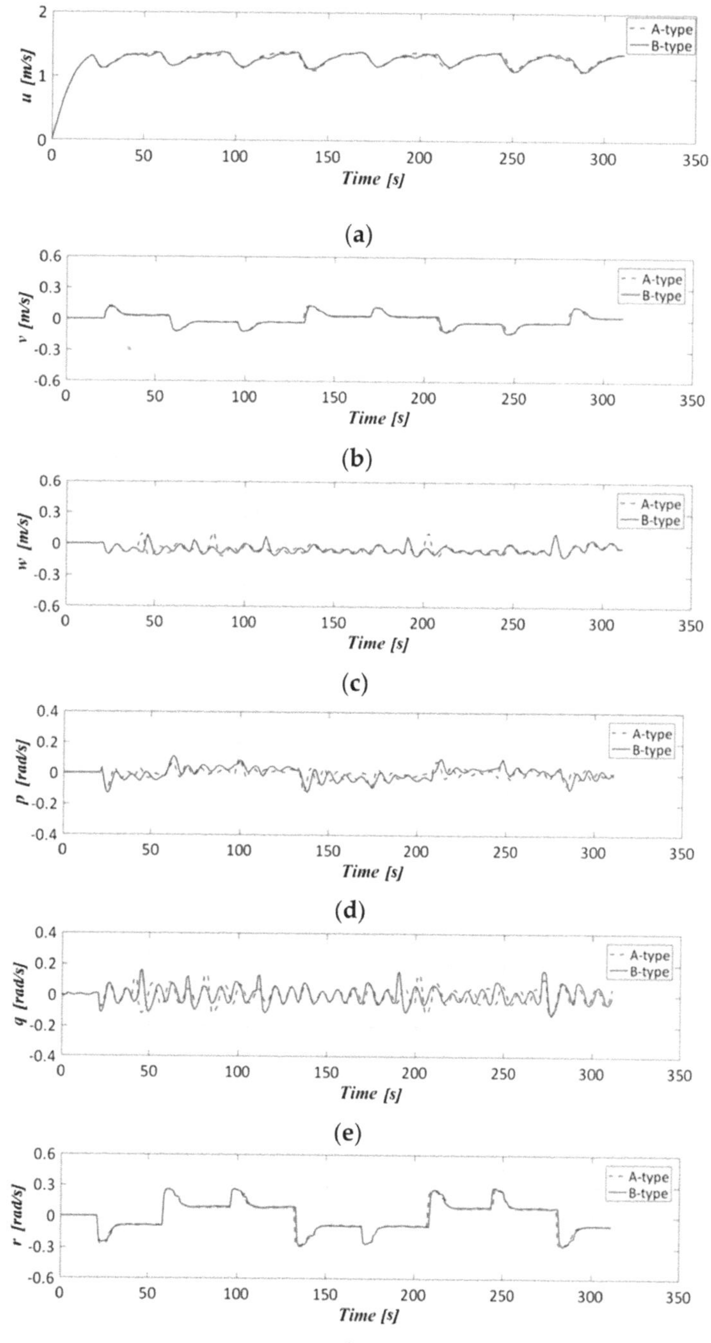

Figure 27. *Time series of velocity components. (a) Surge rate:* $u[\mathrm{m/s}]$*, (b) sway rate:* $v[\mathrm{m/s}]$*, (c) heave rate:* $w[\mathrm{m/s}]$*, (d) roll rate:* $p[\mathrm{rad/s}]$*, (e) pitch rate:* $q[\mathrm{rad/s}]$*, and (f) yaw rate:* $r[\mathrm{rad/s}]$ *for the A-type and B-type controllers, respectively.*

Author Contributions

Y.-H.L. is the principal investigator and Y.-T.L. and Y.-J.C. are the research assistants in this project. All authors have read and agreed to the published version of the manuscript.

REFERENCES

1. Bettle, M. Unsteady Computational Fluid Dynamics Simulations of Six Degrees-of-Freedom Submarine Manoeuvres. Ph.D. Thesis, University of New Brunswick, Fredericton, NB, Canada, 2013.
2. Itard, X. Recovery procedure in case of flooding. In Proceedings of Warship '99: Naval Submarines; RINA: London, UK, 1999.
3. Issac, M.T.; Adams, S.; He, M.; Bose, N.; Williams, C.D.; Bachmayer, R.; Crees, T. Manoeuvring Experiments Using the MUN Explorer AUV. In Proceedings of the 2007 Symposium on Underwater Technology and Workshop on Scientific Use of Submarine Cables and Related Technologies, Tokyo, Japan, 17-20 April 2007; pp. 256-262.
4. Jun, B.H.; Park, J.-Y.; Lee, F.-Y.; Lee, P.-M.; Lee, C.-M.; Kim, K.; Lim, Y.K.; Oh, J.H. Development of the AUV 'ISiMI'and a free running test in an Ocean Engineering Basin. Ocean Eng. 2009, 36, 2-14.
5. Overpelt, B.; Nienhuis, B.; Anderson, B. Free Running Manoeuvring Model Tests on a Modern Generic SSK Class Submarine (BB2). In Proceedings of the Pacific International Maritime Conference, Sydney, Australia, 18-20 August 2015; pp. 1-14.
6. Gertler, M.; Hagen, G.R. Standard Equations of Motion for Submarine Simulation; David w Taylor Naval Ship Research and Development Center: Bethesda, MD, USA, 1967.
7. Feldman, J. Dtnsrdc Revised Standarrd Submarine Equations of Motion; David w Taylor Naval Ship Research and Development Center: Bethesda, MD, USA, 1979.
8. Yoon, H.K.; Rhee, K.P. Identification of hydrodynamic coefficients in ship maneuvering equations of motion by estimation-beforemodeling technique. Ocean Eng. 2003, 30, 2379-2404.
9. Lin, Y.-H.; Li, X.-C. The Investigation of a Sliding Mesh Model for Hydrodynamic Analysis of a SUBOFF Model in Turbulent Flow Fields. J. Mar. Sci. Eng. 2020, 8, 744.
10. Kim, H.; Ranmuthugala, D.; Leong, Z.Q.; Chin, C. Six-DOF simulations of an underwater vehicle undergoing straight line and steady turning manoeuvres. Ocean Eng. 2018, 150, 102-112.
11. Lin, Y.-H.; Tseng, S.-H.; Chen, Y.-H. The experimental study on maneuvering derivatives of a submerged body SUBOFF by implementing the Planar Motion Mechanism tests. Ocean Eng. 2018, 170, 120-135.
12. Yildiz, Ö; Gökalp, R.B.; Yilmaz, A.E. A review on motion control of the underwater vehicles. In Proceedings of the 2009 International Conference on Electrical and Electronics Engineering-ELECO 2009, Bursa, Turkey, 5-8 November 2009; pp. II-337-II-341.
13. Herman, P. Decoupled PD set-point controller for underwater vehicles. Ocean Eng. 2009, 36, 529-534.

14. Soylu, S.; Buckham, B.J.; Podhorodeski, R.P. A chattering-free sliding-mode controller for underwater vehicles with fault-tolerant infinity-norm thrust allocation. Ocean Eng. 2008, 35, 1647-1659.

15. Qi, X. Adaptive coordinated tracking control of multiple autonomous underwater vehicles. Ocean Eng. 2014, 91, 84-90.

16. Mohan, S.; Kim, J. Indirect adaptive control of an autonomous underwater vehicle-manipulator system for underwater manipulation tasks. Ocean Eng. 2012, 54, 233-243.

17. Jun, S.W.; Lee, H.J. Design of TS fuzzy-model-based controller for depth control of autonomous underwater vehicles with parametric uncertainties. In Proceedings of the 2011 11th International Conference on Control, Automation and Systems; ICROS, Gyeonggi-do, Korea, 26-29 October 2011; pp. 1682-1684.

18. Steenson, L.V.; Phillips, A.B.; Turnock, S.R.; Furlong, M.E.; Rogers, E. Effect of measurement noise on the performance of a depth and pitch controller using the model predictive control method. In Proceedings of the 2012 IEEE/OES Autonomous Underwater Vehicles (AUV), Southampton, UK, 24-27 September 2012; pp. 1-8.

19. Subudhi, B.; Mukherjee, K.; Ghosh, S. A static output feedback control design for path following of autonomous underwater vehicle in vertical plane. Ocean Eng. 2013, 63, 72-76.

20. Xu, B.; Pandian, S.R.; Sakagami, N.; Petry, F. Neuro-fuzzy control of underwater vehicle-manipulator systems. J. Frankl. Inst. 2012, 349, 1125-1138.

21. Cooney, L.A. Dynamic Response and Maneuvering Strategies of a Hybrid Autonomous Underwater Vehicle in Hovering; Massachusetts Institute of Technology: Cambridge, MA, USA, 2009.

22. Fossen, T.I.; Fjellstad, O.-E. Nonlinear modelling of marine vehicles in 6 degrees of freedom. Math. Model. Syst. 1995, 1, 17-27.

23. Perrault, D.; Bose, N.; O'Young, S.; Williams, C.D. Sensitivity of AUV response to variations in hydrodynamic parameters. Ocean Eng. 2003, 30, 779-811.

24. Coe, R.G. Improved Underwater Vehicle Control and Maneuvering Analysis with Computational Fluid Dynamics Simulations; Virginia Tech: Blacksburg, VA, USA, 2013.

25. Boger, D.; Davoudzadeh, F.; Dreyer, J.; McDonald, H.; Schott, C.; Aierke, W.; Arabshahi, A.; Briley, W.R.; Busby, J.A.; Chen, J.P.; et al. A Physics-Based Means of Computing the Flow around a Maneuvering underwater Vehicle; Applied Research Laboratory: State College, PA, USA, 1997.

26. Healey, A.J.; Lienard, D. Multivariable sliding mode control for autonomous diving and steering of unmanned underwater vehicles. IEEE J. Ocean. Eng. 1993, 18, 327-339.

27. Roddy, R.F. Investigation of the Stability and Control Characteristics of Several Configurations of the DARPA SUBOFF Model (DTRC Model 5470) from Captive-Model Experiments; David Taylor Research Center Bethesda MD Ship Hydromechanics Dept: Bethesda, MD, USA, 1990.

28. Antonelli, G. On the use of adaptive/integral actions for six-degrees-of-freedom control of autonomous underwater vehicles. IEEE J. Ocean. Eng. 2007, 32, 300-312.

29. McGookin, E.W.; Murray-Smith, D.J.; Li, Y.; Fossen, T.I. The optimization of a tanker autopilot control system using genetic algorithms. Trans. Inst. Meas. Control 2000, 22, 141-178.

30. Xinghua, C.; Juan, L. AUV Planner Tracking Control based on the Line of Sight Guidance Method. In Proceedings of the 2014 IEEE International Conference on Mechatronics and Automation, Tianjin, China, 3-6 August 2014; pp. 1204-1208.
31. McGookin, E.W.; Murray-Smith, D.J.; Li, Y.; Fossen, T.I. Ship steering control system optimisation using genetic algorithms. Control Eng. Pract. 2000, 8, 429-443.
32. Wang, X.; Wu, G. Modified LOS Path Following Strategy of a Portable Modular AUV Based on Lateral Movement. J. Mar. Sci. Eng. 2020,8, 683.
33. Yadav, M.; Tayal, V.K. Performance Enhancement of Induction Motor Using PID Controller with PID Tuner. In Advances in Interdisciplinary Engineering; Springer: Berlin/Heidelberg, Germany, 2019; pp. 783-793.
34. McQuilkin, K.; Powers, A. Cinema 4D: the Artist's Project Sourcebook; Taylor & Francis: New York, NY, USA, 2011.
35. Renilson, M. Submarine Hydrodynamics; Springer: Berlin/Heidelberg, Germany, 2015.
36. Molland, A.F. The Maritime Engineering Reference Book: A Guide to Ship Design, Construction and Operation; Elsevier: Amsterdam, The Netherlands, 2011.

CHAPTER
6

Augmented Reality Lights for Compromised Visibility Navigation

Doupadi Bandara[1], Michael Woodward[1], Christopher Chin[1] and Danchi Jiang[2]

[1] *Australian Maritime College, University of Tasmania, Launceston 7250, TAS, Australia*
[2] *School of Engineering, University of Tasmania, Hobart 7005, TAS, Australia*

ABSTRACT

This paper considers the feasibility of using augmented reality (AR) as a tool for enhancing visualization in maritime operations to avoid collision in different environmental conditions. According to the International Maritime Organization (IMO 2010), 90% of maritime accidents due to collisions at sea are caused in part by human error. This study investigates the new technology (AR) used to superimpose holographic images onto the real world; now reaching a state of readiness for commercial application. This paper demonstrates the competence of AR technology to serve as a maritime navigational aid. The research explores the viability of improving navigational safety in low visibility by projecting holograms of real-world objects in the same geo-location as the real object to make them "visible". The paper presents the logical deconstruction of the technical problems and identified solutions, together with results of experiments used to validate the concept and technology readiness for real word maritime application. The paper presents a verified demonstrator; a proposed holographic bridge interface with an innovative way of presenting information using AR technology. Furthermore, it identifies that new technologies offer the opportunity for enhanced operator performances, with the expectation being that this should lead to reduce risk to persons, property, and the environment.

Keywords: *augmented reality (AR); compromised visibility; microsoft HoloLens; navigation lights; virtual objects*

1. INTRODUCTION

Recent decades have seen the appearance of a plethora of digital technologies on the ship bridge. Many are developed to assist navigation and collision avoidance in poor visibility, but presented on a range of screens and different formats, it is possible that operators may experience information overload. Navigation lights, for night operation, but not visible in fog condition have evolved over centuries and communicate relevant information to the seafarer in a very efficient way. The aim of this study is to evaluate the technical viability of using augmented reality navigation lights that are placed at the physical location of real obstacle to make them 'visible' in poor visibility conditions. The objective is to communicate critical navigational information to the seafarer in an effective and intuitive way, thus reducing the possibility of cognitive load.

To address the task this paper will first review the navigation data, coming from various navigation aids and examine them under different categories. The categories are capacity of information; priority rank of the available information; type of information process; extraction formats such as direct extraction methods, human interaction needed for data process, data extraction from other software. Then the identified data will be projected via holographic objects using software packages including Google SketchUp; simulations with Microsoft HoloLens; Bridge Simulator experiments to develop exploratory case-studies. An example with 'foggy condition operations' is used to evaluate the impact on operational performance of the technology limitations and reliability when applied to the Augmented Reality (AR) platform. This paper presents a verified demonstrator for seafarers using their common language 'navigation light'; a proposed holographic object that can be projected in an innovative way of presenting information using AR technology. Furthermore, it identifies that new technologies offer the opportunity for enhanced operator performances, with the expectation being that this should lead to reduce risk to persons, property, and the environment.

To address the aims and objectives this paper will:

- Review the technology and its application as a maritime navigation aid;
- Identify the technical and practical limitations and opportunities with the chosen proprietary technology Microsoft HoloLens and select a practical solution;
- Validate and calibrate the chosen optimal solution using a dedicated hologram experimentation case study;
- Map out the functional opportunities presented by the findings.

The paper is organized as follows. Section 2 presents a brief introduction on background information about AR technology and current problems in ship operations at sea. Section 3 discusses how the AR can be used as a tool to overcome said problems in a more inventive way. Section 4 provides the AR aided visualization illustrated using artificial navigational light objects. Experimentation and field work of the study and a discussion on the results are presented in Section 5 followed by the conclusions presented in Section 6. Section 7 presents how to integrate the identified data coming from various navigation aids into a common platform using AR technology.

2. BACKGROUND INFORMATION

The maritime traffic environment has consistently seen an ongoing increase in traffic and trends towards larger and faster ships, with the consequent increase in the risk of maritime accidents [1,2]. This could result in threats to the safety of seafarers, the economic performance of shipping industries, and the environment. Aiming to address this issue, enormous efforts have been made to develop measures to improve maritime safety, through improving technology involved in ship building and ship management, advanced navigation technology, and better crew training. International organizations such as the International Maritime Organization (IMO) and national Maritime Safety Administrations (MSAs) have published extensive rules and regulations on safety standards on the safe operations at sea. It is noted that most of the maritime accidents were due to collisions, contacts, and groundings [2-8]; making it important to try to mitigate factors that may cause these accidents. Furthermore, accident analyses have shown that human error is a dominant factor [9-11].

Over the last few decades, ship navigation and safety operations continue to push the limits of existing technology to overcome the current navigation problems using conventional technologies. With the development of new navigation equipment with different technology for safe navigation, the reduction of maritime accidents has not occurred as expected. It is postulated that one aspect may be due to the large amount of information provided to the seafarer on the ship bridge. Running into a 'big data' problem, the ever-increasing volume of information that is also not well organized may result in information overload for the operator. The information is displayed on a variety of different screens and monitors and to read it, the crew members must turn their attention away from the surrounding physical environment. This becomes even more difficult when a ship enters or leaves a harbor. Navigation may become an increasingly difficult task due to obstacles in the proximity of the ship, especially in different environmental conditions. Issues such as transfer of control, communication errors, lack of training, mechanical failure, and poor procedures may be exacerbated [12]. This becomes even more difficult with low visibility due to low light, fog, snow conditions, and rough sea conditions. These findings show that it is important to use new technology to represent the navigation information more effectively to facilitate understanding and enable more accurate decision making for seafarers.

The Latin word "Augere" has the meaning of "to increase" or "to add", which is used to create the word "Augmented". Virtual reality (VR) visually immerses the user totally in a digital view. The new technologies that can be utilized to present information in a holographic form are reaching a state of technology readiness for possible application in supporting maritime operations and training. Therefore, it is important to investigate how AR may be employed to provide a complementary technology for enhanced safety in maritime operations in compromised visibility environments. If maritime accidents are to be reduced, it is important to consider if the information provided to crew is adequate for their decision-making in critical situations. It is also crucial to consider that the information provided to them and the way it is provided is making best use of available and emerging technologies.

The understanding of the navigation light operations at sea is a very familiar language for seafarers. Introducing new and different techniques however may cause more cognitive load. This paper will explore the common language of seafarers "navigation lights" and integrate it into the AR platform using the chosen AR hardware platform. This AR based new concept can be achieved using the international standard navigation lights as the holographic 'language' for representing objects in compromised visibility conditions. The idea of using navigation light to make objects more clearly visible is already known to seafarers; in fact, it is part of their cadet training under International Regulations for Preventing Collisions at sea (COLREGs)_Part C-Rules 20-31. Therefore, it is assumed that using AR navigation lights imposes no additional cognitive load for seafarers. Microsoft HoloLens include a Central Processing Unit (CPU) and Graphical Processing Unit (GPU), and the unit features a Holographic Processing Unit (HPU) that is responsible for the processing that integrates the real world and digitalized holographic data. It is important, therefore, to reduce the highly detailed projections to overcome the limitations on processing power and storage capabilities. To overcome this problem, this paper will focus on using lights instead of images of the whole ship or light house. This will increase the effectiveness of the headset for its users, allowing it to produce higher frame per second, thus reducing latency, lag, and delay [13].

When using AR, the user can still fully see the real world around them; with the addition of digital holographic objects as shown in Figure 1a. This figure is considered to be a daytime ship operation at sea and is difficult to visualize in compromised visibility conditions at night or in fog as illustrated in Figure 1b,c. Figure 1b,d shows a representation of the real-world situation, in that the navigation lights are visible at night (b) but not visible in fog (c). Figure 1d shows as an example that there is a possibility to overcome the problem by using augmented reality navigation lights; locked to the position of the known object in this case a holographic ship but in practice a real one.

AR has a history dating back to the 1950s [14], with a range of ever evolving proprietary platforms coming into the market in recent years. A literature review of the history and more recent contemporary technologies is provided in [15]. From this review, the Microsoft HoloLens was ultimately selected based on both its functionality and on account of being readily available. The unit used for the study is a first generation Microsoft HoloLens with the full specifications in [16].

Over the past few years, AR technology has become an interesting topic in various industries such as training and education, medical, engineering, architecture, defense operations [17-21]. The common AR device called Head-Up-Display (HUD) is being used to enhance the capability of pilots and reduce their time in fighter jets and commercial aircrafts [22]. In terms of developments lending to applying AR to maritime navigation, there are a number of studies published in the open literature.

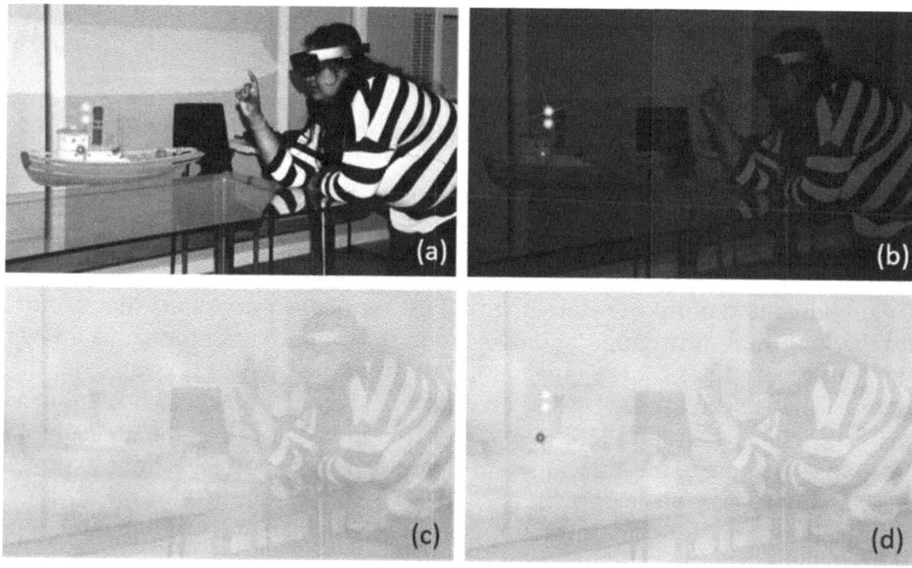

Figure 1. *The user sees the real world as well as the virtual world (Augmented Reality (AR) vessel with navigation lights) in (a) normal condition; (b) night condition with lights visible; (c) fog condition where lights are not visible; (d) fog condition with holographic lights (giving a demonstration/visualization of concept).*

Holder and Pecota [23] postulate that the 'head-down' position when using electronic navigation aids may make individuals reluctant to use them. These results suggested that there are some standard information requirements across situations that could be augmented with task and vessel specific information.

Hugues, et al. [24] integrated a vision system with a thermal camera together with augmentation. Their study finds and concludes that the functionalities provided by AR must differ according to people and weather conditions, requiring a need to provide contextual information.

Oh, et al. [25] investigate 'Advanced Navigation Aids Systems based on AR'. They present work implementing the representation of information using AR; including conducting a survey of the outcomes. In their conclusion they find that, "Although various types of navigation equipment with advanced technology are available to help enhance operation safety for navigators, further research on the method of efficiently displaying and servicing information rather than indiscriminately providing a bounty of information is necessary".

Radu [26] explores the use of AR in education, finding both positive and negative effects on learning. Further, Wu, et al. [27] argue that, from an educational perspective, AR should be viewed as a concept, rather than a technology.

AR can project the information from display devices in bridge equipment to appear as a head-up display. Using a more advanced and contemporary application, this paper proposes that AR can also be used to create holographic representations of real or enhanced objects. In this way, complex information from multiple sources, may be presented to the viewer as if it were part of the real world such as seeing a real/virtual object coming towards you; rather than needing to interpret information from a digital display.

3. DECONSTRUCTION OF THE TECHNICAL PROBLEM (RESEARCH PROBLEM)

The safety of passengers and vessel is one of the major concerns in maritime industry. Despite advances in navigational technology, and improvements in human, legal, and social aspects, ship accidents continue to occur. Of all the maritime accidents reported to date, the highest number of accidents that occurred are due to errors during operations. Among these accidents, ship collision has a high portion and has drawn much attention from the public [2]. Moreover, these accidents have been caused by human error due to inadequate watch keeping and mistakes in ship handling [25]. Therefore, it is important to understand the factors that contribute to these accidents and implement the correct actions to prevent these accidents as soon as possible.

The bridge is the brain of the ship and in recent years has become more modernized; with advanced equipment and tools. It is important for bridge crew members to have an extensive understanding of bridge systems; in addition to navigation skills [28]. There is a large amount of information available on a modern ship bridge, and not all of that information is required for the navigator. In fact, Hareide [29] notes that it is important for the navigator to sort out the information for planning a safe navigation path and that this is difficult when attention is divided between digital displays and the outside world.

As a physical computing and display technology, AR devices are quickly becoming widely used in many industries. This wide adoption implies that it cannot be reasonably argued that the technology is totally useless. Nevertheless, it is only a tool and it is 'what it is used for' that both provides value and provides the foundation of this research. Addressing the contemporary problems associated with 'big-data', the very large amount of data given to the seafarer in a variety of different forms and formats as discussed earlier, this research proposes to investigate 'how' the AR platform can be exploited for organizing and presenting bridge information. Though the aspiration would always be that better organized information should lead to better performance and thus enhanced safety, this will remain implicit within this study. This research paper develops the initial requirements for expected functionality in the final solution and formally address the verification and validation of this functionality at the technical level. It therefore strives to provide a solution for a situation where seafarers may struggle to cope with the very large amount of information provided by the current digital displays. It aims to improve the safety, through collision avoidance, in compromised visibility by organizing and rationalizing information and by exploring 'how' to present that data using a holographic object but with existing data technology. In particular, the integration of data in later stage from growing sources and visualization of that data can improve the safety of navigation in the following aspects:

1. Visualization of normally nonvisible or not clearly visible risky objects;
2. Visualization of risky patterns that can only be captured by expert navigators and;
3. Facilitate quick processing and visualization of information that are distributed in multiple sources which require extensive manual processing.

By making it possible for the seafarer to make navigational decisions based on their view of real and virtual objects, rather than having to interoperate digital display information, it is hoped that the process will become more intuitive for the user. Furthermore, it identifies that new technologies offer the opportunity for possible enhanced operator performances, with the expectation being that this should lead to reduction of risk to persons, property, and the environment.

The main reason to represent holographic objects in terms of navigation light operations at sea is due to the common and familiar language for the seafarers. Furthermore, holographic navigation lights have the power to communicate most of the visual clues that disappear under compromised visibility environment conditions discussed in the next section.

4. AR AS A VIABLE NAVIGATION AID IN THE MARITIME INDUSTRY

Navigators on modern ships work in a technologically advanced environment compared to older times. Therefore, it is important to understand how to identify/distinguish important data from different data sources for example but not limited to: Electronic Chart Display Information System (ECDIS), Radar, sonar/-depth-sounder integrate into the AR platform in a more intuitive way to improve operations in compromised visibility, while addressing the 'big-data' problem. This problem is commonly known as the very large amount of data given to the seafarer in a variety of different forms and formats.

4.1. Review of Ambient Data Sources

It is important to understand the ambient information available to the seafarer. This may include visual clues apart from the navigation aids, and the conscious or unconscious usage of the information during operations at sea for safe navigation. It is postulated here that these visual clues may disappear in compromised visibility conditions. To address this possibility, this study proposes to review and critique ambient information sources; for possible inclusion in an AR model. Table 1 represents the identified visual clues used in vessel navigation in daylight. Table 1 also strives to suggest/offer possible ways of replacing the 'clues' in an AR platform.

A conclusion that can be derived from Table 1 is the importance of the visual clues apart from the navigation aids during operations at sea for safe navigation such as seeing the water flow around a buoy or bank. Furthermore, it has been shown that the disappearance of these visual clues at night can lead to maritime accidents which damage human lives, property, and environment. It is therefore very important to develop a new concept of safe navigation with the help of AR

Table 1. *Identifying objects in night environment conditions.*

Navigation Aid Type:	**Fixed Features**	
Information:	**Example Feature/Object:**	
Marked (lit) land feature, lighthouse, leading line marks	**Lighthouse**—fixed light of known position and height, identifiable by unique light flashing pattern (attributed to that particular lighthouse). May include colored light sectors (red, white, green) indicating approximate position line (bearing to mark).	
Transmutation:	**Leading Marks**—usually display a triangle shape. The front lead has its apex pointing up and the rear or back lead has its apex pointing down. When leads in line, observer (vessel) is in middle of channel. At night, major leads are usually lit with blue lights.	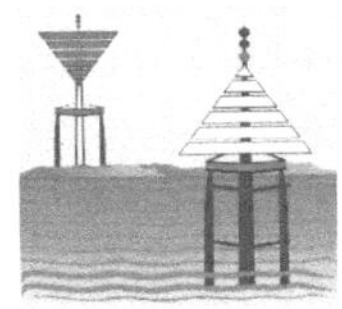
ECN, RADAR, (AIS)	**Blue Middle Channel Marks**—fixed blue lights that mark the middle of the channel for vessels passing under a bridge. Directional and sector lights have a similar purpose to leading lights at channel entrances and inshore waters.	
Leads to know:		
Position line with respect to fixed object(s)		
Suggested AR indicator		
Int. Standard navigation light (light, color, flashing pattern)	**Directional Lights**—used as steering marks and sector lights display a light of different colors (usually green, white, and red). **Port Traffic Signal Lights**—three red vertical flashing lights that are remotely controlled by Maritime Safety traffic services to tell others in the area that large commercial ships are moving in the port, harbor, marina, or other confined waterway. Vessels must not enter or depart the port or harbor area when the port traffic signal lights are flashing.	
Navigation Aid Type:	**Buoys, Marks, and Beacons**	
Information:	**Example Feature/Object:**	
Waterway floating device system display information need for navigation at day-time and night-time	**Lateral Marks**—show the port (left) and starboard (right) sides of navigable waters or channels. A port mark is red with a can-like shape. A starboard mark is green with a cone-like shape. At night, a port buoy shows a red flashing light and a starboard buoy shows a green light.	
Transmutation:		
ECN, RADAR, COMPASS	**Cardinal Marks**—show where the deepest and safest water to travel is. Marks have black and yellow bands with black double cones on top showing the different compass direction that identifies the safest and deepest water to navigate according to each direction (North, East, South, and West). At night, each type of cardinal mark has a flashing white light with different groupings of flashes (continuous, or groups of 3, 6, or 9).	
Leads to know:		
Position line with respect to fixed object(s)		
Suggested AR indicator		
Int. Standard navigation light (light, color, flashing pattern)		

(Cont.)

Table 1. *Cont.*

Navigation Aid Type:	**Buoys, Marks, and Beacons**	
Information:	**Example Feature/Object:**	
	Isolated Danger Marks—show where there is an isolated danger that has navigable water all around it (for example, an isolated shoal, rock, or wreck). Marks are black with one or more red horizontal bands and two spheres as the top mark. At night, the white light flashes in groups of two.	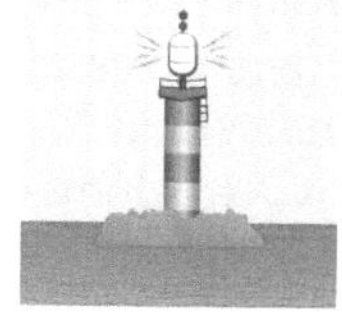
	Safe Water Marks—show that there is navigable water all around the mark. For example, fairway, mid-channel, or landfall mark. Marks have red and white vertical stripes with a single red sphere as the top mark. At night, a single white light shows one long flash every 10 s.	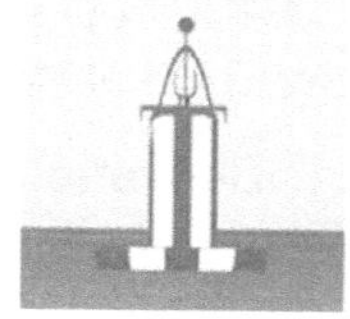
	Special Marks—show a special area or feature. For example, to show that a channel divides or to mark cables or pipelines. Special marks are yellow and sometimes have a yellow X as the top mark. At night, the flashing light is yellow.	
	Emergency Wreck Marking Buoys—used to identify new dangers or wrecks. They have blue and yellow vertical stripes and are a pillar or spar shape with a yellow cross as the top mark. At night, the flashing light alternates between 1 s of blue light and 1 s of yellow light, with 0.5 s of darkness in between.	

Navigation Aid Type:	**Identify Speed through Water**	
Information:	**Example Feature/Object:**	
Speed through water and depth of the water **Transmutation:** Visual **Leads to know:** Speed or depth of the water with respect to fixed object(s)	**Speed through Water**—speed through water and speed over the ground are independent. Watching water near the bank going by may give one impression whereas watching the flow around a buoy, post, or rock may show that the speed through water is different. **Speed/Depth through water**—the characteristic Kelvin wake of a ship changes angle in shallow water, and is sometimes observed by the pilot to indicate water depth.	

technology to enhance visualization of seafarers in compromised visibility conditions. This can be achieved while presenting lots of visual clues information in the form of holographic objects in the AR platform at compromised visibility conditions such as night, fog, or snow. This paper evaluates the technical viability of using navigation lights as AR objects which are geo-locked to the real object, and the capability to represent information that is normally unavailable in poor visibility.

4.2. Navigation Light Requirements

According to the IMO, COLREGs, navigation lights should appear steady and non-flashing. The COLREGs specify the required number, location, position, color, luminosity, number of spare lights, of navigation lights that must be present on different vessel types and in different operations [30]. Therefore, it is important to understand the regulations on COLREGs for developing holographic objects in an AR environment to symbolize the real light intensity in a reasonable and realistic way.

5. AR AIDED VISUALIZATION ILLUSTRATED USING ARTIFICIAL NAVIGATIONAL LIGHT OBJECTS

5.1. Derivation of Holographic Signals

Having identified the international standards for navigation lights, the task of identifying a suitable holographic display is undertaken. The aim of this section is to determine the method of holographic navigation light display and calibrate the derived objects lights. The objective is to define the baseline object light that is meaningfully displayed to a degree of realism for further development of the platform case study. This will be addressed in the following section by, first, presenting the approach/method for calibrating the holographic signals; second, presenting the model design and experimental methodology, then presenting the results and discussion.

5.2. Calibration of Holographic Signal

It is first important to note that the HoloLens technology is designed and calibrated to operate in the near environment such as with a physical room. To verify the functionality and calibrate the outputs on typical scales of kilometers, case studies were used. A case study was conducted to examine how holographic navigation lights may be used for visual awareness of objects/obstacles when in compromised visibility. Holographic spheres are used to represent different virtual lights, for vessels at different distances. The experiment used a sphere as a 'virtual light', which represented the other vessel's navigation light to the seafarers as the view is the same from all directions and is just a way of modeling 'brightness' in a practical way. The main purpose is to place a hologram 'light' in the same place as a navigational light on another vessel so that it appears as it would at night but in the intended case that it will appear this way on a foggy day to aid the user to intuitively account for the object with the least cognitive load.

Graphical software was used to generate arbitrary models of navigation lights, and to compare their holographic appearance at different distances with real lights. The SketchUp software was used for the modeling as it has a readily usable interface for the HoloLens. Relevant tools that are available in the software were used to prepare models and then projected using a head-mounted holographic visualization tool in a Microsoft HoloLens device.

5.3. Model Design and Experiment Methodology

The study considers spherical virtual lights of different radii (0.5, 0.75, 1, 2, 3, 4, 5, 6, 7, 8, 9, 10 m). The different radii are explored in this case to simulate a visual approximation of luminosity of the holographic light. A model was developed and subsequently used to identify and calibrate the luminosity of the virtual navigation light at different distances.

The images of different light luminosity were calibrated for eight sets, representing eight selected distances (150, 500, 1000, 1500, 2000, 2400, 2800, 3600 m) covering a variety of terrain scenarios including land, open water, and various background lights.

The designed model projected into the Microsoft HoloLens Sketchup Viewer through Google SketchUp "SketchUp Viewer AR/VR" uses the same Wi-Fi network on both devices. The projected model is then displayed on a 2.3 Megapixel widescreen stereoscopic head-mounted display.

The Microsoft HoloLens is the first commercially available AR Head-Mounted Device (HMD) to reach the market as a consumer version which was first released as a development release in 2016. Since this device is so new, and the HoloLens is the first to market in this area, there are little to no competitors for consumer grade wireless AR HMDs. The Microsoft HoloLens features four Intel Atom x5-Z8100 1.04 GHz Intel Airmont Logical Processors, an HPU/GPU Holographic Processing Unit, 64 GB Flash, 2 GB RAM, 2.3 Megapixel widescreen stereoscopic head-mounted display. HoloLens runs with Windows 10 operating system. Holograms enable the user to apply five fundamental features of HoloLens which are gaze control, gesture control, voice controls, spatial mapping, and spatial sound.

Using the defined 3-D models, projected through the holographic interface, the visual appearance of the virtual lights was compared with real lights of known distance and brightness. The images were captured and used to assess the 'best-fit' solution.

6. EXPERIMENTATION AND FIELD WORK

6.1. Initial Field Work

Before moving into the results of the case study, it is important to look into how a projected holographic boat/ship appears when viewed from the bridge of a real ship. The objective was to see how the HoloLens technology and derived holograms look and function in the real environment and to identify any practical/technical issues that may not be apparent from desk studies.

In this field study, a holographic fishing boat is examined, when viewed from the bridge of the vessel Bluefin (Australian Maritime College Research Vessel) as illustrated in Figures 2 and 3 . The holographic boat highlighted in Figure 2 by the white arrow is found to be fully visible in daylight conditions. In addition, it was possible to make the holographic image appear to be outside the ship bridge; despite being viewed through the glass windows on the bridge.

Figure 2. *Projected view of fishing boat at AMC Surface Vessel Bluefin at angle of* 115.4°.

Figure 3. *Projected view of fishing boat at AMC Surface Vessel Bluefin at angle of* 75.6°.

Figure 3 shows the same holographic model being viewed from a different location on the ship bridge including a second viewer, wearing a HoloLens. It is worth noting that the technology used is capable of pairing, so that multiple bridge personnel can view the same virtual object at the same time. This may help the bridge crew member's perception, especially in confirmation bias which can be a serious threat to safety. The confirmation biasing is when the crew members only understand the information which can confirm their assumption and reject any information that falsifies their assumption. The observations from Figures 2 and 3 show the exploration of holographic objects with a real world environment. These results are used as proof-of-concept; that AR technology can be implemented for maritime industry to enhance visualization of the operations at sea.

From the field experiments it was identified that projecting images of real ships added 'visual clutter' that was not information rich. Additionally, projecting the hologram of a full ship uses a significant amount of CPU availability. By deconstructing the problem, and in discussion with seafaring experts/instructors, the processes of projecting only navigation lights were identified. The justification for

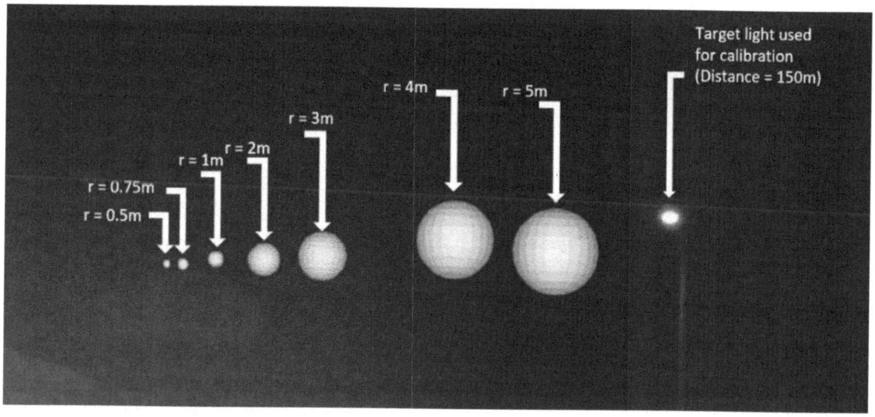

Figure 4. *Matching spheres (size and luminosity) with the actual light at a distance of* 150 m.

this is based on the assumption that seafarers are already familiar with navigation lights as a 'language'; resulting in less computation time for the device and an information rich output.

6.2. Field Studies for Navigation Lights

The aim of this experiment was to demonstrate the view of the projected holographic object appearance when viewed in night vision, especially displaying navigation light. The holographic technology is typically calibrated for indoor use, so the second aim of the experiment was to identify if the technology functioned adequately in a real outdoor environment; on scales of hundreds of meters rather just a few meters. The objective was to calibrate the visual objects so that they appear as close to reality in terms of size and luminosity as is practicable. There are two sets of surveys that were conducted: one in Newnham Tasmania, Australia and one at Lake St. Clare Tasmania, Australia. The first stage occurred during the first stage of COVID-19 lockdown; requiring some lateral thinking. To progress, a practical solution was identified, by calibrating against a streetlight at a known distance and the outcomes are presented in Figure 4. The result demonstrates that the virtual object can be created in an outdoor environment. The distance between the street-light pole and operator or observer is 150 m and the holographic objects are placed in the same distance from the operator or observer. The same experiment was conducted using different colors of spheres such as red, green, and blue as shown in Figure 5; which are commonly used in navigation lights.

The next set of experiments were conducted in Newnham Tasmania, at 2400, 2800, and 3600 m across the Tamar River. Figure 6 gives the example with a 2400 m distance.

The survey was conducted in Newnham across Tamar River, a metropolitan area with lots of lights. This made it difficult to determine the light reflection of a single light on the water. To add an additional reference point, and using more controlled conditions, extra tests were conducted in a rural area with very low light

(a) (b) (c)

Figure 5. *Matching spheres of different colors (size and luminosity) with the actual light at a distance of* 150 m. *(a) Red color spheres; (b) blue color spheres; (c) green color spheres.*

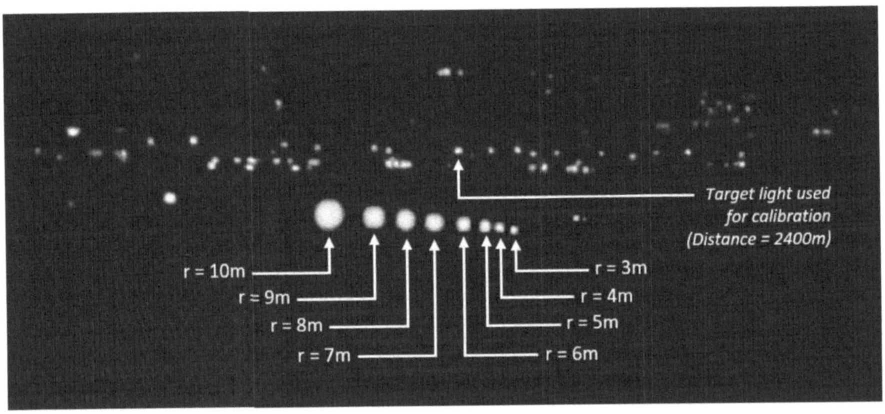

Figure 6. *Comparison of light and sphere at a distance over water of* 2400 m *in Newnham Tasmania.*

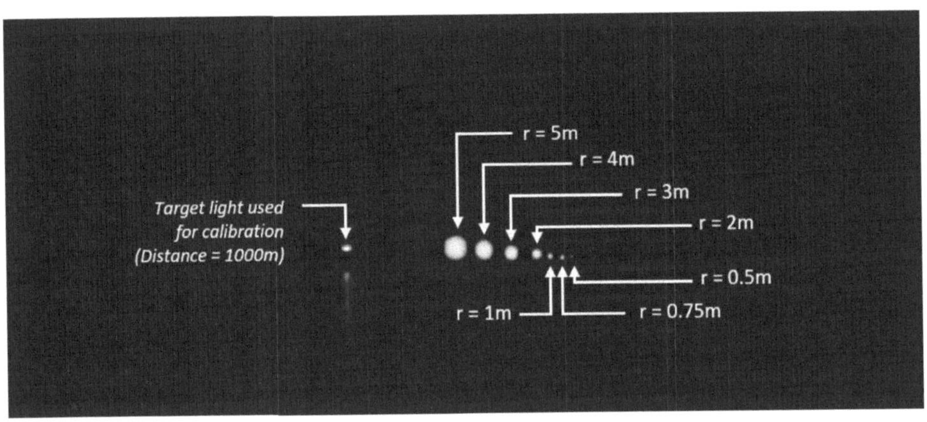

Figure 7. *Comparison of light and spheres at a distance over water of* 1000 m *(Lake St Clair).*

pollution, which is assumed to be more representative of an ocean environment. This was undertaken at Lake St. Clair beach where a pump house is located 1000 m away from the beach. The same experiment was conducted, using the same model; set to 1000 m as shown in Figure 7.

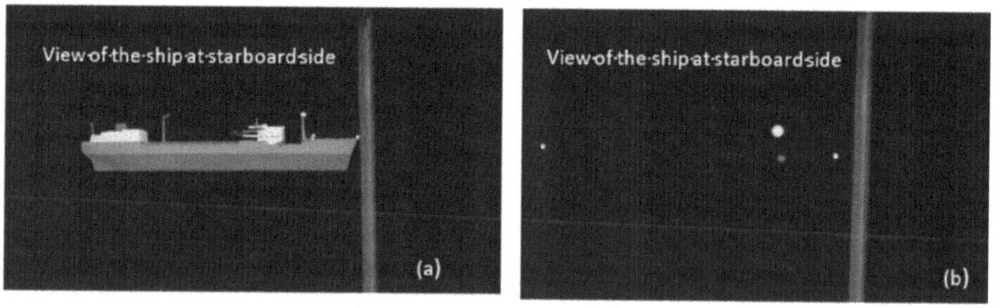

Figure 8. *(a) The projected holographic ship through the physical window and (b) the navigation light projection of the same ship through the ship bridge.*

Figure 7 shows the navigation light projection visualization on the river or sea with the reflection on the water. This image provided an additional concept information that will be considered at a later stage of the research.

The results presented in this section show the exploration of holographic objects in a real-world environment. These results provide proof-of-concept, that AR technology can be implemented for projecting virtual navigation lights that resemble visual appearance and luminosity at a distance between 150 and 3600 m in a dark environment. These results were conducted using a small datasets during the night. Ongoing research will investigate visual appearance of the derived holograms in other light conditions, especially in fog condition. The final outcomes of the proposed augmented reality navigation light can be visualized through the physical window as shown in Figure 8 . This result showed that without the redundancy of information in terms of without having full shape of the object, it is still possible to give the proper dimensions and the direction of the ship correctly. Moreover, this result revealed the enhancement of visualization of AR device Microsoft HoloLens for ship operations at night.

6.3. Data Process and Data Analysis

Data is collected by projecting holographic lights at various distances and comparing them to known lights at the same distance. To model the holographic lights, a set of spheres of different sizes are generated in a 3-D modeling software and imported into the holographic projection hardware. The holograms are then projected in a night environment over a body of water to lights of known size and distance. The calibration light used is standard type of street-light and is used throughout so that all samples are of the same size and lumen.

6.3.1. Data Processing

Images of the target light and closest fit holographic lights are captured using the same scale as shown in Figure 9, Row 1. To account for the dispersion of the light over the image and the fuzziness, an image processing software is used to convert the image into three distinct regions of brightness as illustrated in Figure 9, Row 2. The area of each of the three regions is then estimated by comparison with a 10-by-10 grid; giving a relative percentage for each region. From this an

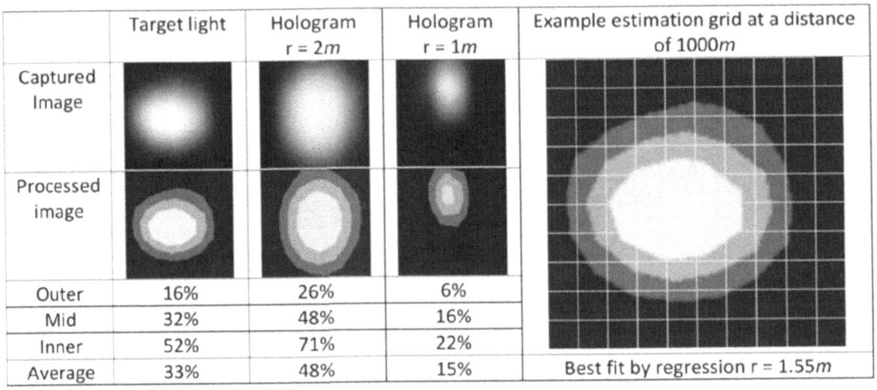

Figure 9. *Example of identified best holographic spheres to represent navigation light in the real-world environment.*

average is taken for each image and regression analysis is used to identify the best match size. A similar process was carried out for all the data sets used in both case studies to identify appropriate holographic spheres for best matching size to represent navigation lights in the real environment.

6.3.2. Data Analysis

The best match size is then plotted against distance and the relationship estimated as illustrated in Figure 10. The data is somewhat spread so there is no justification to use any higher order curve than a linear fit. This results in the estimated representative holographic radius, r, as a function of distance D, given by Equation (1):

$$r = 0.0010 + 0.4308 \tag{1}$$

with a standard regression of 0.7. Standard experimental uncertainty methods are used to identify the variable uncertainty and sensitivity of each input and are found to give a combined uncertainty of $+/-0.08$ m for the estimated radius.

Analysis of the results shown in Figure 10 confirms that holographic images of suitably sized spheres can reasonably be used to represent navigation lights in a real-world environment. The calibration process identifies that a simple linear relationship between sphere radius and distance from objects can be used to give a reasonable approximation to size and luminosity.

6.4. Discussion

On examination of the available information and processor capacity, navigation lights are identified as the most intuitive candidate for a holographic language. This conclusion depends on the efficiency of the processor which is used to project the objects for holographic display when compared to, for example, projecting the whole ship or lighthouse. In this paper it is postulated that seafarers are already familiar with navigation lights and as such would not have to learn a new skill.

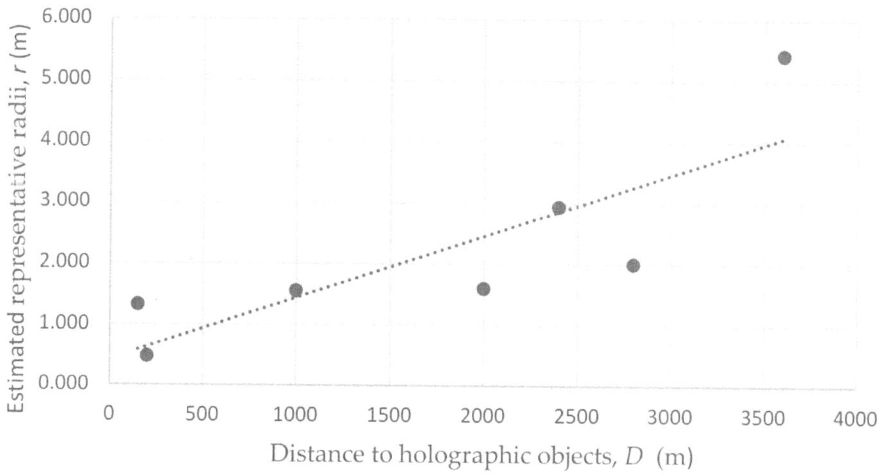

Figure 10. *Linear regression estimation curve for the representative hologram (sphere) radii as a function of distance.*

This research paper has identified the possibility of using virtual lights to mimic the navigation lights in other vessels and operations at sea in compromised visibility conditions. Proof-of-concept is provided that holographic navigation lights can be achieved with existing holographic technologies and in environments including distance over water comparable to those in real navigational situations.

Other visual clues such as light reflection in the water are noted. Their importance for realism such as cognitive acceptance will be considered, and outcome will inform future representation of AR objects.

This research paper contributes to the area of Holographic AR based new algorithm for advanced ship navigation, safety, and operations at sea using Microsoft HoloLens for the first time in the maritime industry and for maritime operators. This ongoing research proposed to develop further the concept in vision-based navigation using AR as a tool for enhancing information propagation; with the expectation of improved situational awareness. The next stage of the study will formalize the process for identifying and extracting information such as location of other navigational hazards. This will then be used to inform the geo-location of the AR navigation light presented in this paper.

7. CONCLUSIONS

This paper outlined the viability of representing AR holographic navigation lights for ship operations to enhance the visualization of navigational hazards; in terms of safety in compromised visibility. It is identified that there is a significant amount of research work being conducted to realize safety operations at sea, and that safety operations at sea is the major challenge in maritime industry especially in compromised visibility. It is further identified that AR is reaching a stage of technology readiness suitable for commercial application. Holographic navigation

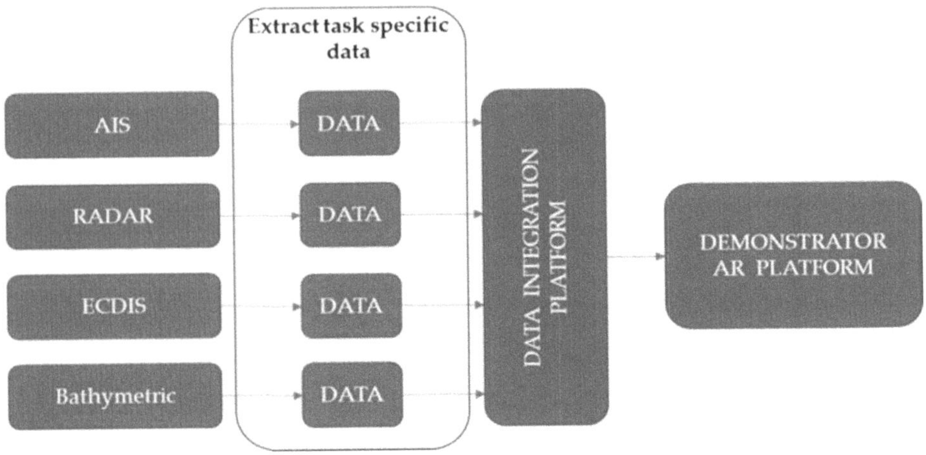

Figure 11. *Overall design process.*

lights that are geo-fixed at the location of real objects are identified as an information rich way of presenting hazards. Field studies are conducted and demonstrate the functionality of an AR technology platform in a real maritime environment. Moving beyond the calibrated and intended scope of the device used, a practical way of representing realistic lights is derived and verified at representative distances in km scales. View size and luminosity of the AR objects are calibrated using regression analysis, and a suitable scaling factor identified.

The study identifies that using AR navigation lights that could be geo-locked to real physical hazards is a viable way of presenting navigational information in compromised visibility.

8. FUTURE WORK

Future work will deliver a real-time AR integrated and interactive user interface to demonstrate the advantages of using AR technology while operating in compromised visibility such as snow, fog, rain, and rough seas. This can be done by developing a new algorithm to import data into the AR platform to identify objects in different light and environment conditions and use it to run the application as illustrated in Figure 11. Future research aims to:

- Identify the key data sources, their purpose, and their data output format;
- Identify ways to extract the necessary data;
- Unify the various data and data protocols into an integrated platform;
- Develop a case study in AR demonstrator platform.

Though this study has a focus on the technology, it is also important to recognize the need to verify the user interface using human subjects. It is recommended that future studies explore the response and opinion of seafarers; who will ultimately use the technology. Future work would also strongly recommended field research with examples given by Procee et al. [31,32].

Author Contributions

D.B.-PhD researcher; M.W.-Primary supervisor; C.C. and D.J.-Co-supervisors. All authors have read and agreed to the published version of the manuscript.

Acknowledgments

Our appreciation and special thanks go to Damien Freeman for his great support and guidance. We also would like to thank all the staff members in the bridge simulator, AMC and University of Tasmania.

REFERENCES

1. Clinch, S. MAIB Annual Report 2014; Marine Accident Investigation Branch: Southampton, UK, 2015.
2. Huang, Y.; van Gelder, P.; Mendel, M. Imminent ships collision risk assessment based on velocity obstacle. In Risk, Reliability and Safety: Innovating Theory and Practice, Proceedings of the ESREL 2016, Glasgow, Scotland, 25-29 September 2017; Walls, L., Revie, M., Bedford, T., Eds.; Taylor and Francis Group: London, UK, 2017.
3. Bye, R.J.; Aalberg, A.L. Maritime navigation accidents and risk indicators: An exploratory statistical analysis using AIS data and accident reports. Reliab. Eng. Syst. Saf. 2018, 176, 174-186.
4. Chauvin, C.; Lardjane, S.; Morel, G.; Clostermann, J.-P.; Langard, B. Human and organisational factors in maritime accidents: Analysis of collisions at sea using the HFACS. Accid. Anal. Prev. 2013, 59, 26-37.
5. Luo, M.; Shin, S.-H. Half-century research developments in maritime accidents: Future directions. Accid. Anal. Prev. 2019, 123, 448-460.
6. Uğurlu, Ö.; Köse, E.; Yıldırım, U.; Yüksekyıldız, E. Marine accident analysis for collision and grounding in oil tanker using FTA method. Marit. Policy Manag. 2015, 42, 163-185.
7. MAIB. Annual Overview of Marine Casualties and Incidents 2018; Marine Accident Investigation Branch: Southamton, UK, 2018.
8. Agency, E.M.S. Annual Overview of Marine Casualties and Incidents 2019; European Maritime Safety Agency: Lisboa, Portugal, 2019.
9. Squire, D. The Human Element in Shipping. Retrieved July 2005, 1, 2008. Available online: https://www.healert.org/filemanager/root/site_assets/-standalone_pdfs_0355-/HE00350.pdf (accessed on 11 December 2020).
10. Gauss, B.; Rötting, M.; Kersandt, D. Naridas-evaluation of a risk assessment system for the ship's bridge. In Proceedings of the Human Factors in Shop Design, Safety and Operation, RINA-The Royal Institution of Naval Architects, International Conference, London, UK, 21-22 March 2007.

11. Acar, U.; Ziarati, R.; Ziarati, M. Collisions and Groundings-Major Causes of Accidents at Sea; Uola, K., Ed.; Satakunta University of Applied Sciences: Pori, Finland, 2012; pp. 40-47.
12. Woodward, M.; Landamore, M.; Rees, G.; Allen, N.; De Graauw, A.; Mejer, A.; Hutchins, J.; Irvine, M.; Short, S.; Trodden, D.; et al. On the safe pilotage of ship's with propulsion that can azimuth for steering. Trans. R. Inst. Nav. Archit. Part A Int. J. Marit. Eng. 2015, 157, 282.
13. Evans, G.; Miller, J.; Pena, M.I.; MacAllister, A.; Winer, E. Evaluating the Microsoft HoloLens through an augmented reality assembly application. In Degraded Environments: Sensing, Processing, and Display, Proceedings of the SPIE Defense + Security, Anaheim, CA, USA, 9-13 April 2017; SPIE: Bellingham, WA, USA, 2017.
14. Λαχανας, B. Augmented Reality for Simulation Based Training and Assessment in Minimal Invasive Surgery. Ph.D. Thesis, National and Kapodistrian University of Athens (EKPA), Athens, Greece, 2015.
15. Bandara, D.; Woodward, M.; Chin, C.; Jiang, D. The application of Augmented Reality Technology in Maritime Operations and Training. In Proceedings of the International Marine Simulation Forum AGM, Johore Bahru, Malaysia, 23-25 September 2019.
16. Microsoft. HoloLens (1st Generation) Hardware. Available online: https://docs.-microsoft.com/en-us/ hololens/hololens1-hardware (accessed on 11 July 2020).
17. Akçayır, M.; Akçayır, G. Advantages and challenges associated with augmented reality for education: A systematic review of the literature. Educ. Res. Rev. 2017, 20, 1-11.
18. Guo, P.; Kaplan, D. Augmented Reality Based Component Replacement and Maintenance. U.S. Patent US10,198,620 B2, 5 February 2019.
19. Pratt, P.; Ives, M.; Lawton, G.; Simmons, J.; Radev, N.; Spyropoulou, L.; Amiras, D. Through the HoloLens ™ looking glass: Augmented reality for extremity reconstruction surgery using 3D vascular models with perforating vessels. Eur. Radiol. Exp. 2018, 2, 1-7.
20. Slocombe, G. Augmented reality is a useful technique for defence. Asia Pac. Def. Report. 2002 2017, 43, 94-95.
21. Wang, W.; Wu, X.; Chen, G.; Chen, Z. Holo3DGIS: Leveraging Microsoft HoloLens in 3D Geographic Information. ISPRS Int. J. Geoinf. 2018, 7, 60.
22. Kim, G.-H.; Jo, S.-I.; Park, J.-H. Application of head-up-display technology to telematics. Electron. Telecommun.Trends 2008, 23, 153-162.
23. Holder, E.; Pecota, S.R. Maritime Head-Up Display: A Preliminary Evaluation. J. Navig. 2011, 64, 573-594.
24. Hugues, O.; Cieutat, J.-M.; Guitton, P. An experimental augmented reality platform for assisted maritime navigation. In Proceedings of the 1st Augmented Human International Conference, Megève, France, 2-4 April 2010; p. 12
25. Oh, J.; Park, S.; Kwon, O.-S. Advanced Navigation Aids System based on Augmented Reality. Int. J. eNavig. Marit. Econ. 2016, 5, 21-31.

26. Radu, I. Augmented reality in education: A meta-review and cross-media analysis. Pers. Ubiquitous Comput. 2014, 18, 1533-1543.
27. Wu, H.-K.; Lee, S.W.-Y.; Chang, H.-Y.; Liang, J.-C. Current status, opportunities and challenges of augmented reality in education. Comput. Educ. 2013, 62, 41-49.
28. Hareide, O.S.; Ostnes, R. Maritime Usability Study by Analysing Eye Tracking Data. J. Navig. 2017, 70, 927-943.
29. Hareide, O. Improving Passage Information Management for the Modern Navigator. In Proceedings of the 19th IALA Conference, Incheon, Korea, 27 May-2 June 2018.
30. International Maritime Organization. MSC.253(83) Adoption of the Performance Standards for Navigation Light, Navigation Light Controllers and Associated Equipment; International Maritime Organization: London, UK, October 2007; p. 8 .
31. Procee, S.; Borst, C.; van Paassen, R.; Mulder, M.; Bertram, V. Using Augmented Reality to Improve Collision Avoidance and Resolution. In Proceedings of the 17th International Conference on Computer and IT Applications in the Maritime Industries, Pavone, Italy, 14-16 May 2018; pp. 14-16.
32. Procee, S.; Borst, C.; van Paassen, M.; Mulder, M.; Bertram, V. Toward functional augmented reality in marine navigation: A cognitive work analysis. In Proceedings of the 17th International Conference on Computer and IT Applications in the Maritime Industries, Pavone, Italy, 14-16 May 2018; pp. 298-312.

CHAPTER 7

Adaptive Formation Control of Unmanned Underwater Vehicles with Collision Avoidance under Unknown Disturbances

Zheping Yan, Anzuo Jiang and Chonglang Lai

College of Intelligent Systems Science and Engineering, Harbin Engineering University, Harbin 150001, China

ABSTRACT

In this paper, the formation control problem for a group of unmanned underwater vehicles (UUVs) is investigated considering collision avoidance and environment disturbances. To address the external force effect of the environment, such as waves and currents, a sliding mode disturbance observer is designed to compensate for the unknown dynamic disturbances in finite time. A bounded artificial potential field is incorporated into the control law to ensure collision avoidance among UUVs. The form of an artificial potential function is much simpler and convenient for engineering applications. A controller is devised to guarantee all the error signals are bounded, and the formation pattern can be achieved in finite time after collision avoidance. The stability of UUV formation with collision avoidance is proven by using the Lyapunov theorem, and the scheme has been shown to be convergent using Barbalat's lemma. Comparative simulations are presented to demonstrate the effectiveness of the proposed method in 2-D and 3-D environments.

Keywords: *multiple unmanned underwater vehicles; formation control; collision avoidance; adaptive sliding mode disturbance observer*

1. INTRODUCTION

In recent years, formation control for multiple agents has become a popular area of research for scholars, due to high efficiency, a wide searching area, and so on [1-6]. In the research process, there are a lot of challenges to address, such as nonlinearity in parameters, communication constraints, and dynamic environmental disturbance. Various methods have been proposed for formation control of autonomous underwater vehicles groups in a decentralized manner, which is also known as high-level control and examples of which include artificial potential field or methods based on agreement protocol such as leaderfollower, virtual structure, and behavioral approaches [7]. Ghommam [8] investigated a coordinated adaptive controller for a group of unmanned underwater vehicles (UUVs) to cope with uncertain dynamics and bounded unknown disturbances, and coordinate transformation was introduced to overcome the control difficulty caused by underactuation of UUVs. In [9], an adaptive neural network formation controller was developed for multiple autonomous underwater vehicles with unknown model coupling terms and unknown disturbances. A sliding mode formation controller was proposed for multiple autonomous underwater vehicles to compensate the environmental disturbances [10]. Gao proposed a fixed-time sliding control scheme with a disturbance observer to solve the compound disturbance, which included both external environment disturbance and parameter uncertainties [11]. Based on the minimal learning parameter algorithm, Lu proposed a robust adaptive formation controller to achieve the formation control of multiple underactuated surface vessels [12]. Due to the complicated underwater situation, the research progress of UUV formation has been relatively slow, but the research on unmanned aerial vehicle formations, autonomous ground vehicles, and multiagents has advanced. In [13], an adaptive neural network control scheme was investigated to address the formation control problem of multiple unmanned surface vessels with data dropout and time delays [14].

The research on the collision avoidance strategy of multiagents has been deepened. In [15-17], types of potential functions were designed to achieve connectivity preservation and collision avoidance, and the Lyapunov function was used for stability analysis. For multiagent systems, the virtual potential force of the artificial potential field (APF) was treated as an external disturbance, which proved to be robust based on $H\infty$ analysis in [18]. Bong investigated an unmanned surface vessel formation system, which suffered from heterogeneous limited communication ranges and designed a novel nonlinear transformed formation error to achieve the initial connectivity preservation and collision avoidance [19]. In [20], a distributed control strategy was presented for the multiagent system to reach the target plane with predesigned orientation, circulate around the target with prescribed radius, and avoid collisions among multiagents. He and Wang proposed a formation control method with collision avoidance for actuated multiple unmanned surface vessels to address the parametric and nonparametric uncertainties and external disturbances by imposing proper prescribed performance [21]. By integrating the gradient of repulsive APF with a fast terminal sliding mode surface, a novel sliding mode surface-like variable was improved for a position control scheme of an unmanned surface vessel formation [22].

In practical engineering experiments, external disturbance cannot be ignored, which can influence the stability of the multiagent system. Then, many control methods have been proposed such as the $H\infty$ method, disturbance observer, and adaptive sliding mode control theory to compensate for the external disturbance [23-25]. Nair [26] designed an adaptive gain higher order sliding mode observer to approximate the noisy position and velocity measurements, and the adaptive tuning algorithms could cope with the unexpected state jitter. Hua [27] integrated the finite-time control, sliding mode control, and the super-twisting algorithm to solve the large chattering of the control inputs for a high-order multiagent system with mismatched disturbance and unknown leader input.

Motivated by the above observations, this paper presents an adaptive control scheme for UUV formation with collision avoidance under an unknown disturbance. The specific contributions of this paper can be summarized as follows:

1. An adaptive formation control law of UUVs with collision avoidance under unknown bounded disturbance is proposed. The stability of the proposed method is proved, and simulation results are presented to verify its effectiveness;
2. Based on sliding mode control theory, a novel sliding mode disturbance observer was designed to approximate the unknown, nonlinear, and bounded disturbance. The simulation results show the observer achieves a good performance;
3. The studies on collision avoidance for multi-UUV formation are mostly on the three degree of freedom (DOF)F model in the horizontal plane. The state feedback linearization method is used to transfer the nonlinear and coupling mathematical model of UUVs into a second-order system model in 5-DOF. The artificial potential field theory is applied to cope with the collision avoidance among the UUVs. The form of the potential function is much simpler;
4. Based on the Lyapunov theory, the stability of the formation system is proven, and the proposed controller is valid and performs well, which can be seen in the 3-D simulation figures.

2. PRELIMINARIES AND PROBLEM FORMULATION

2.1. Feedback Linearization of UUV Model

The kinematic and dynamic models of a UUV are described in two coordinate frames, which are the Earth-fixed frame $\{E\}$ and the body-fixed frame $\{B\}$ as shown in Figure 1.

According to the structure of the UUV studied in the engineering application in our lab, there is no thruster to control the angular velocity in roll. Meanwhile, the rolling has little influence on the translational motion, so the roll speed can be ignored. Then, the kinematic and dynamic equations of UUVs can be described as:

$$\begin{aligned} \dot{\eta} &= J(\eta)v \\ M\dot{v} &= \bar{\tau} + \bar{\omega} - D(v)v - C(v) \end{aligned} \tag{1}$$

where $\eta = [x, y, z, \theta, \psi]^T$ denotes the position and the attitude angle of the UUV in the Earth-fixed frame; $v = [u, v, w, q, r]^T$ is the velocity vector of the UUV in the body-fixed frame. $\bar{\omega} \in \mathbb{R}^5$ is an unknown time-varying disturbance due to currents and waves. $\bar{\tau} \in \mathbb{R}^5$ is the control input acting on the UUV in the body-fixed frame. In addition, $J(\eta)$ is the transformation matrix; M, $D(v)$, and $C(v)$ denote the

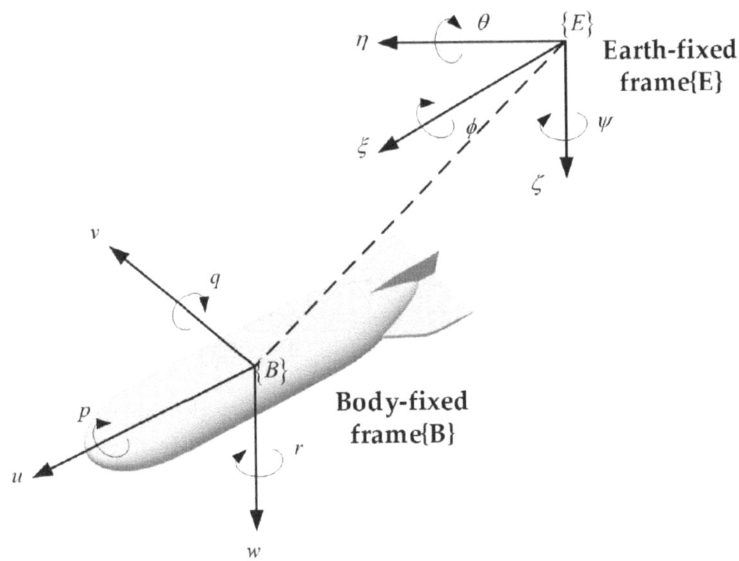

Figure 1. *Coordinate system of a UUV.*

inertia matrix, damping matrix, and the matrix of Coriolis and centrifugal terms, respectively. The structures of $J(\eta)$ and M are as follows:

$$J(\eta) = \begin{bmatrix} \cos\psi\cos\theta & -\sin\psi & \cos\psi\sin\theta & 0 & 0 \\ \sin\psi\cos\theta & \cos\psi & \sin\psi\sin\theta & 0 & 0 \\ -\sin\theta & 0 & \cos\theta & 0 & 0 \\ 0 & 0 & 0 & 1 & 0 \\ 0 & 0 & 0 & 0 & 1/\cos\theta \end{bmatrix}$$
$$M^{-1} = \begin{bmatrix} m_{11} & 0 & 0 & 0 & 0 \\ 0 & m_{22} & 0 & 0 & m_{25} \\ 0 & 0 & m_{33} & m_{34} & 0 \\ 0 & 0 & m_{43} & m_{44} & 0 \\ 0 & m_{52} & 0 & 0 & m_{55} \end{bmatrix} \tag{2}$$

The mathematical model of UUV is nonlinear and strong coupling. To solve the problem, the feedback linearization method is adopted to simplify the UUV model. The standard double integrator dynamic model can be described as:

$$\begin{aligned} \dot{p} &= v, \\ \dot{v} &= \tau, \\ p \in \mathbb{R}^5&, v \in \mathbb{R}^5 \end{aligned} \tag{3}$$

The specific linearization process can be obtained from [28]. In (3), the external disturbances are not considered. Since the disturbance is unknown and nonlinear, no matter how the conversion is performed in the linearization process, the final form of disturbance is still unknown and nonlinear. Then, a modified linearization model is proposed in this paper as:

$$\begin{aligned} \dot{p}_i &= v_i \\ \dot{v}_i &= \tau_i + \omega \end{aligned} \tag{4}$$

where $p_i \in \mathbb{R}^5, v_i \in \mathbb{R}^5$ is the position and velocity of the UUV_i in UUV formation with $i = 1, 2, \ldots, n. \tau_i \in \mathbb{R}^5$ is the control input of $UUV_i . \omega \in \mathbb{R}^5$ is an unknown time-varying disturbance.

Due to the complexity of the underwater environment, if faults occur to the leader in leader-follower method, the mission of UUV formation will not be able to be completed. To enhance the fault tolerance ability of the formation control, a virtual leader is introduced and defined as:

$$\begin{aligned} \dot{p}_l &= v_l \\ \dot{v}_l &= g_l(t) \end{aligned} \tag{5}$$

where $p_l \in \mathbb{R}^5$ is the position of the virtual leader, and $v_l \in \mathbb{R}^5$ is the velocity of the virtual leader. $g_l(t)$ is a given bounded and time-varying function, and $\varnothing$ is a positive constant and $\|g_l(t)\| < \eth$. Define the error variable of the system as:

$$\begin{aligned} e_{pi} &= p_i - p_l - \varepsilon_i \\ e_{vi} &= v_i - v_l \end{aligned} \tag{6}$$

where $\varepsilon_i \in \mathbb{R}^5$ is the desired relative position between UUV_i and the virtual leader. The desired velocity of UUV_i is the same as the velocity of the virtual leader, such that $\dot{\varepsilon}_i = 0$. Further, we assume e_{pi} and e_{vi} are bounded. Let

$$\begin{aligned} E_p &= \left[e_{p1}^T, e_{p2}^T, \ldots, e_{pn}^T \right]^T \\ E_v &= \left[e_{v1}^T, e_{v2}^T, \ldots, e_{vn}^T \right]^T \end{aligned} \tag{7}$$

2.2. Graph Theory

Considering a multi-UUV system consisting of n vehicles, we use the graph theory to model the information exchange among UUVs. Let the graph $G = (N, E)$ be an undirected graph, which consists of a node set $N = \{n_1, n_2, \cdots, n_n\}$, an edge set $E \subseteq N \times N$, and the adjacency matrix $A = [a_{ij}] \in R^{n \times n}$. The element $a_{ij} = 1$ denotes that the node i can receive information from the node j, otherwise $a_{ij} = 0$. In addition, for all $i, a_{ii} = 0$. Moreover, G is undirected if $a_{ij} = a_{ji} = 1$. The collection of connected neighbors of n_i is denoted as $N_i = \{n_j \in N : (n_j, n_i) \in E\}$. The in-degree matrix $B = \operatorname{diag}\{b_1, b_2, \cdots b_n\} \in R^{n \times n}$ is a diagonal matrix, where $w_i = \sum_{j=1}^{n} a_{ij}, i = 1, 2, \ldots, n$ is the in-degree of node n_i. Then the Laplacian matrix $L = [l_{ij}] \in R^{n \times n}$ is defined as $L = B - A \in R^{n \times n}$.

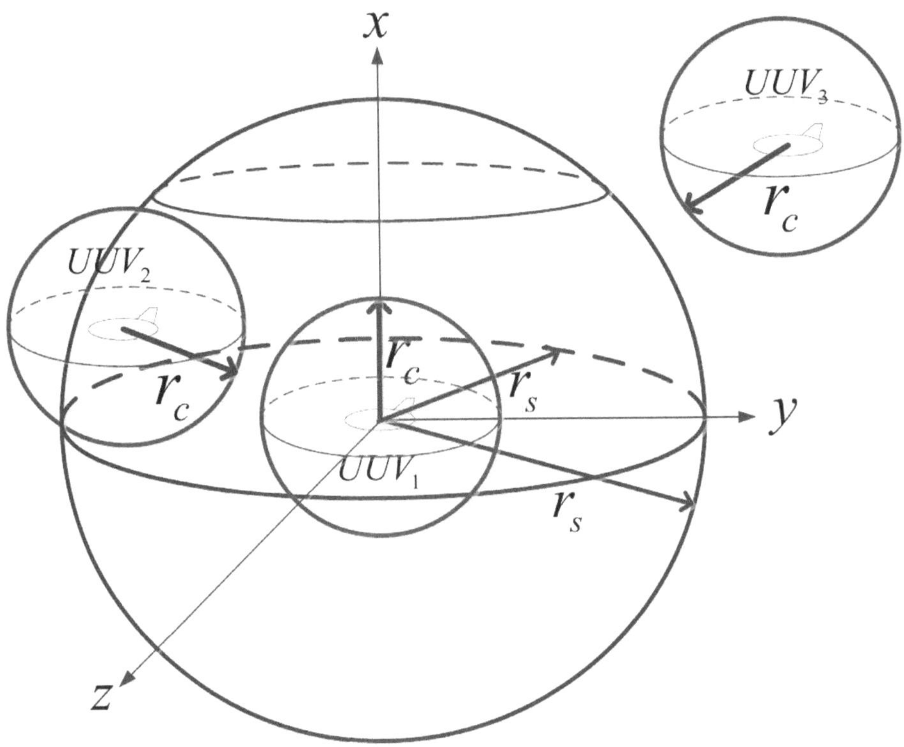

Figure 2. *The collision sphere and the collision avoidance sphere of UUVs.*

2.3. Artificial Potential Field

The APF method regards each UUV as a high-potential field. If any UUV is close to its neighbor, the repulsive force will repel the UUV away from other UUVs' potential field. There are two advantages for collision avoidance by using the APF method. The first one is that the individuals of the multi-UUV systems can be separated from each other to avoid collisions. The second one is that fewer parameters need to debug, and the controller design is much simpler than other collision avoidance methods. In practical engineering, a UUV is a rigid body with volume instead of a particle. Then, we assume that each UUV has the same structure and define the collision sphere and the collision avoidance sphere of a UUV, as shown in Figure 2. The collision avoidance sphere is defined by the black sphere with safe radius r_S. The collision sphere is shown by the red sphere with collision radius r_C. We define $\|d_{ij}\|$ as the distance between UUV_i and UUV_j, where $\|d_{ij}\| = \|p_i - p_j\| = \sqrt{\|x_i - x_j\|^2 + \|y_i - y_j\|^2 + \|z_i - z_j\|^2}$. Shi [29] concludes that UUV_j is a collision avoidance neighbor N_i^c of UUV_i while $\|d_{ij}\| \leq r_S$. When $\|d_{ij}\| \leq 2r_c$, collision occurs between UUV_i and UUV_j.

To ensure that no collision occurs among the UUVs, an artificial potential function $\delta_{ij}(d)$ and an action function $\varsigma(d)$ are defined as:

$$\delta_{ij}(d) = \int_{r_s}^{d} \varsigma(s)ds$$
$$\varsigma(d) = \left\{ \begin{array}{ll} -\frac{\bar{\beta}_i}{d^2} & d \in (2r_c, r_s) \\ 0 & d \in [r_s, \propto) \end{array} \right\} \tag{8}$$

where $\bar{\beta}_i > 0$ is a design parameter. When $\bar{\beta}_i$ is large enough, and $d \to 2r_c$, the potential function $\delta_{ij}(d)$ will tend to infinity. Thus, the repulsive force for collision avoidance is:

$$\tau_i^{ca} = \beta_i \sum_{j \in N_i^c} -\nabla_{x_i} \delta_{ij}(d) = -\beta_i \sum_{j \in N_i^c} \varsigma\left(\|d_{ij}\|\right) \frac{d_{ij}}{\|d_{ij}\|} \tag{9}$$

where $-\nabla_{x_i}$ denotes a negative gradient along x_i.

Assumption 1. *The disturbance is time-varying and bounded. Then, there exists a positive constant σ, such that $\|\omega\| < \sigma$.*

Assumption 2. *The velocity of UUVs and the virtual leader are not zero, e.g., $v_i \neq 0$ and $v_l \neq 0$. According to the characteristics of UUV, the velocity of UUVs is bounded, e.g., $\|v_i\| < \gamma_i$ and $\|v_l\| < \gamma_l$, where γ_i and γ_l are positive constants.*

Assumption 3. *At least one UUV can receive the information from the virtual leader, e.g., $\vartheta \neq 0$, where ϑ is the communication matrix between the virtual leader and followers. In addition, the communication between the UUVs is connected, e.g., $A \neq 0$ and $L \neq 0$.*

Assumption 4. *The initial error in relative position and relative velocity of any two UUVs are bounded. e.g., $\|e_{pi} - e_{pj}\| < \alpha_p$, $\|e_{vi} - e_{vj}\| < \alpha_v.\alpha_p$ and α_v are positive finite values.*

Lemma 1 ([30]). *Suppose $a_1, a_2, \ldots, a_n$ and $0 < p < 2$ are all positive numbers; then, the following inequality holds:*

$$\left(a_1^2 + a_2^2 + \cdots + a_n^2\right)^p \leq \left(a_1^p + a_2^p + \cdots + a_n^p\right)^2 \tag{10}$$

Lemma 2 ([31]). *$S(t) > 0$ is a continuous function for any times, and the initial state of $S(0)$ is bounded. If the inequality holds, $\dot{S}(t) > \lambda S(t)$ for $t > 0, \lambda > 0$, then, we have the following inequality:*

$$S(t) > e^{q(t-t_0)} S(t_0) \tag{11}$$

Lemma 3 ([29]). *$\vartheta = \operatorname{diag}\{\theta_1, \theta_2, \ldots, \theta_n\}$ and $\theta_i \geq 0$. If the Laplacian matrix L of the undirected graph is irreducible, then the eigenvalues of the matrix $L + \vartheta$ are positive definite, e.g., $L + \vartheta > 0$.*

3. ADAPTIVE FORMATION CONTROL SCHEME DESIGN

3.1. Adaptive Sliding Mode Disturbance Observer Design

In the process of performing underwater tasks, UUVs will be subject to environmental disturbances from ocean currents and waves. These disturbances are unknown and nonlinear. In this section, we design an adaptive sliding mode disturbance observer to estimate the disturbance ω. Firstly, an auxiliary state estimation error $e_0 \in \mathbb{R}^5$ is defined as:

$$e_0 = z - v \tag{12}$$

where $z \in \mathbb{R}^5$ is the auxiliary variable vector and its dynamic equation is designed as:

$$\dot{z} = \tau + v_S \tag{13}$$

where $v_s \in \mathbb{R}^5$ is the switching term to be designed. Substituting (4) and (14) into the result of differentiating (13), it yields:

$$\dot{e}_0 = v_s - \omega \tag{14}$$

To guarantee that the system can deduce the sliding motion, we design the switching term as:

$$v_s = -\Lambda_1 e_o - \Lambda_2 e_0^{\frac{m}{n}} - \mathrm{Ksgn}\left(e_0\right) \tag{15}$$

where $\Lambda_1 = \mathrm{diag}\left\{\lambda_{11}, \lambda_{12}, \lambda_{13}, \lambda_{14}, \lambda_{15}\right\}, \Lambda_2 = \mathrm{diag}\left\{\lambda_{21}, \lambda_{22}, \lambda_{23}, \lambda_{24}, \lambda_{25}\right\}$, and $K_i = \mathrm{diag}\left\{k_1, k_2, k_3, k_4, k_5\right\}$ are all positive definite diagonal matrices. m and n are odd positive integers, and $m < n$.

We design the Lyapunov function as follows:

$$V_O = \frac{1}{2} e_o^T e_o \tag{16}$$

Substituting (15) and (16) into the time derivative of V_O, it yields:

$$\begin{aligned} \dot{V}_O &= e_o^T e_o = e_o^T\left(v_S - \omega\right) = e_o^T\left(-\Lambda_1 e_o - \Lambda_2 e_o^{\frac{m}{n}} - \mathrm{Kgnn}\left(e_0\right) - \omega\right) \\ &= \sum_{i=1}^{5} -\lambda_{1i} e_{oi}^2 - \lambda_{2i} e_{oi}^{\frac{m+n}{n}} - k_i\left|e_{oi}\right| - e_{oi}\omega_i \end{aligned} \tag{17}$$

According to Assumption 1, ω is bounded. Choose $k_i > |\omega_i|$, such that $-k_i\left|e_{oi}\right| - e_{oi}\omega_i \leq -k_i\left|e_{oi}\right| + \left|e_{oi}\right|\left|\omega_i\right| < 0$. Thus,

$$\dot{V}_O \leq \sum_{i=1}^{5} -\lambda_{1i} e_{oi}^2 - \lambda_{2i} e_{oi}^{\frac{m+n}{n}} \tag{18}$$

According to Lemma 1, we have

$$\left(\sum_{i=1}^{5} e_{oi}^{2}\right)^{\frac{1}{2}} < \left(\sum_{i=1}^{5} e_{oi}^{\frac{m+n}{n}}\right)^{\frac{n}{m+n}} \Rightarrow -(2V_O)^{\frac{m+n}{2n}} > -\sum_{i=1}^{5} e_{oi}^{\frac{m+n}{n}} \tag{19}$$

Substituting (16) into (14), we obtain

$$\dot{V}_O \leq -2^{\frac{m+n}{2n}} \lambda_{1\min} (V_O)^{\frac{m+n}{2n}} - \lambda_{2\min} V_O \tag{20}$$

where $\lambda_{1\min} = \min\{\lambda_{11}, \lambda_{12}, \lambda_{13}, \lambda_{14}, \lambda_{15}\} > 0$, and $\lambda_{2\min} = \min\{\lambda_{21}, \lambda_{22}, \lambda_{23}, \lambda_{24}, \lambda_{25}\} > 0$. According to [31], since $\frac{1}{2} < \frac{m+n}{2n} < 1, V_O \to 0$ in the finite time, such that $e_O \to 0$. Then, v_s can be used to estimate the disturbance ω, i.e.,

$$v_s = \hat{\omega} \tag{21}$$

where $\hat{\omega}$ is the estimation of the ω.

3.2. Adaptive Formation Control with Collision Avoidance under Unknown Disturbances

The formation control scheme of the multi-UUVs are defined as:

$$\tau_i^f = -\mu_1 \sum_{j \in N_i} a_{ij} \left(e_{pi} - e_{pj}\right) - \mu_2 \sum_{j \in N_i} a_{ij} \left(e_{vi} - e_{vj}\right) - \theta_i \left(c_1 e_{pi} + c_2 e_{vi}\right) - v_s \tag{22}$$

where μ_1, μ_2, c_1, and c_2 are parameters to be designed; $\theta_i = 1$ means that UUV_i can receive the information from the virtual leader, and $\theta_i = 0$ otherwise. The formation controller with collision avoidance under unknown disturbances is designed as

$$\tau_i = \tau_i^f + \tau_i^{ca} \tag{23}$$

Theorem 1. *Under Assumptions 1-3, consider N UUVs with dynamics (1), control input (24) with adaptive sliding mode disturbance observer (16), and artificial potential function (8). If the control parameters are selected such that $k_i > |\omega_i|, m < n$, then the disturbance estimation error $m < n$ converges to zero in finite time, and $\dot{V}_C < 0, \|d_{ij}\| > 2r_c$ holds, which means that (a) the tracking errors converge to a small neighborhood around zero; (b) collision avoidance can be guaranteed for each UUV.*

Proof of Theorem 1. We define the Lyapunov function as follows:

$$V_C = \frac{1}{2} \sum_{i=1}^{5} \left(\theta_i e_{pi}^T e_{pi} + e_{vi}^T e_{vi}\right) + \frac{1}{2} E_p^T \left(L \otimes I_5\right) E_p \tag{24}$$

where $\otimes$ represents the Kronecker product. Taking the time derivative of the Lyapunov function, we obtain

$$\begin{aligned}\dot{V}_C &= \sum_{i=1}^{5}\left(\theta_i e_{pi}^T \dot{e}_{pi} + e_{vi}^T \dot{e}_{vi}\right) + E_p^T \left(L \otimes I_5\right) \dot{E}_p \\ &= \sum_{i=1}^{5}\left(\theta_i e_{pi}^T e_{vi} + e_{vi}^T \tau_i + e_{vi}^T \omega\right) + E_p^T \left(L \otimes I_5\right) E_v\end{aligned} \tag{25}$$

Substituting (23) and (25) becomes

$$\begin{aligned}\dot{V}_C &= \sum_{i=1}^{5}\theta_i e_{pi}^T e_{vi} + \sum_{i=1}^{5} e_{vi}^T \omega + E_p^T \left(L \otimes I_5\right) E_v - \mu_1 \sum_{i=1}^{5} e_{vi}^T \sum_{j\in N_i} a_{ij}\left(e_{pi} - e_{pj}\right) \\ &- c_1 \sum_{i=1}^{5}\theta_i e_{vi}^T e_{pi} - c_2 \sum_{i=1}^{5}\theta_i e_{vi}^T e_{vi} - \mu_2 \sum_{i=1}^{5} e_{vi}^T \sum_{j\in N_i} a_{ij}\left(e_{vi} - e_{vj}\right) - \sum_{i=1}^{5} e_{vi}^T v_s \\ &- \beta_i \sum_{i=1}^{5} e_{vi}^T \sum_{j\in N_i^c} -\nabla_{p_i}\delta_{ij}(d)\end{aligned} \tag{26}$$

We define that $\widetilde{\omega} = \hat{\omega} - \omega$ is the error between the real disturbance and the estimated disturbance. For an undirected graph, the repulsive forces between UUVs are equal and opposite in direction [32]. Therefore, $\beta_i \sum_{i=1}^{5} e_{vi}^T \sum_{j\in N_i^c} -\nabla_{p_i}\delta_{ij}(d) = 0$. Then, we obtain

$$\begin{aligned}\dot{V}_C &= \sum_{i=1}^{5} e_{vi}^T(\omega - \hat{\omega}) + E_p^T \left(L \otimes I_5\right) E_v - \mu_1 \sum_{i=1}^{5} e_{vi}^T \sum_{j\in N_i} a_{ij}\left(e_{pi} - e_{pj}\right) \\ &+ (1 - c_1) \sum_{i=1}^{5}\theta_i e_{vi}^T e_{pi} - c_2 \sum_{i=1}^{5}\theta_i e_{vi}^T e_{vi} - \mu_2 \sum_{i=1}^{5} e_{vi}^T \sum_{j\in N_i} a_{ij}\left(e_{vi} - e_{vj}\right) \\ &= \sum_{i=1}^{5} e_{vi}^T \widetilde{\omega} + (1 - \mu_1) E_v^T \left(L \otimes I_5\right) E_p + (1 - c_1) \sum_{i=1}^{5}\theta_i e_{vi}^T e_{pi} \\ &- c_2 E_v^T \left(\vartheta \otimes I_5\right) E_v - \mu_2 E_v^T \left(L \otimes I_5\right) E_v\end{aligned} \tag{27}$$

where $\vartheta = \text{diag}\left\{\theta_1, \theta_2, \theta_3, \theta_4, \theta_5\right\}.v_S$ can approximate the disturbance ω in finite time; thus, $\widetilde{\omega} \to 0$. Let $\mu_1 = c_1 = 1$, then

$$\begin{aligned}\dot{V}_C &= -c_2 E_v^T \left(\vartheta \otimes I_5\right) E_v - \mu_2 E_v^T \left(L \otimes I_5\right) E_v \\ &= -E_v^T \left(\left(\mu_2 L + c_2 \vartheta\right) \otimes I_5\right) E_v\end{aligned} \tag{28}$$

Since $c_2 > 0, \mu_2 > 0, L$ and ϑ are semi-definite. Thus, $\dot{V}_C$ is negative semi-definite, which proves that V_C is bounded. Then, the second derivative of V_C is found as

$$\ddot{V}_C = -2E_v^T \left(\left(\mu_2 L + c_2 \vartheta\right) \otimes I_5\right) \left(T_i + W\right) \tag{29}$$

where $T_i = [\tau_1, \tau_2 \ldots, \tau_5]^T$, and $W_i = [\omega_1, \omega_2 \ldots, \omega_5]^T$. Using polygon inequality, we obtain

$$\|\tau_i\| \le \mu_1 \sum_{j\in N_i} a_{ij} \left\|(e_{pi} - e_{pj})\right\| + \mu_2 \sum_{j\in N_i} \left(a_{ij} \left\|(e_{vi} - e_{vj})\right\| + \theta_i \left\|(c_1 e_{pi} + c_2 e_{vi})\right\|\right) + \|v_s\| + \beta_i \sum_{j\in N_i^c} \varsigma\left(\|d_{ij}\|\right) \frac{d_{ij}}{\|d_{ij}\|} \tag{30}$$

By choosing the proper $\bar{\beta}$, the function $\zeta(d)$ is bounded. According to Assumption 1 and 4 , we can see that the control law τ_i is bounded. This implies $\ddot{V}_C$ is bounded. Combining this with the fact that V_C is bounded and $\dot{V}_C$ is negative semidefinite, and applying the Barbalat's lemma, we can obtain $\dot{V}_C \to 0$, as $t \to \infty . \mu_2 L + c_2 \Pi$ is positive definite, from (29), which indicates $t \to \infty, e_v \to 0$. In addition, $\dot{e}_v = \tau_i + \omega \to 0$, which is only possible when all individual control parts become zero, and v_S can estimate the disturbance ω [33]. Then, we can obtain

$$\begin{aligned}
\lim_{t\to\infty} \sum_{i=1}^{5} e_{pi}^T \dot{e}_{vi} &= \lim_{t\to\infty} \sum_{i=1}^{5} e_{pi}^T \left(\tau_i + \omega_i\right) \\
&= -\lim_{t\to\infty} \sum_{i=1}^{5} e_{pi}^T \left\{ \mu_1 \sum_{j\in N_i} a_{ij} \left(e_{pi} - e_{pj}\right) + \mu_2 \sum_{j\in N_i} a_{ij} \left(e_{vi} - e_{vj}\right) \right. \\
&\quad \left. + c_1 \theta_i e_{pi} + c_2 \theta_i e_{vi} - \beta_i \sum_{j\in N_i^c} -\nabla_{p_i} \delta_{ij}(d) + v_s - \omega_i \right\} \\
&= -\lim_{t\to\infty} \sum_{i=1}^{5} \left\{ \sum_{j\in N_i} a_{ij} \left(e_{pi}^T e_{pi} - e_{pi}^T e_{pj}\right) + c_1 e_{pi}^T \theta_i e_{pi} \right\} \\
&= -\lim_{t\to\infty} E_p^T \left((W - A + \vartheta) \otimes I_m\right) E_p \\
&= -\lim_{t\to\infty} E_p^T \left((L + \vartheta) \otimes I_m\right) E_p \\
&= 0
\end{aligned} \tag{31}$$

where $L + \vartheta > 0$ can be obtained from Lemma 3. Therefore, $\lim_{t\to\infty} e_{pi} = 0$. Then, the adaptive UUV formation system with disturbance achieves performance.

We define an energy function:

$$S(t) = \frac{1}{2} d_{ij}^T(t) d_{ij}(t) + \frac{1}{2} v_i^T(t) v_i(t) \tag{32}$$

where $d_{ij}(t)$ denotes the relative position variable between the *UUVi* and its collision avoidance neighbor UUVj. Because all the UUVs in the formation have the same structure and characteristics of kinematic and dynamic, each UUV's collision avoidance performance is analyzed in the same way. The specific proof can be obtained from [26]. Taking the time derivative of (29) yields:

$$
\begin{aligned}
\dot{S}(t) &= d_{ij}^T(t)\dot{d}_{ij}(t) + v_i^T(t)\dot{v}_i(t) \\
&= d_{ij}^T(t)\left(\dot{p}_i(t) - \dot{p}_j(t)\right) + v_i^T(t)\left(\tau_i(t) + \omega_i\right) \\
&= d_{ij}^T(t)\left(v_i(t) - v_j(t)\right) + v_i^T(t)\left(-\mu_1 \sum_{j\in N_i} a_{ij}\left(e_{pi} - e_{pj}\right) - \sum_{j\in N_i} c_1\theta_i e_{pi}\right. \\
&\left. - \sum_{j\in N_i} c_2\theta_i e_{vi} - \mu_2 \sum_{j\in N_i} a_{ij}\left(e_{vi} - e_{vj}\right) - \hat{\omega}_i - \beta_i \sum_{j\in N_i^c} -\nabla_{p_i}\delta_{ij}(d) + \omega_i\right)
\end{aligned} \tag{33}
$$

From (30), it is easy to obtain $\widetilde{\omega} = \hat{\omega} - \omega \to 0$, because of the disturbance observer we designed. The error terms e_{vi} and e_{pi} tend to zero. $v_i(t) \neq 0, v_j(t) \neq 0$, and $d_{ij}(t) \to 2r_c$ are bounded. However, the potential function δ_{ij} approaches infinity, if $d_{ij}(t) \to 2r_c$. Then, we can obtain the inequality by choosing the appropriate parameter as follows:

$$
\begin{aligned}
v_i^T(t)\nabla_{p_i}\delta_{ij}\left(d_{ij}(t)\right) > &\frac{1}{2}d_{ij}^T(t)d_{ij}(t) + \frac{1}{2}v_i^T(t)v_i(t) \\
&- \frac{1}{\beta_i}d_{ij}^T(t)\left(v_i(t) - v_j(t)\right) + \frac{1}{\beta_i}v_i^T(t)\left(-\mu_1 \sum_{j\in N_i} a_{ij}\left(e_{pi} - e_{pj}\right)\right. \\
&\left. - \sum_{j\in N_i} c_1\theta_i e_{pi} - \sum_{j\in N_i} c_2\theta_i e_{vi} - \mu_2 \sum_{j\in N_i} a_{ij}\left(e_{vi} - e_{vj}\right) - \widetilde{\omega}_i\right)
\end{aligned} \tag{34}
$$

Then,

$$
\dot{S}(t) > \beta_i S(t) \tag{35}
$$

According to Lemma 2, we can obtain the following inequality

$$
d_{ij}^T(t)d_{ij}(t) > 2e^{\beta_i(t-t_0)}S(t) - v_i^T(t)v_i(t) \tag{36}
$$

By designing the parameter β_i, we have $2e^{\beta_i(t-t_0)}S(t) - v_i^T(t)v_i(t) > {r_c}^2$. Substituting it into (34), we obtain $\left\|d_{ij}(t)\right\| > 2r_c$. Therefore, the UUV formations can avoid collisions among them by the proposed control law.

4. EXPERIMENTAL RESULTS AND SIMULATION

To illustrate the effectiveness of the proposed algorithms, the simulation results are presented. The verification is mainly divided into two parts. The first one analyzes the effect of the sliding mode disturbance observer. The other part is about the effectiveness of the collision avoidance. The UUV formation consists of four

Table 1. *Parameters.*

Entry	Value	Entry	Value	Entry	Value
ϑ_1	0	ϑ_2	0	c_1	1
ϑ_3	1	ϑ_4	1	c_2	2
r_c	0.3	μ_1	1	$\underline{\beta}$	5
r_s	0.8	μ_2	2	$\overline{\beta}$	2

UUVs in the simulation experiment, and UUV_0 is the virtual leader. Before showing the simulation results, some parameters used in the system are listed in Table 1.

The adjacency matrix and Laplacian matrix are

$$A = \begin{bmatrix} 0\,0\,1\,1 \\ 0\,0\,0\,1 \\ 1\,0\,0\,1 \\ 1\,1\,1\,0 \end{bmatrix}, L = \begin{bmatrix} 2 & 0 & -1 & -1 \\ 0 & 1 & 0 & -1 \\ -1 & 0 & 2 & -1 \\ -1 & -1 & -1 & 3 \end{bmatrix} \tag{37}$$

The initial velocities of the UUVs are designed as zero, and the initial position of the UUVs is given as

$$p_i = \begin{bmatrix} p_1 \\ p_2 \\ p_3 \\ p_4 \end{bmatrix} = \begin{bmatrix} -10 & 10 & 0 & 0\,0 \\ 0 & -10 & 0 & 0\,0 \\ 0 & 0 & -10 & 0\,0 \\ -5 & -5 & 0 & 0\,0 \end{bmatrix} \tag{38}$$

The position of the virtual leader is given as $p_l = [20\cos(0.1t), 20\sin(0.1t), 0.1t, 0, 0]$.

4.1. Disturbance Observer Simulation Result

Since the disturbance we considered is nonlinear and bounded, ω_i is given as:

$$\omega_i = \begin{bmatrix} (\sin(0.2t))^2 \times (\cos(1+0.2t)) \\ (\sin(0.2t))^2 \times (\cos(2+0.2t)) \\ (\sin(0.2t))^2 \times (\cos(1+0.2t)) \\ (\sin(1+0.2t))^2 \times (\cos(0.2t)) \\ (\sin(2+0.2t))^2 \times (\cos(0.2t)) \end{bmatrix} \tag{39}$$

The disturbance observer we designed is

$$v_S = -\Lambda_1 e_o - \Lambda_2 e_o^{\frac{m}{n}} - \mathrm{Kggn}\,(e_o) \tag{40}$$

The parameters are selected as $m = 3, n = 5, \Lambda_1 = \mathrm{diag}\{1,1,1,1,1\}, \Lambda_2 = \mathrm{diag}\{1,1,1,1,1\}$, and $K_i = 1.1 \times \mathrm{diag}\{1,1,1,1,1\}$.

In this section, our goal is to illustrate the effectiveness of the sliding mode disturbance observer. Thus, the control law has included the part of collision avoidance in simulation. The virtual leader trajectory and the UUVs trajectories are plotted in Figure 3 to show the formation control performance; the initial and

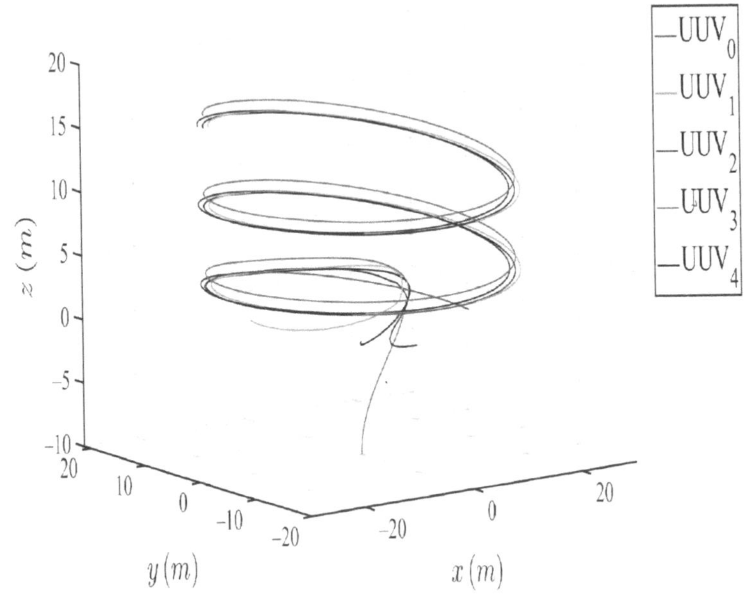

Figure 3. *3D trajectories of UUVs with disturbance observer.*

final position of the UUVs are also shown. The 3D trajectory in Figure 3 shows the controller with the disturbance observer. The tracking trajectory is smooth and steady. Figure 4 displays the 3D trajectory results from the controller without the disturbance observer. It is obvious that the tracking trajectory has jitter, and the disturbance has a great influence on the formation maintenance. The simulation results show that the adaptive sliding mode disturbance observer achieves good performance.

To verify the performance of the disturbance observer, an adaptive nonlinear disturbance observer (ANDO) proposed in [34] is compared with the observer proposed by us. The results of comparation are shown in Figures 5 and 6 . The red line is the real disturbance value curve. The blue line is the estimated value produced by the observer proposed in this paper. In addition, the green line is the estimation of the ANDO observer. From the enlarged views in Figures 5 and 6, compared with the green line, the blue one is always closer to the red line. It means the estimation of SMDO is more precise than the ANDO. The SMDO achieves a better performance.

4.2. Collision Avoidance Simulation Result

To show the performance of the controller of collision avoidance, we used the proposed disturbance observer to estimate the disturbance. Since the UUV will not stop after a collision, the difference of the 3D trajectories between the control law with collision avoidance and the controller without collision avoidance is small. So, the 3D trajectories are not displayed in this section. The distance between any two UUVs is more clearly to show the collision. The $2r_c$ line is set at the bottom of the Figures 7 and 8 . If the distance $\|d_{ij}\| < 2r_c$, the collision occurs. Figure 7 shows

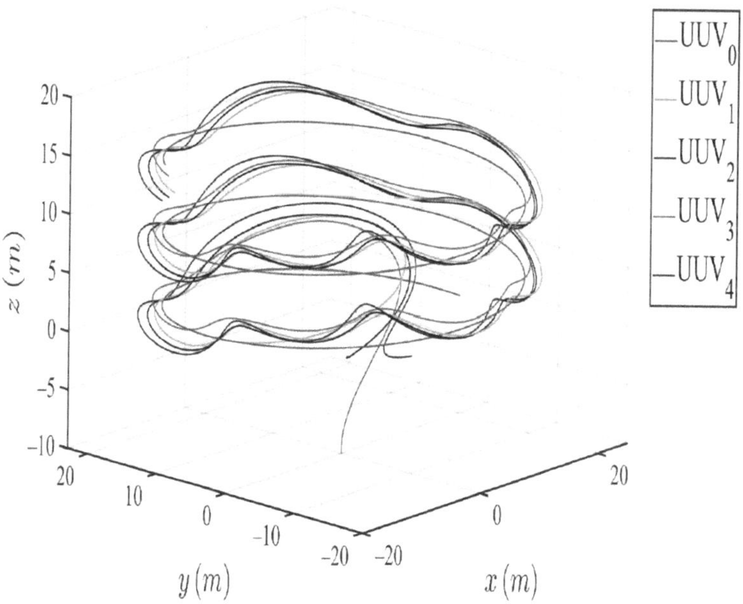

Figure 4. *3D trajectories of UUVs without disturbance observer.*

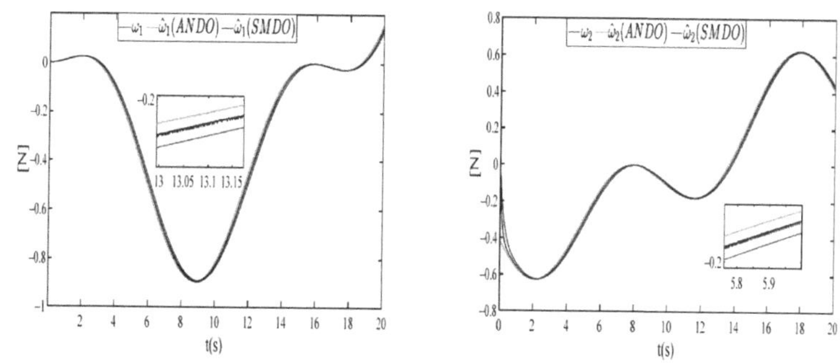

Figure 5. *The disturbance estimation comparation in* ω_1 *and* ω_2*.*

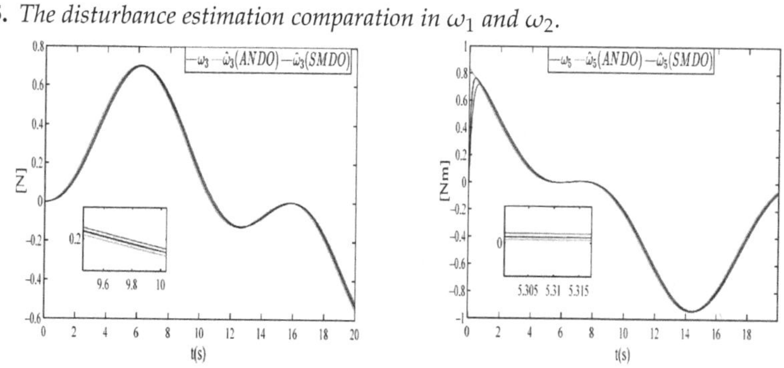

Figure 6. *The disturbance estimation comparation in* ω_3 *and* ω_5*.*

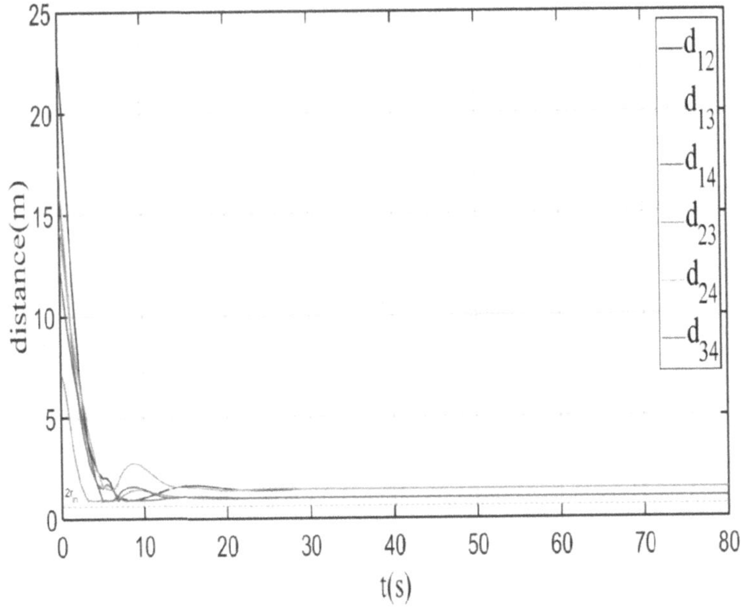

Figure 7. *The distance between any two UUVs of the formation with collision avoidance.*

the distance between any two UUVs of the formation with considering collision avoidance. There is no line under the $2r_c$ line. Due to the spiral trajectory, there are some fluctuations in the distance between UUVs. It has no influence on the stability of system. Figure 8 shows the distance between any two UUVs of the formation without considering collision avoidance. It is obviously that collision occurs several times from 4 s to 6 s. The distance between UUVs is below $2r_C$. Without the collision avoidance control, the formation performance cannot be guaranteed.

The position error and velocity error performance are validated as shown in Figures 9-11. All the errors converge to zero in finite time, which means that the controller works satisfactorily and the system is stable. The simulation results show the four UUVs can track the virtual leader well without collision and maintain a good formation.

5. CONCLUSIONS

The adaptive formation control for multi-UUVs with collision avoidance under unknown disturbance was discussed. Graph theory was utilized to model the communications between UUVs. Feedback linearization was used to simplify the 5-DOF mathematical model. The overall structure of the control law was composed of a formation controller and a collision avoidance controller. The adaptive sliding mode disturbance observer was adopted to approximate the nonlinear and unknown dynamic disturbance term. By integrating the artificial potential field

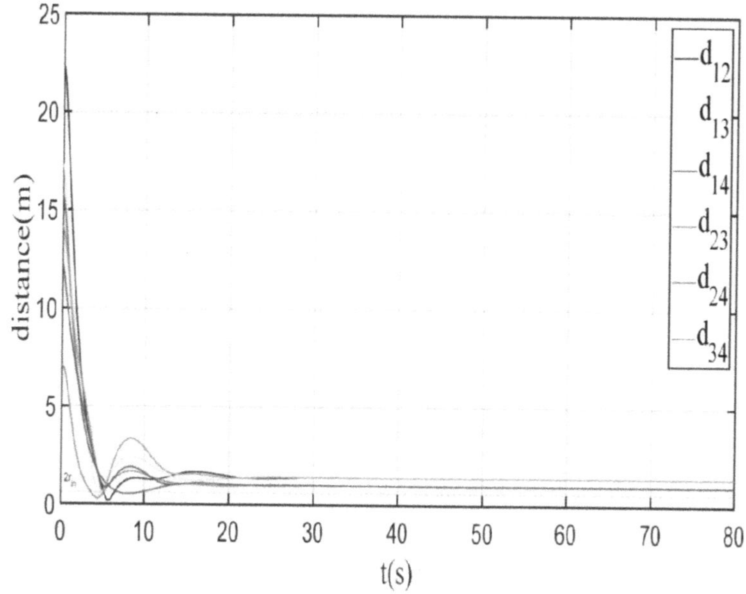

Figure 8. *The distance between any two UUVs of the formation without collision avoidance.*

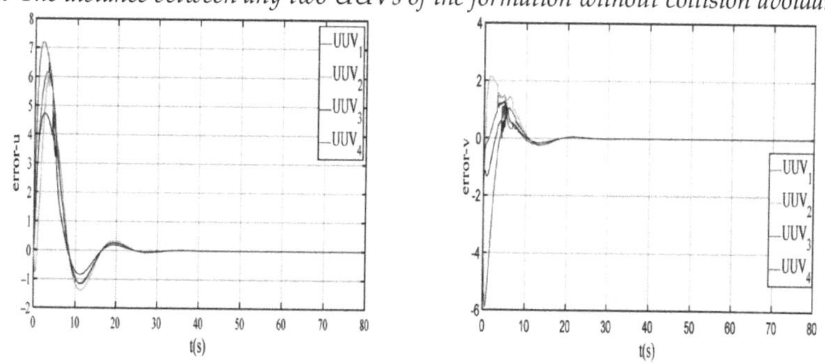

Figure 9. *Velocity errors of UUVs in u and v.*

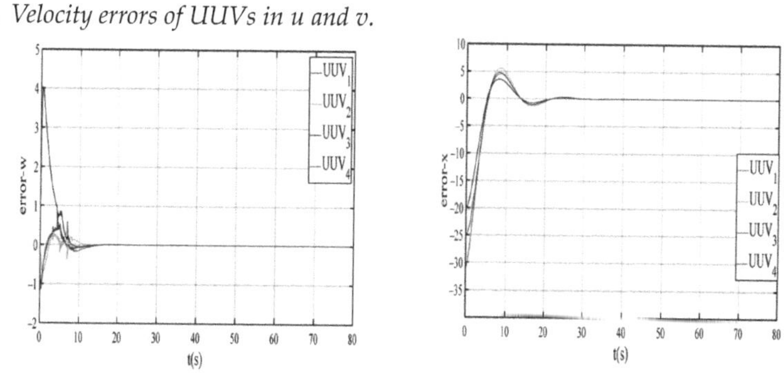

Figure 10. *Velocity errors of UUVs in w, and position errors of UUVs in x direction.*

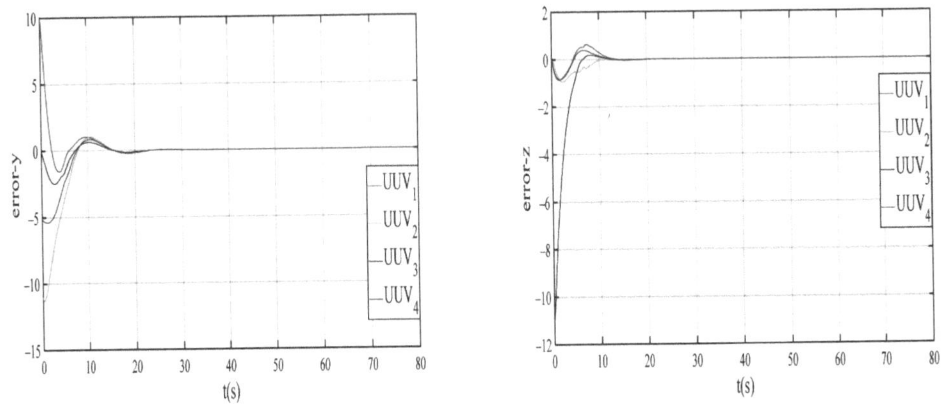

Figure 11. *Position errors of UUVs in y and z directions.*

method into the virtual leader-following formation strategy, the problem of collision among UUVs was solved. The system stability was proven using the Lyapunov theory. Finally, we compared the proposed disturbance observer with the observer in another paper. The result shows the observer proposed in this paper performed better. The simulation results on formation control with collision avoidance have demonstrated the controller designed in this paper is valid and performs well. In the future, we will work on improving the observer to eliminate the jitter caused by the sliding mode and formation control subject to constraints.

Author Contributions

Conceptualization, A.J. and Z.Y.; methodology, A.J.; software, A.J. and C.L.; validation, A.J.; formal analysis, Z.Y.; investigation, A.J. and C.L.; resources, Z.Y.; data curation, A.J.; writing-original draft preparation, A.J.; writing-review and editing, Z.Y. and C.L.; supervision, Z.Y. All authors have read and agreed to the published version of the manuscript.

REFERENCES

1. Liang, H.; Zhang, L.; Sun, Y.; Huang, T. Containment Control of Semi-Markovian Multiagent Systems with Switching Topologies. IEEE Trans. Syst. Man Cybern. Syst. 2019, 51, 3889-3899.
2. Liu, Y.; Shi, P.; Lim, C.-C. Collision-Free Formation Control for Multi-Agent Systems with Dynamic Mapping. IEEE Trans. Circuits Syst. II Express Briefs 2020, 67, 1984-1988.
3. Li, X.; Wen, C.; Chen, C. Adaptive Formation Control of Networked Robotic Systems with Bearing-Only Measurements. IEEE Trans. Cybern. 2021, 51, 199-209. [PubMed]
4. Zhou, K.-B.; Wu, X.-K.; Ge, M.-F.; Liang, C.-D.; Hu, B.-L. Neural-Adaptive Finite-Time Formation Tracking Control of Multiple Nonholonomic Agents with a Time-Varying Target. IEEE Access 2020, 8, 62943-62953.

5. Damasceno, B.C.; Xie, X. Deadlock-free scheduling of manufacturing systems using petri nets and dynamic programming. IFAC Proc. Vol. 1999, 32, 4870-4875.
6. Foumani, M.; Gunawan, I.; Smith-Miles, K. Resolution of deadlocks in a robotic cell scheduling problem with post-process inspection system: Avoidance and recovery scenarios. In Proceedings of the 2015 IEEE International Conference on Industrial Engineering and Engineering Management (IEEM), Singapore, 6-9 December 2015; pp. 1107-1111.
7. Chen, W.; Wu, X.; Lu, Y. An Improved Path Planning Method Based on Artificial Potential Field for a Mobile Robot. Cybern. Inf. Technol. 2015, 15, 181-191.
8. Ghommam, J.; Saad, M. Backstepping-based cooperative and adaptive tracking control design for a group of underactuated AUVs in horizontal plan. Int. J. Control 2014, 87, 1076-1093.
9. Park, B.S. Adaptive formation control of underactuated autonomous underwater vehicles. Ocean Eng. 2015, 96, 1-7.
10. Huang, H.; Zhang, G.-C.; Li, Y.-M.; Li, J.-Y. Fuzzy sliding-mode formation control for multiple underactuated autonomous underwater vehicles. In Proceedings of the International Conference on Swarm Intelligence, Bali, Indonesia, 25-30 June 2016; pp. 503-510.
11. Gao, Z.; Guo, G. Fixed-time sliding mode formation control of AUVs based on a disturbance observer. IEEE/CAA J. Autom. Sin. 2020, 7, 539-545.
12. Lu, Y.; Zhang, G.; Sun, Z.; Zhang, W. Robust adaptive formation control of underactuated autonomous surface vessels based on MLP and DOB. Nonlinear Dyn. 2018, 94, 503-519.
13. Park, B.S.; Kwon, J.-W.; Kim, H. Neural network-based output feedback control for reference tracking of underactuated surface vessels. Automatica 2017, 77, 353-359.
14. Franze, G.; Casavola, A.; Famularo, D.; Lucia, W. Distributed Receding Horizon Control of Constrained Networked LeaderFollower Formations Subject to Packet Dropouts. IEEE Trans. Control Syst. Technol. 2018, 26, 1798-1809.
15. Ajorlou, A.; Aghdam, A.G. A bounded distributed connectivity preserving aggregation strategy with collision avoidance property. Syst. Control Lett. 2013, 62, 1098-1104.
16. Sakai, D.; Fukushima, H.; Matsuno, F. Leader-Follower Navigation in Obstacle Environments While Preserving Connectivity without Data Transmission. IEEE Trans. Control Syst. Technol. 2018, 26, 1233-1248.
17. Dong, Y.; Su, Y.; Liu, Y.; Xu, S. An internal model approach for multi-agent rendezvous and connectivity preservation with nonlinear dynamics. Automatica 2018, 89, 300-307.
18. Wen, G.; Chen, C.L.P.; Liu, Y.-J. Formation Control with Obstacle Avoidance for a Class of Stochastic Multiagent Systems. IEEE Trans. Ind. Electron. 2018, 65, 5847-5855.
19. Park, B.S.; Yoo, S.J. An Error Transformation Approach for Connectivity-Preserving and Collision-Avoiding Formation Tracking of Networked Uncertain Underactuated Surface Vessels. IEEE Trans. Cybern. 2019, 49, 2955-2966.
20. Zhong, H.; Wang, Y.; Miao, Z.; Tan, J.; Li, L.; Zhang, H.; Fierro, R. Circumnavigation of a Moving Target in 3D by Multi-agent Systems with Collision Avoidance: An Orthogonal Vector Fields-based Approach. Int. J. Control Autom. Syst. 2019, 17, 212 – 224.

21. He, S.; Wang, M.; Dai, S.-L.; Luo, F. Leader-Follower Formation Control of USVs With Prescribed Performance and Collision Avoidance. IEEE Trans. Ind. Inform. 2019, 15, 572-581.

22. Huang, Y.; Liu, W.; Li, B.; Yang, Y.; Xiao, B. Finite-time formation tracking control with collision avoidance for quadrotor UAVs. J. Frankl. Inst. 2020, 357, 4034-4058.

23. Pan, L.; Bao, G.; Xu, F.; Zhang, L. Adaptive robust sliding mode trajectory tracking control for 6 degree-of-freedom industrial assembly robot with disturbances. Assem. Autom. 2018, 38, 259-267.

24. Chen, Z.; Yang, C.; Liu, X.; Wang, M. Learning control of flexible manipulator with unknown dynamics. Assem. Autom. 2017, 37, 304-313.

25. Yang, C.; Jiang, Y.; He, W.; Na, J.; Li, Z.; Xu, B. Adaptive Parameter Estimation and Control Design for Robot Manipulators with Finite-Time Convergence. IEEE Trans. Ind. Electron. 2018, 65, 8112-8123.

26. Nair, R.R.; Behera, L. Robust adaptive gain higher order sliding mode observer based control-constrained nonlinear model predictive control for spacecraft formation flying. IEEE/CAA J. Autom. Sin. 2016, 5, 367-381.

27. Hua, Y.; Dong, X.; Han, L.; Li, Q.; Ren, Z. Finite-Time Time-Varying Formation Tracking for High-Order Multiagent Systems with Mismatched Disturbances. IEEE Trans. Syst. Man Cybern. Syst. 2019, 50, 1-9.

28. Yan, Z.-P.; Liu, Y.-B.; Yu, C.-B.; Zhou, J.-J. Leader-following coordination of multiple UUVs formation under two independent topologies and time-varying delays. J. Cent. South Univ. 2017, 24, 382-393.

29. Shi, Q.; Li, T.; Li, J.; Chen, C.P.; Xiao, Y.; Shan, Q. Adaptive leader-following formation control with collision avoidance for a class of second-order nonlinear multi-agent systems. Neurocomputing 2019, 350, 282-290.

30. Yu, S.; Yu, X.; Shirinzadeh, B.; Man, Z. Continuous finite-time control for robotic manipulators with terminal sliding mode. Automatica 2005, 41, 1957-1964.

31. Wen, G.; Chen, C.L.P.; Dou, H.; Yang, H.; Liu, C. Formation control with obstacle avoidance of second-order multi-agent systems under directed communication topology. Sci. China Inf. Sci. 2019, 62, 1-14.

32. Pang, Z.-H.; Zheng, C.-B.; Sun, J.; Han, Q.-L.; Liu, G.-P. Distance- and Velocity-Based Collision Avoidance for Time-Varying Formation Control of Second-Order Multi-Agent Systems. IEEE Trans. Circuits Syst. II Express Briefs 2021, 68, 1253-1257.

33. Mondal, A.; Bhowmick, C.; Behera, L.; Jamshidi, M. Trajectory Tracking by Multiple Agents in Formation with Collision Avoidance and Connectivity Assurance. IEEE Syst. J. 2018, 12, 2449-2460.

34. Li, J.; Du, J.; Hu, X. Robust adaptive prescribed performance control for dynamic positioning of ships under unknown disturbances and input constraints. Ocean Eng. 2020, 206, 107254.

CHAPTER

8

Capturing Expert Knowledge to Inform Decision Support Technology for Marine Operations

Jennifer Smith, Fatemeh Yazdanpanah, Rebecca Thistle, Mashrura Musharraf and Brian Veitch

Faculty of Engineering and Applied Science, Memorial University of Newfoundland, St. John's, NL A1B 3X5, Canada

ABSTRACT

The digital transformation of the offshore and maritime industries will present new safety challenges due to the rapid change in technology and underlying gaps in domain knowledge, substantially affecting maritime operations. To help anticipate and address issues that may arise in the move to autonomous maritime operations, this research applies a human-centered approach to developing decision support technology, specifically in the context of ice management operations. New technologies, such as training simulators and onboard decision support systems, present opportunities to close the gaps in competence and proficiency. Training simulators, for example, are useful platforms as human behaviour laboratories to capture expert knowledge and test training interventions. The information gathered from simulators can be integrated into a decision support system to provide seafarers with onboard guidance in real time. The purpose of this research is two-fold: (1) to capture knowledge held by expert seafarers, and (2) transform this expert knowledge into a database for the development of a decision support technology. This paper demonstrates the use of semi-structured interviews and bridge simulator exercises as a means to capture seafarer experience and best operating practices for offshore ice management. A case-based reasoning (CBR) model is used to translate the results of the knowledge capture exercises into an early-stage ice management decision support system. This paper will describe the methods used and insights gained from translating the interview data and expert performance from the bridge simulator into a case base that can be referenced by the CBR model.

Keywords: *knowledge capture; decision support; artificial instruction; case-based reasoning*

1. INTRODUCTION

The digital transformation of the offshore and maritime industries will present new safety challenges for operations in ice-covered waters. A main safety concern with the rapid advancement of automation lies in its inherent goal to replace human operators with fully autonomous ships. The forecasted replacement of crews' responsibilities with automated functions is problematic. Many researchers have argued that humans are vital to the safe operation of complex socio-technical systems [1-3]. In response to the technology-driven transformation, offshore and maritime regulators and classification societies, such as the International Maritime Organization (IMO) and Lloyd's Register, have developed regulations and guidelines to shape the safe implementation of automation for offshore and maritime operations [4,5].

Researchers in the maritime domain have advocated for a more human-centered approach to the development and integration of autonomous operations [6,7]. This is especially important for offshore ice management operations as the safety of these operations relies heavily on the judgement and decision making of experienced captains and their bridge teams, and the team's ability to routinely perform task adaptations. The expectations of seafarers and the current need for routine adaptations and customized ice management plans highlight two underlying knowledge gap in the domain: (1) a time-consuming and informal on-the-job training of new mates and cadets, and (2) a lack of consensus on operating procedures for the safe management of pack ice offshore. These gaps highlight the need to capture expert knowledge and to standardize the distribution of ice management training. Therefore, emphasis should be placed on improving the transfer of skills from experienced operators to new mates and cadets. To do this, we need a better understanding of how seafarers perform current pack ice management operations and how they adapt to address problems.

A human-centered approach to developing a decision support system can help with the needed knowledge capture and mobilization. Our approach is to use simulation technology to capture expert knowledge on offshore pack ice management operations. We plan to develop an expert-informed decision support system from the information gained from the simulation exercises to foster better knowledge exchange between seafarers. The expert knowledge can be embedded into a decision support system using case-based reasoning (CBR). CBR is well suited to capture the expertise of captains and provide standardized assistance and learning opportunities to cadet and other seafarers.

This paper focuses on the early development stages of a CBR decision support system, which includes (1) gathering contextual knowledge from expert seafarers (e.g., data collection), and (2) indexing the seafarers experiences from a variety of scenarios into a CBR case base. This paper will describe the methods used and the insights gained from translating interview data and expert performance from a bridge simulator into a case base that can be referenced by the CBR model.

2. BACKGROUND

2.1. Offshore Ice Management Operations

The year-round activity of offshore oil operations combined with the harsh ice conditions off the coast of Newfoundland present an ongoing need for ice management operations to protect the oil platforms in the area and ensure safe operations. In particular, the Grand Banks commonly experience annual seasons of icebergs and variable seasons of pack ice [8-10]. Ice management operations offshore involve a broad scope of activities that include (1) monitoring and forecasting ice conditions, (2) detecting and tracking icebergs and pack ice, (3) reporting ice conditions, (4) avoiding, breaking, and deflecting ice threats [9, 11].

Focusing on the physical interventions of ice management, there is a variety of operations, such as towing icebergs, deflecting ice or icebergs with water cannons and prop wash, as well as breaking up and dispersing sea ice [9]. This research focuses on offshore operations used to deflect or disperse encroaching pack ice from the operational area of an oil platform. Dunderdale and Wright in their review of pack ice management in offshore Newfoundland outlined five common techniques used by offshore support vessels to physical move pack ice, the techniques include linear, sector, circular, stationary/prop wash and pushing [8]. Combinations of these techniques are used to protect offshore platforms against approaching ice. Overall, operations for managing pack ice are complex and performed in dynamic circumstances (e.g., unpredictable maritime winds and currents, and considerable time constraints to respond to danger). Consequently, it is difficult to anticipate how a situation will unfold due to the uncertainty in ice management operations. Since each pack ice situation offshore is different, many operations require customized strategies and approaches.

The majority of ice management training is provided through on-the-job experience in which inexperienced personnel learn informally by supporting experienced crewmembers in performing ice management operations [12]. Informal on-the-job training is both a regulatory and operational problem. From a regulatory perspective, it has taken time to put new regulations into practice. From an operational perspective, much of the competence requirements and training guidelines associated with ice management operations have been developed for navigating Arctic waters, which may not be directly applicable to the specific context of pack ice management around offshore oil platforms. In Canada, much of the regulatory guidance on ice management training stems from two model courses (Basic and Advanced Training for Ships Operating in Polar Waters) that where developed by IMO in collaboration with the Marine Institute's Centre for Marine Simulation and the professional organization Master Mariners of Canada [13,14]. Other regulatory guidance focuses on certification requirements of 'ice navigators', which are based on: (1) IMO code for Ships Operating in Polar Waters (Polar Code) [15], and (2) the Canadian Arctic Shipping Safety and Pollution Prevention Regulations (ASPPR) that were developed for navigating in Arctic waters [16]. According to Canada's ASPPR regulations, a seafarer receives the designation of an 'ice navigator' after serving on a vessel as master for at least 50 days, where 30 days of this time must have been in Arctic waters when the vessel is in ice conditions requiring ice avoidance or ice breaking maneuvers [12, 16]. Conversely, to receive certification as an 'ice advisor', the Joint Industry-Government Guidelines for the Control vessels in Ice Control Zones of Eastern Canada (JIGs) requires master seafarers to complete at least 90 days navigating in ice-covered waters over the course of 5 years (e.g., six trips each with a minimum of 15 days in ice conditions) [17].

Applying the regulations and guidelines for an ice advisor to the context of offshore ice management, it will take seafarers many ice seasons to develop the needed proficiency and expertise to respond to situations where offshore platforms are impeded by sea ice. As such, the conventional on-the-job form of ice management training has created an inefficient knowledge exchange from captains to new mates and training cadets. To compound this issue further, pack ice management operations offshore rely on the inherited experience of captains and their bridge teams, and the team's joint ability to develop ad-hoc strategies and perform task adaptations. Consequently, the informal nature of on-the-job practice for ice operations has created a skill disparity in the domain. This is evident from Veitch et al. research, which tasked experienced seafarers with performing ice management operations in emergency conditions in a simulator and found a high level of variability in performance [18]. This variability in ice management skills amongst experienced seafarers has likely propagated down the hierarchy to mates and cadets.

2.2. Case-Based Reasoning Theory

Case-based reasoning (CBR) is a constructivist paradigm to solving problems that is useful for machine learning and artificial intelligence applications. The aim of *CBR* is to mimic how experienced people make decisions based on using their experiences to form plausible approaches for new situations. The CBR method is suitable for domains that require regular interpretation of open-ended or poorly understood problems [19]. Over the last 30 years, CBR has been employed or combined with other methods (e.g., rule-based reasoning) to develop artificially intelligent applications for decision support and learning in education [20], medical diagnosis [21,22], and environmental emergency decision making [23].

Figure 1 provides the general CBR model. The CBR model builds a case base of previously established knowledge and specific examples from past situations, known as a case base, to help understand new problems and offer solutions [20,24,25]. Cases can take the form of successfully solved problems as well as failed problems [24]. The solved problems create a basis of features and examples used for matching cases to the conditions of new situations. The failed problems form reminders of what approaches did not work and how to avoid repeating mistakes when presented with new problems.

As depicted in Figure 1, there are four main stages of the CBR model: retrieval, reuse, revision, and retention. The following is an explanation of the stages based on Aamodt and Plaza [24]. Case retrieval involves identifying features of a new problem, searching for a feature match within the existing case base and selecting the matched or partially matched case for use. Case reuse is a closer examination

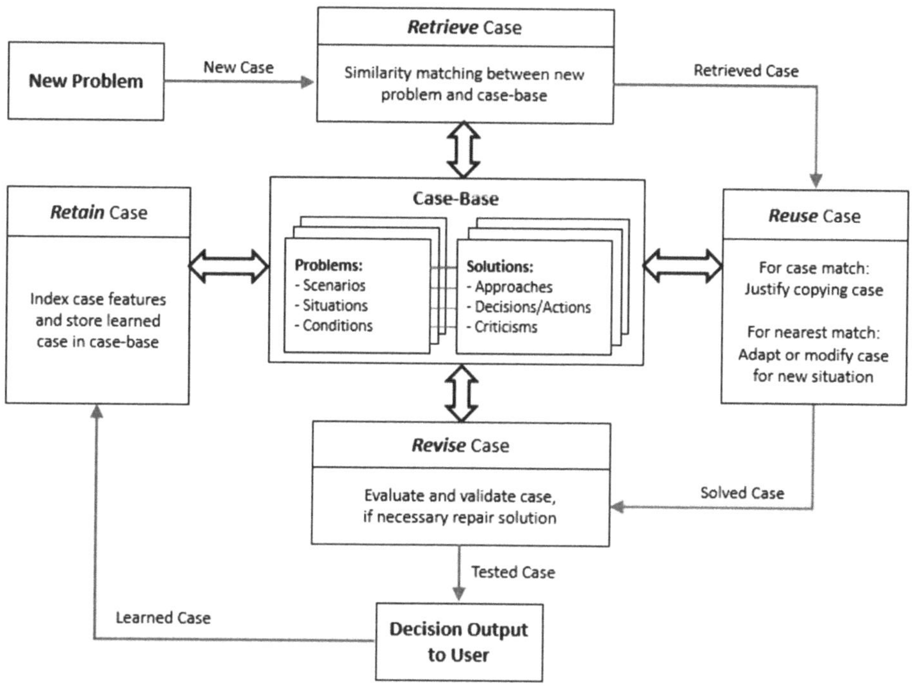

Figure 1. *Case-based reasoning model for decision support, showing the retrieval, reuse, revision, and retention of cases (adapted from [24] with permission from IOS Press, 1994).*

of the reusable aspects and differences between the new and retrieved cases. For a matched case, a copy of the retrieved case is reused as the solution for the new situation. For a partially or an inferred matched case, adaptations are made to the case solution before it can be transferred for reuse in the new situation. Case revision is a form of validation that involves testing the solution by asking a subject matter expert or by applying the decision in real life or in a simulation. Decisions that resulted in failed solutions are repaired by creating explanations for the errors and modifying the cases to avoid making the same errors again. Case retention is how the CBR model learns by working through new cases. Useful aspects of the successful and repaired cases from the revision stage are indexed and integrated into the case base.

For decision support applications, the CBR model provides an incremental learning method for modeling complicated decisions by representing and organizing expert knowledge in a way that enables decision support systems to quickly retrieve similar cases and adapt them to new situations [20, 26]. Gathering subject matter expert knowledge is vital for the development of two aspects of the CBR model: (1) specific domain knowledge, needed to generate example solutions for the case base; and (2) general domain knowledge, needed to develop the reasoning structure of the CBR model. The general domain knowledge guides the case feature matching, retrieval, and indexing by explaining the semantic similarities and contextual links in the case-base network [24].

The CBR model requires an embedded case memory to support the case matching, retrieval, and retention functionality. Case memory indexing can be based on flat organizational models, or more detailed hierarchal models (e.g., 3-layer, dynamic memory, decision tree induction based, and object based) and network-based models (e.g., concept retrieval nets, category exemplar, and ontology based) that are informed by domain knowledge [27]. A common network-based memory model for the reasoning structure of the CBR is the category-exemplar model [27,28]. The category-exemplar model creates contextual meaning in the CBR model by using three indices to link and store case data (e.g., case features, domain informed categories, and index pointers that link the data by feature, case, and differences [24]. The semantic indexing helps identify and compile case features that are similar to the conditions of a new situation and offers the nearest matching case to generate the new solution.

2.3. Case-Based Reasoning for Ice Management Decision Support

The CBR is appropriate for the first-degree automation in which seafarers are provided with onboard decision support. CBR is a good fit for providing decision support in the maritime domain because it offers advice from an embedded expert that is similar to how seafarers currently provide training on-the-job. Seafarers offer words of wisdom (or 'rules of thumb') by sharing their experiences on-the-job, often through analogies and storytelling. The CBR model replicates this information exchange. To illustrate the similarities in problem solving, the following is a simple example (modelled after examples from [24]):

A captain is actively clearing pack ice in close proximity to an offshore oil platform when they are reminded of a relevant past experience. Some weeks earlier, this same captain experienced a situation where encroaching pack ice quickly enveloped an offshore platform. Due to the unpredictable environmental conditions, the drift speed of the ice floe abruptly changed, complicating their ice management operations. In particular, they remember a mistake they made of not orienting their vessel away from the platform in order to have a bailout plan. The current ice management task is very similar to the situation from their memory (e.g., within the operational zone of the platform and a sudden change in weather conditions). They use this memory to avoid repeating the mistake in the current situation.

Applying the CBR model to an onboard decision support system could assist a cadet in a similar way to how a captain would use their experience to solve a problem. Collecting a group of expert seafarers' experiences in ice management can inform the reasoning structure of the CBR decision support system and populate an initial case base of solved problems. The CBR model allows the decision support system to incrementally learn (e.g., build its case base) and increase its accuracy (e.g., improved indexing) as it assists cadets in problem solving.

3. METHODS

To gather seafarers' experiences in managing pack ice offshore, similar to the example above, we conducted a pilot study that used a combination of semi-structured interviews and bridge simulator exercises. The pilot study employed a three-phased knowledge elicitation plan: (i) interview seafarers on their approach to ice management scenarios and what factors they considered during ice management, (ii) ask the seafarers to evaluate the performance of cadets in the same ice management scenarios, and (iii) have the seafarers complete the same ice management scenarios in a bridge simulator. This section describes the participants, the ice management scenarios used in this study, and explains the methods used to collect information from the participants on their approach, criticism, and performance to each of the scenarios. An explanation of the data analysis and the integration of the data into the CBR case base is also provided.

3.1. Participants

Experienced seafarers were recruited to participate in this study. Participants were given the option of attending one or both sessions (e.g., interview or simulation exercises). Information was collected from four participants in the interview session and from four participants in the simulator session. Three participants completed both sessions while two participants attended one session. In total, five participants completed this study.

All participants were experienced seafarers. Participants ranged in experience operating at sea from 10 to 30 years ($M = 17.6; SD = \pm 7.9$). Participants also ranged in experience operating in the presence of ice from 2 to 7 years ($M = 4.8; SD = \pm 1.9$). Participants also had a range of experience operating in different regions and with different types of vessels. The regions the seafarers operated in included coastal Newfoundland and Labrador, the Arctic/ North of 60, the Gulf of Saint Lawrence, and the Great Lakes. The different vessels included ice breakers, offshore supply vessels (OSVs) and anchor handling tug supply (AHTS), tanker/ bulk/ cargo vessels, and coastal ferries. In relation to the types of ice operations, two participants had only performed watch keeping during transit in the presence of ice. Three participants had experience watch keeping in the presence of ice as well as performing the following operations in ice: ice management in open water and confined water, maneuvering a ship to escort another vessel, towing or emergency response, and maneuvering a ship while being escorted.

In terms of the training the participants had in ice operations, two seafarers had no formal training in ice operations, and the other three seafarers had completed advanced training in ice operations.

3.2. Offshore Ice Management Scenarios

An anchor handling tug supply (AHTS) vessel was selected for this pilot study as it is commonly used for offshore support and handling pack ice management operations in offshore Newfoundland. The AHTS was modeled in the bridge simulator and described to the participants during the interviews and simulator exercises. The following vessel parameters were explained to the participants. The virtual ship was modeled after a 75 m twin-screw AHTS vessel with an ice class of ICE-C [29]. The ship had two 5369 kW engines and was equipped with 896 kW tunnel thrusters in the bow and stern.

Three scenarios that required ice management support from the AHTS vessel were used to gather information from the seafarers. Table 1 provides a brief description of the tabletop scenarios. For comparison purposes, the scenarios were based on two previous simulator experiments [18, 29].

3.3. Seafarers' Approach to Ice Management

The interview phase used a semi-structured approach to enquire about two aspects: (i) identifying what factors the participants prioritized in different ice management situations, and (ii) capturing the experienced participants' approach to the different ice management tabletop scenarios. Participants completed the interview individually and their responses were audio recorded.

To assist the researchers in conducting the interview, participants were first asked to complete an experience questionnaire designed to gain background information about the participant and their experience at sea. Then participants were asked to list all the factors they consider during ice management to understand what information seafarers prioritized in their decision making. For each of the factors listed, the researchers asked further questions about how these factors were considered during ice management. If the participant had difficulty generating a list, the researchers provided a list of factors for the participant to comment on. Once the list was generated, the participant was tasked with ranking the factors based on overall importance and again for specific ice management situations. To capture the participants' approaches to the ice management scenarios, the participants were provided with one of the three ice management tabletop scenarios (shown in Table 1) and were asked to describe how they would approach each situation. They were asked to draw their approach on a schematic and to denote their decision points. The researchers asked the participants to elaborate on the reasoning for their choices at each decision point. Each of the ice management scenarios were discussed individually.

3.4. Critiquing Cadet Training Examples

Once the participants explained their strategy, they were asked to review prerecorded examples of cadets performing in the same scenarios in a bridge simulator. The objective of this phase was to gather the experienced participants' advice, recommendations, and feedback on cadets' performance in the same ice management situations.

The cadet examples used in this phase came from a previous study [29] that trained cadets in different pack ice management techniques using simulation exercises. Six cadet examples were used for the participants' evaluation (two cadet examples for each of the three ice management scenarios).

The participants, who were experienced seafarers, were shown anonymized replay files of the cadets' performance in the ice management scenarios. This was

Table 1. *Pack ice management tabletop exercises and simulation scenario.*

Sc. #	Description	Schematic
Sc.1	Objective: Stand-by vessel support is required to clear a zone mid-ship of the tanker to allow researchers to launch equipment and reduce the risk of ice-related damage. Environmental Conditions: Current: 1 knot drift Winds: Light Ice: 0.3–0.7 m first year ice, five-tenths concentration Intended Cadet Training Technique: Leeway Time: 15 min	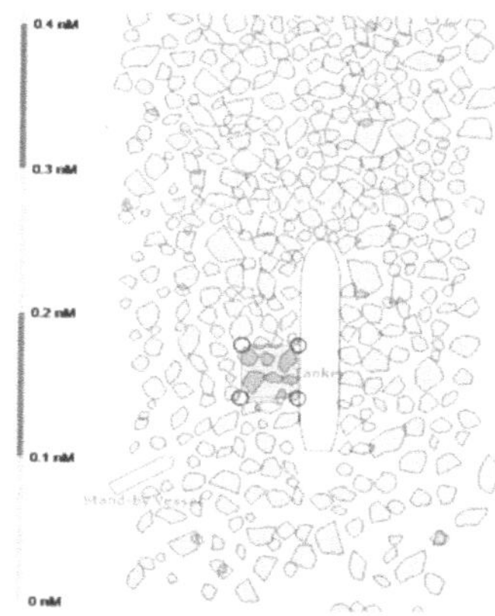
Sc.2	Objective: Stand-by vessel support is required to clear encroaching pack ice from around a platform. Environmental Conditions: Current: 0.4 knot drift Winds: Light Ice: 0.3–0.7 m first year ice, four-tenths concentration Intended Cadet Training Technique: Pushing Time: 15 min	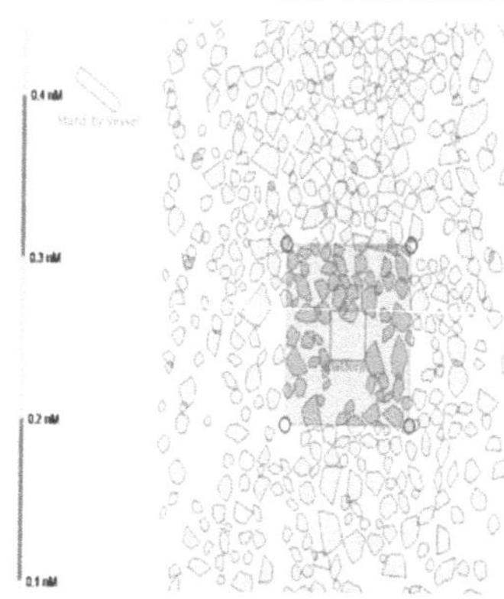
Sc.3	Objective: Stand-by vessel support is required to clear the build up of ice on the port side of the Floating Production Storage and Offloading (FPSO) to create an ice-free lifeboat launch zone. Environmental Conditions: Current: 0.5 knot drift Winds: Light Ice: 0.3–0.7 m first year ice, seven-tenths concentration. Intended Cadet Testing Scenario: Emergency Time: 30 min	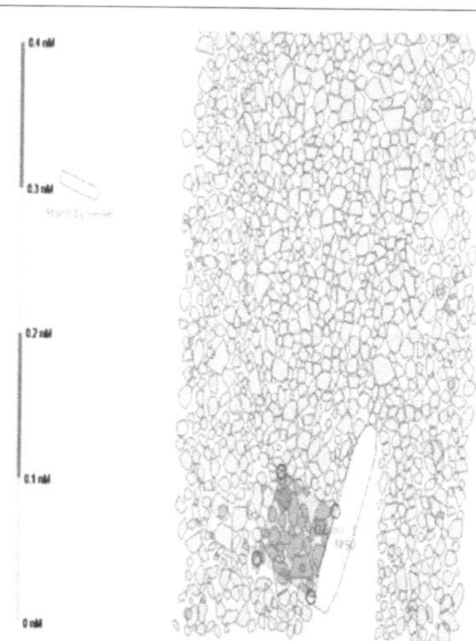

in the form of top-down replay videos displaying an offshore supply vessel managing sea ice. The participants were asked to review the cadets' performance and comment on the efficacy of the cadet examples. The participants provided their observations about the cadets' performance as well as advice at critical stages in the ice management operations (e.g., how to guide the cadet to maintain course or how to provide intervening instructions). Then the participants were asked to comment on the cadets' compliance or violation of rules (both regulatory and common-practice procedures). Participants completed this phase of this study individually and their observations and interventions for the cadets were audio recorded.

3.5. Ice Management Bridge Simulator and Exercises

The bridge simulator was used to gather more detailed information about the pack ice management techniques and actions taken by experienced seafarers. Prior to the simulation exercises, participants completed a series of short habituation scenarios that were designed to familiarize the seafarers with the simulator and the control configuration (to reduce any influence the simulator would have on their performance). Then the participants were tasked with completing the three ice management scenarios in the bridge simulator (described in Table 1). After completing each scenario, they were asked a series of debriefing questions related to their approach to the exercise and the overall efficacy of their performance. Participants completed the simulation exercises individually and their responses to the debriefing questions were audio recorded.

3.6. Data Analysis and Integrating Information into the CBR Case Base

The participants' ice management approaches, reviews of the cadet examples, and debriefs from the simulation exercises were transcribed from the audio recordings. The transcriptions were used to develop the initial reasoning structure of the CBR and to populate cases in the CBR case base.

The general domain knowledge gathered from the participants (e.g., the participants' factor rankings and comments on ice management strategies) was used to inform the reasoning structure of the CBR (e.g., similarity matching, retrieval, and indexing). The specific domain knowledge gathered from the participants (e.g., the participants' verbal explanations of their ice management strategies and their observations from the cadets' ice management performance) was parsed into example cases for the CBR case base.

4. RESULTS

The transcribed data from the three-phased knowledge elicitation study was used to develop cases to populate the CBR case base and to develop the reasoning structure of the CBR. This section describes the knowledge capture process and explains the CBR knowledge representation.

4.1. Knowledge Capture-Gathering Cases to Populate the CBR Case Base

Domain knowledge gathered from the participants was parsed into cases for the CBR case base. The specific domain knowledge included (i) the participants'

Table 2. *Ice management techniques as described by participants*

Pushing Technique	Sector Technique	Prop-Wash Technique	Leeway Technique	Circular Technique
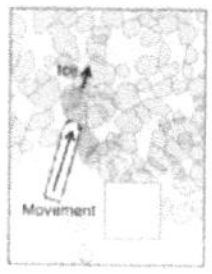	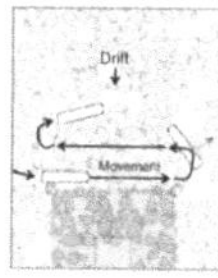	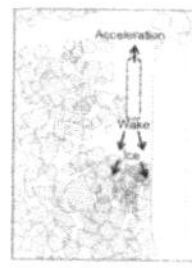	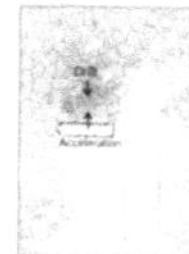	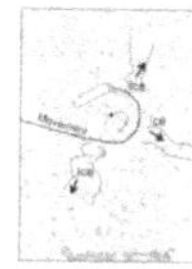
The pushing technique involves displacing pack ice using the bow or broadside of the vessel.	The sector technique is similar to the pushing technique and involves moving the vessel in a back and forth motion to disperse ice up current from the offshore platform.	The prop-wash technique involves maintaining a stationary position ahead of the zone and using the vessel's propeller wake wash to flush pack ice from the offshore platform.	The leeway technique involves holding position and using the side of the vessel to deflect and block pack ice from drifting into the area near an offshore platform.	The circular technique involves moving the vessel in a circular motion and using a combination of pushing and prop wash to disperse the pack ice upstream of the offshore platform.

verbal explanations of their approaches, decision points and instructions to the ice management scenarios; (ii) the participants' observations, criticisms, and interventions for the cadets' performances; and (iii) the participants' actions in the simulator scenarios and their reflections on their own performance. From the three sources, a list of techniques (and corresponding features) employed by participants for offshore ice management operations was recorded.

4.1.1. Ice Management Techniques Described by Participants

Table 2 outlines the main ice management techniques that were described in the pilot study. For each scenario, participants suggested and/or performed in the simulator a distinct technique or a combination of techniques. For example, a common approach to scenario Sc.3 was to use the leeway and prop-wash techniques to manage the build up of ice on the FPSO.

4.1.2. Translating Participant Data into Cases

The knowledge captured from the pilot study resulted in 34 cases. Table 3 shows a summary of the cases collected and highlights the techniques suggested or employed for each scenario. Specifically, we extracted 18 cases from the interviews, 6 cases from reviewing and advising the cadet examples, and 10 cases from the simulation exercises. Each case linked the pre-programmed attributes of the scenario (e.g., the scenario objective, the weather and ice conditions, and the properties of the target vessel) to the participants' data (e.g., attributes of their described approaches to the scenario). As such, each case contained details about the ice management technique and the series of decision points (from the approaches interview), interventions (from the cadet examples), or actions (from the simulation exercises) described or performed by the participants. The Navicat Data Modeler software was used as a tool to design the cases and the CBR case base.

4.2. Knowledge Representation-Developing the CBR Reasoning Structure

The knowledge representation involved using general domain knowledge gathered from the participants in the three-phased pilot study to establish the reasoning structure of the CBR (e.g., how the case-base data was indexed for similarity matching and case retrieval). The general domain knowledge included (i) the

Table 3. *Summary of ice management cases collected from each phase of this study.*

Scenario	Phase	Technique	Total Cases
Sc.1 Ice management alongside vessel to allow lowering of research equipment	Interview Approach	• Prop wash • Leeway Combinations: • Leeway + prop wash (x3)	5
	Cadet Example Interventions	Combinations: • Leeway + prop wash (x2)	2
	Simulator Exercises	• Prop wash • Leeway (x2)	3
Sc.2 Ice management support for pack ice from around a platform	Interview Approach	• Pushing • Prop wash • Circular • Sector Combinations: • Pushing + prop wash (x2) • Leeway + prop wash	7
	Cadet Example Interventions	• Circular • Pushing	2
	Simulator Exercises	• Pushing • Prop wash Combinations: • Sector + prop wash • Leeway + pushing	4
Sc.3 Emergency ice management of FPSO create ice-free lifeboat launch zone	Interview Approach	• Pushing • Leeway Combinations: • Leeway + pushing • Leeway + prop wash (x3)	6
	Cadet Example Interventions	Combinations: • Leeway + prop wash • Pushing + leeway	2
	Simulator Exercises	• Pushing • Combinations • Leeway + prop wash (x2)	3
		Total Cases:	**34**

participants' generated list of important ice management factors and subsequent rankings, (ii) their comments about general ice management strategies, and (iii) their comments on the cadets rule adherence or violations.

4.2.1. Key Ice Management Factors and Rankings

Participants were asked to generate a list of factors that influence their decision making during ice management operations. They were also asked to rank the

Table 4. *Important ice management factors generated and ranked by participants.*

Category	Factors	Average Ranking			
		Initial	Sc.1	Sc.2	Sc.3
Task/Objective	Area to be Cleared	4.0	4.0	4.0	4.5
	Level of Urgency	4.3	4.5	4.3	5.0
	Acceptable Level of Risk	3.8	3.8	4.0	4.0
Vessel Properties	Location Relative to Target Vessel	3.5	3.8	3.8	4.3
	Vessel Heading	2.8	4.0	4.3	4.5
	Vessel Speed	4.0	4.0	4.0	4.0
	Vessel Ice Class	4.0	4.0	4.0	4.0
Weather Conditions	Visibility and Weather Conditions	4.3	4.3	4.5	3.8
	Drift Speed (Current)	4.0	4.0	4.3	4.0
Ice Conditions	Floe Size	3.3	3.3	3.3	3.5
	Ice Type	3.5	3.5	3.5	3.5
	Ice Thickness	4.0	4.7	4.0	4.3
	Ice Loads	4.3	4.3	4.3	4.0
	Ice Concentration	4.3	4.8	4.3	4.0
Operator Characteristics and Actions	Experience	3.3	3.8	3.8	3.8
	Strategy	3.8	3.8	3.8	3.8

importance of each factor and how they relate to each of the tabletop scenarios. Table 4 lists the factors the participants discussed in the pilot study.

The list of factors was used to develop the CBR knowledge representation. Figure 2 depicts the resulting CBR knowledge representation modeled using Navicat Data Modeler software. Additional factors were added based on the participants' comments throughout the interviews and simulation exercises (e.g., separate categories for 'scenario attributes', 'target vessel properties' and splitting the 'operator characteristics and actions' category into two: operator characteristics and the operator's actions represented by the label 'ice management technique').

4.2.2. Indexing Cases for Matching and Retrieval Using the Category-Exemplar Model

The category-exemplar model [28] was used for the case memory knowledge representation. Indexing features and categorizing common themes that occurred in the case base help to match similar cases with the conditions of future scenarios. To illustrate the indexing, two different ice management techniques described by participants were used. Figure 3 shows the case links for the pushing technique using the interview and cadet example data. Similarly, Figure 4 shows an example of case indexing for the leeway technique using the interview, cadet examples and simulation data. These examples show the relationships between features of the ice management techniques for the seafarer and cadet example cases. The participant or cadet informed cases that used the given technique were compared to identify their commonalities and differences.

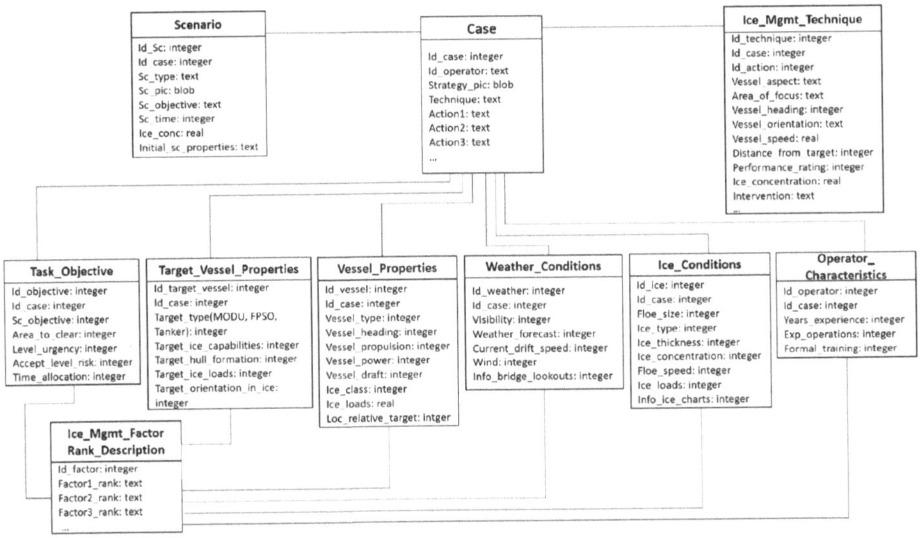

Figure 2. *Class diagram of the CBR for ice management.*

In Figure 3, the participants' verbal explanations from the interviews and reviewing the cadet examples provided details on the following features: (F1) setting the aspect, (F2) area of focus for the ice clearing, (F3) approximate vessel heading, (F4) orientation of vessel to the target vessel, (F5) the specific vessel maneuvers for the technique, (F6) an estimate of the vessel speed (F7) an approximate distance from the platform, and (F8) whether a combination of techniques were required. A proxy measure of the technique's efficacy was also added as a feature (F9): the participant's priority ranking of the technique for the situation and their rating of the cadets' performance in the examples.

Depending on the source of the data (i.e., interviews or simulation exercises), some cases contained more detailed information on the case features than others. For example, the approaches the participants described in the interviews (Figure 3) represent the techniques under static and ideal circumstances. However, as shown in Figure 4, the simulator data provides more granularity on parameters that cannot be described in detail during the interviews. Such features include (* F3) vessel heading, (* F5) vessel maneuvers, (* F6) vessel speed, and (* F7) distance from target during clearing and the size of the clearing zone. Objective measures of the scenario outcomes were also added as features for the simulator exercises: ($*F10$) the change in ice concentration in the target zone and ($*F11$) estimate of the ice loads endured by the vessel.

The indexing process is repeated for all other themes that were identified as important factors by the seafarers. Such categories include the ice management task itself, the weather and ice conditions, and other aspects of the ice management technique. The network of case indices forms the basis of the case memory for the CBR-model.

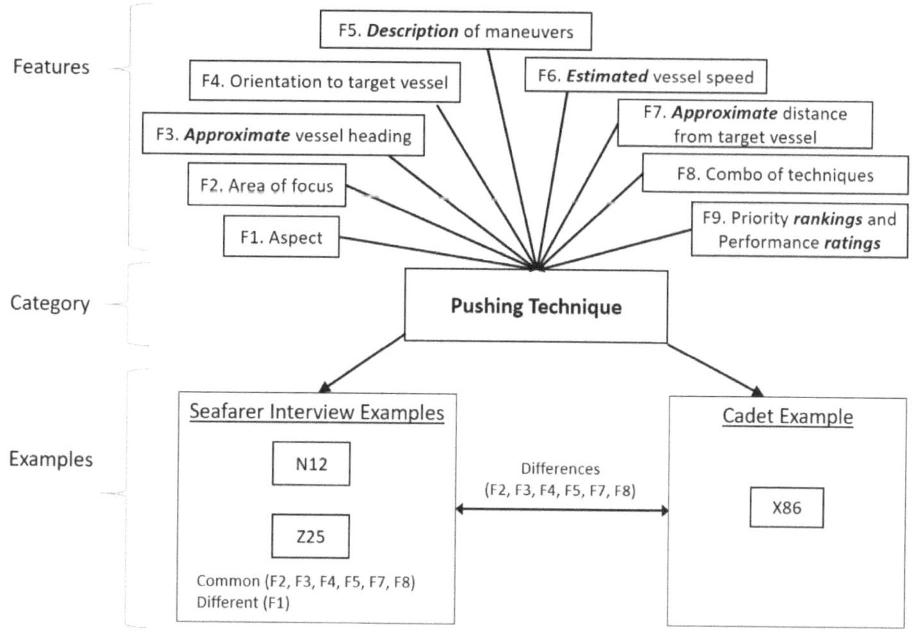

Figure 3. *Case indexing for the pushing technique using the interview and cadet data.*

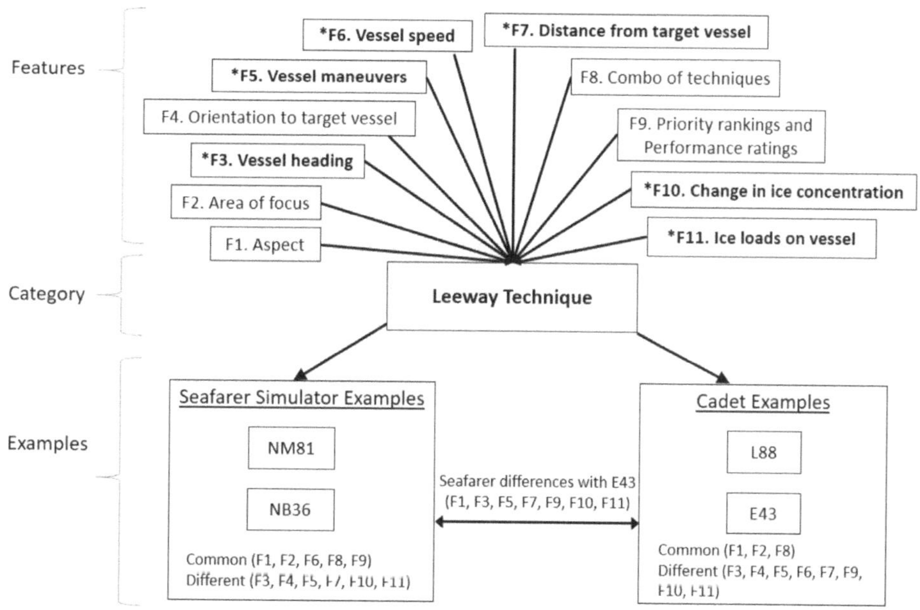

Figure 4. *Case indexing for the leeway technique using the interview, cadet examples, and simulator data.*

5. DISCUSSION

5.1. Ice Management Interviews to Establish the CBR Model and Populate Cases

Data captured from the participant interviews was used to develop an early-stage decision support system with a CBR structure and case base. The process of transcribing the audio recordings from the interviews (e.g., seafarer approaches and evaluations of the cadet examples) and converting the data into cases was tedious and time consuming. However, the information captured using this method was useful for classifying ice management strategies and identifying important ice management factors and the relationships between these factors. This information was beneficial for gathering domain knowledge used in establishing the CBR structure and case feature indices.

During the interviews, participants communicated their ice management strategies in a combination of anecdotal descriptions of approaches and highlighted the aspects to consider while conducting ice management operations (e.g., common practices and rules of thumb). In particular, participants described the flexibility of their strategies. They explained that their plan was not fixed and that it was common to approach the situation with multiple strategies or techniques, often referring to it as a trial and error process. This is reflected in the lower ranking of the strategy factor in comparison to other factors in Table 4 . When describing the step-by-step decision points of their approach, participants had a tendency to position their vessel in such a way to safely abandon and attempt alternative approaches (e.g., if the current strategy was not working or the conditions of the situation changed, the participants would adjust their strategy and try something different).

It is important to note that the participants' descriptions of their strategies were in ideal circumstances and resulted in somewhat static strategies. As a result, the interviews were limited in the amount of information they could provide on how to execute the techniques and the resulting outcome of the prescribed approach (e.g., Was the approach successful? How proficient in the technique does a captain need to be to reach the intended outcome?). These limitations were addressed with the data collected from the cadet examples and simulator exercises.

5.2. Evaluating Cadet Examples to Define the Scope of the Ice Management Operations

The CBR model requires a collection of successful and unsuccessful cases to capture the scope of the ice management problem and to offer solutions on how to correct or avoid making mistakes. As such, the participants reviewed examples of successful and unsuccessful ice management operations performed by cadets and provided observations and interventions. The participants' advice to correct or improve the cadet examples was added as cases to the case base. The cadet examples provided opportunities for the participants to comment on the ways in which ice management techniques can go wrong or indicate which situations require higher proficiency to perform the technique effectively. This review process also prompted the participants to elaborate on why it was necessary to perform the techniques in a certain way or under different conditions. The information captured from evaluating the cadet examples formed the basis of comparison for the feature indexing (e.g., linking the commonalities and differences among the seafarer and cadet cases as represented in Figures 3 and 4).

5.3. Using Simulation Exercises to Refine the CBR Model and Add Details to Cases

In order for the CBR decision support system to be capable of recommending ice management techniques or providing adjustments during the execution of a technique, the CBR model must include a measure of the ice management efficacy. The simulation exercises from the pilot helped improve the CBR model by adding feature details such as the scenario outcome to the case base. For example, cases that were developed from the participant interviews did not contain an objective measure of the technique's success or failure due to the static circumstances of the tabletop scenarios (as shown in Figure 3). Instead, these cases used subjective measures of the scenario outcomes such as the participants' priority rankings of their approaches and the participants' performance ratings of the cadet examples.

Conversely, the simulation data captured the dynamic aspects of ice management approaches. As a result, cases from the simulation exercises provided continuous information on parameters that could not be described in detail in the interviews and outlined the detailed consequences of the participants' actions in the scenario (as shown in Figure 4). The objective performance metrics from the simulation exercises provided a measure of how well the ice management technique was executed. The performance metrics include (1) the change in ice concentration in the target zone (a measure of ice management effectiveness) and (2) an estimate of the ice loads endured by the vessel (whether the technique is feasible, safe and within the operating capabilities of the vessel). The metrics recorded in the simulator exercises drew out the subtleties of the techniques that were needed to help novice operators, such as cadets, improve their proficiency in the ice management tasks.

6. CONCLUSIONS

A human-centered approach is necessary for successful digital transformation for the offshore and maritime industries. We envision that experienced seafarers will play an important role in facilitating the transition from fully crewed vessels to remotely operated vessels and fully autonomous operations. Since one of the goals of first-degree automation is to address the deficiencies of conventional on-the-job training and foster better knowledge exchange between seafarers, it is essential that experienced seafarers be involved in the development of onboard decision support technology. As such, this paper presented a human-centered approach to designing an early-stage CBR decision support system for offshore ice management operations.

This pilot study employed three different data collection methods (e.g., semi-structure interviews on ice management approaches, reviewing cadet examples, and performing simulation exercises) to gather information from experienced seafarers on offshore pack ice management. Overall, the methods used to capture expert knowledge were suitable for the CBR application. Each form of data collection was useful for different aspects of developing the CBR model and case base.

While the 34 cases documented in the pilot study were sufficient to populate the initial CBR case base, the CBR model (both the case memory and corresponding case base) could be improved by adding more cases. The addition of more expert seafarer cases would increase the accuracy of the model's solution generation, which is important for providing decision support. Cases can be added by repurposing data collected from previous simulation studies [18, 29] or collecting new data using simulator exercises. However, simply adding more cases to the CBR model could compromise the decision support system's efficiency because it takes the system longer to sort through the cases to find a match if the case indexing is deficient. To avoid this problem, it is important to have a well-established and tested case memory to effectively match new situations with the case features. We recommend testing the initial CBR model's similarity matching and solution generation before collecting more data to populate cases.

The next step of this research is to gather more cases, develop the retrieval algorithm for the CBR model, and test the CBR decision support system's similarity matching and solution generation functionality in a simulated environment. To get the most value out of future simulation trials, we recommend corresponding the data collection to the intended decision support functionality so that specific case metrics can be recorded automatically. The data collection methods used in this pilot study required substantial post-processing and interpretation before they were integrated into the CBR model as cases. While this was an adequate process to demonstrate the concept of the CBR decision support system, from a practical perspective, future work should focus on automating the data collection. This can be achieved by embedding the case feature recording, parsing, and indexing into the existing simulation technology. This integration would make the decision system more informative and effective, which would improve the practicality of the CBR decision support system.

Future work will focus on the verification and validation of the CBR decision support system with cadets and seafarers in a simulator setting. Training simulators, such as the bridge simulator used in this pilot study, have been shown to be useful human behaviour laboratories for both knowledge capture and testing training interventions such as a decision support system. This pilot study demonstrated that the information gathered from the simulator exercises improved the design of the initial CBR decision support system and increased the level of detail of the case base. Once the development of CBR decision support system has been completed, testing the CBR model in a simulator setting would verify the decision support system's capabilities of offering onboard guidance on pack ice management techniques.

Author Contributions

Conceptualization and methodology, J.S., F.Y, M.M., and B.V.; data collection and analysis, J.S., F.Y., and R.T.; writing-original draft preparation, J.S.; writing-review and editing, J.S., F.Y., and B.V.; supervision, M.M. and B.V.; project administration and funding acquisition, B.V. All authors have read and agreed to the published version of the manuscript.

Acknowledgments

The authors acknowledge with gratitude the support of the NSERC/Husky Energy Industrial Research Chair in Safety at Sea. Conflicts of Interest: The authors declare no conflict of interest. The funders had no role in the design of this study; in the collection, analyses, or interpretation of data; in the writing of the manuscript, or in the decision to publish the results.

REFERENCES

1. Bainbridge, L. Ironies of automation. Automatica 1983, 19, 775-779.
2. Parasuraman, R.; Wickens, C. Humans: Still vital after all these years of automation. Hum. Factors 2008, 50, 511-520.
3. Hollnagel, E.; Wears, R.; Braithwaite, J. From safety-I to safety-II: A white paper. In The Resilient Health Care Net; University of Southern Denmark: Odense, Denmark; University of Florida: Gainesville, FL, USA; Macquarie University: Sydney, Australia, 2015.
4. IMO Takes First Steps to Address Autonomous Ships. Available online: http://www.imo.org/en/MediaCentre/ PressBriefings/Pages/08-MSC-99-MASSscoping.aspx (accessed on 11 June 2020).
5. Lloyd's Register Cyber-Enabled Ships: ShipRight Procedure Assignment for Cyber Descriptive Notes for Autonomous & Remote Access Ships. Available online: http://info.lr.org/l/12702/2016-07-07/32rrbk (accessed on 11 June 2020).
6. Mallam, S.; Nazir, S.; Sharma, A. The human Element in Future Maritime Operations-Perceived Impact of Autonomous Shipping. Ergonomics 2020, 63, 334-345.
7. Aylward, K.; Johannesson, A.; Weber, R.; MacKinnon, S.; Lundh, M. An evaluation of low-level automation navigation functions upon vessel traffic services work practices. WMU J. Marit. Affairs 2020.
8. Dunderdale, P.; Wright, B. Pack Ice Management on the Southern Grand Banks Offshore Newfoundland Canada; PERD/CHC Report 20-76; Noble Denton Canada Ltd.: St. John's, NL, Canada, 2005; pp. 1-42.
9. McClintock, J.; McKenna, R.; Woodworth-Lynas, C. Grand Banks Iceberg Management; PERD/CHC Report 20-84; Report prepared for PERD/CHC, National Research Council Canada, Ottawa, ON; Report prepared by AMEC Earth & Environmental, St. John's, NL, R.F. McKenna & Associates: Wakefield, QC, Canada; PETRA International Ltd.: Cupids, NL, Canada, 2007; pp. 1-82.

10. Canadian Coast Guard. Ice Navigation in Canadian Waters; Ice Breaking Program Maritime Services Canadian Coast Guard Fisheries and Oceans Canada: Ottawa, ON, Canada, 2012; pp. 1-158. Available online: https: / /www.ccg-gcc.gc.ca / publications / icebreaking-deglacage / ice-navigation-glaces / page01-eng.html (accessed on 19 June 2020).

11. Keinonen, A. Ice management for ice offshore operations. In Proceedings of the Offshore Technology Conference, Houston, TX, USA, 5-8 May 2008. OTC 19275.

12. Østreng, W.; Eger, K.M.; Fløistad, B.; Jørgensen-Dahl, A.; Lothe, L.; Mejlænder-Larsen, M.; Wergeland, T. Shipping and Arctic Infrastructure. In Shipping in Arctic Waters: A Comparison of the Northeast, Northwest and Trans Polar Passages; Springer Praxis Books: Berlin / Heidelberg, Germany, 2013; pp. 177-239.

13. International Maritime Organization. Model Course 7.11: Basic Training for Ships Operating in Polar Waters, 2017th ed.; International Maritime Organization: London, UK, 2017; pp. 1-59.

14. International Maritime Organization. Model Course 7.12: Advanced Training for Ships Operating in Polar Waters, 2017th ed.; International Maritime Organization: London, UK, 2017; pp. 1-37.

15. International Maritime Organization. International Code for Ships Operating in Polar Waters (Polar Code); MEPC 68 / 21 / Add.1; International Maritime Organization: London, UK, 2017; pp. 1-42.

16. Arctic Shipping Safety and Pollution Prevention Regulations SOR / 2017-286. Available online: https: / / lawslois.justice.gc.ca / eng / regulations / SOR-2017-286 / FullText.html (accessed on 6 July 2020).

17. Transport Canada Joint Industry-Government Guidelines for the Control of Oil Tankers and Bulk Chemical Carriers in Ice Control Zones of Eastern Canada (JIGs) TP15163 (2015). Available online: https: / / tc.canada.ca / en / marine-transportation / marine-safety / joint-industry-government-guidelinescontrol-oil-tankers-and (accessed on 6 July 2020).

18. Veitch, E.; Molyneux, D.; Smith, J.; Veitch, B. Investigating the influence of bridge officer experience on ice management effectiveness using a marine simulator experiment. J. Offshore Mech. Arct. Eng. 2019, 141, 1-12.

19. Kolodner, J. An Introduction to Case-Based Reasoning. Artif. Intell. Rev. 1992, 6, 3-34.

20. Kolodner, J.; Camp, P.J.; Crismond, D.; Fasse, B.; Gray, J.; Holbrook, J.; Puntambekar, S.; Ryan, M. Problem-based learning meets case-based reasoning in the middle-school science classroom Putting learning by design (tm) into practice. J. Learn. Sci. 2003, 12, 495-547.

21. Rossille, D.; Laurent, J.; Burgun, A. Modelling a decision-support system for oncology using rule-based and case-based reasoning methodologies. Int. J. Med. Inform. 2005, 74, 299-306.

22. Kumar, K.; Singh, Y.; Sanyal, S. Hybrid approach using case-based reasoning and rule-based reasoning for domain independent clinical decision support in ICU. Expert Syst. Appl. 2009, 36, 65-71.

23. Wang, D.; Wan, K.; Ma, W. Emergency decision-making model of environmental emergencies based on case-based reasoning method. J. Environ. Manag. 2020, 262, 110382.

24. Aamodt, A.; Plaza, E. Case-based reasoning: Foundational issues, methodological variations, and system approaches. AI Commun. 1994, 7, 39-59.

25. Homem, T.; Santos, P.; Costa, A.; da Costa Bianchi, R.; de Mantaras, R.L. Qualitative case-based reasoning and learning. Artif. Intell. 2020, 283, 103258.

26. Han, J.; Kamber, M.; Pei, J. Classification: Advanced methods. In Data Mining: Concepts and Techniques, 3rd ed.; Elsevier: Waltham, MA, USA, 2011; pp. 425-426.

27. Soltani, S.; Martin, P. Case-Based Reasoning for Diagnosis and Solution Planning; Technical Report No. 2013-611; Queen's University: Kingston, ON, Canada, 2013; pp. 1-54.

28. Porter, B.; Bareiss, R.; Holte, R. Concept learning and heuristic classification in weak theory domains. Artif. Intell. 1990, 45, 229-263.

29. Thistle, R.; Veitch, B. An evidence-based method of training to targeted levels of performance. In Proceedings of the SNAME Maritime Convention, Tacoma, WA, USA, 30 October-1 November 2019; SNAME-SMC-2019-030. Available online: https://www.onepetro.org/conference-paper/SNAME-SMC-2019030 (accessed on 2 June 2020).

CHAPTER
9

Application of Radar Image Fusion Method to Near-Field Sea Ice Warning for Autonomous Ships in the Polar Region

Tsung-Hsuan Hsieh, Bo Li, Shengzheng Wang and Wei Liu

Merchant Marine College, Shanghai Maritime University, Shanghai 201306, China

ABSTRACT

Mastering the real-time dynamics of near-field sea ice is the primary condition to guaranteeing the navigation safety of autonomous ships in the polar region. In this study, a radar image fusion process combining marine radar and ice radar is proposed, which can effectively solve the problems of redundant information and spatial registration during image fusion. Then, using the fused radar images, this study proposes a set of near-field sea ice risk assessment and warning processes applicable to both low- and high-sea-ice-concentration situations. The sea ice risk indexes in these two situations are constructed by using four variables: sea ice area, sea ice grayscale, distance between sea ice and the own-ship, and relative bearing of sea ice and the own-ship. Finally, visualization processing is carried out according to the size of the risk index values of each piece of sea ice to achieve a better near-field sea ice risk assessment and warning effect. According to the example demonstration results, through the radar image fusion process and the set of near-field sea ice risk assessment and warning processes proposed in this study, the sea ice risk distribution in the near-field area of the ship can be well obtained, which provides effective support for the assisted decision-making of autonomous navigation in the polar region.

Keywords: *autonomous ships in the polar region; sea ice warning; radar image fusion; ice radar*

1. INTRODUCTION

In recent years, with the continuous warming of the global climate, much research data have shown that the extent of sea ice in the Arctic and Antarctica has decreased dramatically [1-3], and many studies have even predicted that the Arctic Ocean will be ice-free in summer within 15 years [4]. As the sea ice in the polar region gradually decreases, the number of voyages into the polar region also gradually increases. In the Arctic region, the Arctic shipping route has the economic benefit of greatly shortening the route distance between the three major economic circles of Northeast Asia, Europe, and North America. In the Antarctic region, the number of people engaged in scientific research and tourism activities has also increased year by year. Although global warming has increased the possibility of polar navigation, sea ice remains a key factor affecting navigation safety in the polar region. In order to ensure navigation safety, it is of great significance to conduct near-field sea ice perception and warning research, which is also one of the key technologies for the development of autonomous ships in the polar region.

To realize the autonomous and unmanned ship, the technology of perception, decisionmaking, and control must be developed, and the quality of perception information is the source and key to determine the effect of decision-making and control. The sensing equipment of autonomous ships can be divided into two types: shipboard sensors and satellite-based sensors [5]. Shipboard sensors include the very high frequency (VHF) radio-based automatic identification system (AIS), electronic chart display and information system (ECDIS) [6], marine radar, attitude indicator, and depth sounder. Satellite-based sensors include the global navigation satellite system (GNSS), satellite-based AIS [7], the meteorological and oceanographic (METOC) data receiver, synthetic aperture radar (SAR), and the long-range identification and tracking (LRIT) system. In general, different sensors have different detection objects, data types, and applicability, and not all of them are suitable for the ice detection requirements of polar autonomous ships.

This study summarizes and compares the related studies on sea ice risk warning, as shown in Table 1. According to the sensing range of sea ice by the sensing equipment, it can be divided into three categories: large-scale sea ice data, medium-scale sea ice data, and small-scale sea ice data. Among them, the large-scale sea ice data are acquired mainly through a satellite remote sensing method such as NOAA (the National Oceanic and Atmospheric Administration satellite) [8], RADARSAT (Canada's satellite) [9], and MODIS (the moderate resolution imaging spectroradiometer on board NASA's satellite) [10], which is not affected by polar night and weather, is widely used to retrieve the information of sea ice concentration [9,11-15] and thickness / grayscale [8,16-19], and monitors the spatialtemporal changes in sea ice conditions. However, its temporal resolution is usually one day, which is more suitable for grasping the large-scale sea ice range, but poor in grasping near-field and real-time sea ice conditions. The small-scale sea ice data are acquired mainly through the shipboard or airborne camera [20-22], which is the observation method closest to human eyes and can accurately obtain the close-range sea ice image in the near-field area of the ship. However, due to its short viewing distance and inability to obtain extensive and complete sea ice information, it is more suitable for verifying ice conditions. The mediumscale sea ice data are acquired mainly through the airborne electromagnetic induction device (i.e., a so-called EM-bird) [23], marine radar, ice radar, etc. Compared with the other two categories of data, it can better grasp the near-field sea ice conditions in real time.

In fact, radar is sensing equipment suitable for ice detection in the near-field waters [24]. It is clearly stated in the International Code for Ships Operating in Polar Waters (Polar Code) that ships sailing in polar waters should be actively equipped with radars with ice detection capabilities [25]. Compared with the research that only uses a single-sensor marine radar to obtain ice information [26], this study uses the fused image of the marine radar and the ice radar, which can provide more comprehensive sea ice information for the near-field sea ice perception.

In addition, what variables should be considered when conducting near-field sea ice risk warning studies? As can be seen from Table 1, although most studies have taken into account the impact of sea ice physical characteristics such as sea ice concentration and sea ice area on sea ice risk when conducting sea ice risk assessment [8-16,18-23,26,27], they have not considered the relationship between the own-ship and sea ice from the point of view of navigation safety and the perspective of navigators. As for the differences in different sea ice concentration situations, the sea ice risk assessment and warning methods should also be considered and dealt with separately.

Based on this, this study uses the fused radar image, takes into account the risk variables such as sea ice concentration, sea ice area, sea ice grayscale, and the newly added risk variables such as the distance between sea ice and the own-ship and the relative bearing of sea ice and the own-ship, and proposes a set of near-field sea ice risk assessment and warning processes that meet the needs of autonomous ships in the polar region and are suitable for two situations. Ships can choose to adopt nonpartition or partition sea ice risk assessment methods according to the actual sea ice concentration in the near-field waters to ensure effective warning of sea ice risks and guarantee navigation safety.

2. RADAR IMAGE FUSION PROCESS

Radar is an effective equipment for autonomous ships in the polar region to detect the near-field sea ice targets. Among them, marine radar is a necessary navigation equipment for ships. Its detection range is farther, but the detection effect of sea ice is poor, and only part of the edge of sea ice can be delineated. Ice radar can reflect the overall ice texture better, but the detection range is closer and the noise interference is greater. Compared with a single sensor, the fusion of image information of these two radars will help to perceive the near-field sea ice scene more comprehensively. However, even if two radar images at the same time point are selected for radar image fusion, the problems of redundant information and spatial registration must be solved. The problem of redundant information is how to eliminate the interference of various redundant information and correctly extract the target echo information on the radar echo display area. The redundant information includes: (1) various menus, options, words, and numbers on the radar image; (2) the marking lines and frames on the radar echo display area; (3) the echo at the center of the radar distance scale ring may be the hull of the own-ship rather than sea ice. The spatial registration problems include: (1) as the radar display orientation may be head-up or north-up, the heading angle of the two radar images may

Table 1. *Comparison of studies related to sea ice risk warning.*

Related Studies	Sensing Range	Sensing Equipment	Multi-Source Data	SIC [1]	SIA [2]	SIG [3]	Dist [4]	Relative Bearing	Visual Warning
Haverkamp et al. (1995) [19]	Large	Satellite (SAR)	X	X	O	O	X	X	X
Spreen et al. (2008) [11]	Large	Satellite (AMSR-E sensor)	X	O	O	X	X	X	O
Gu et al. (2013) [8]	Large	NOAA satellite	X	X	O	O	X	X	O
Beitsch et al. (2014) [12]	Large	Satellite (AMSR-2 sensor)	X	O	X	X	X	X	O
Posey et al. (2015) [13]	Large	Satellite (AMSR-2 and SSMIS sensors)	O	O	X	X	X	X	O
Hebert et al. (2015) [14]	Large	Satellite (AMSR-2 sensor)	O	O	X	X	X	X	O
Wang et al. (2016) [9]	Large	Satellite (SAR-RADARSAT-2)	X	O	X	X	X	X	O
Zeng et al. (2016) [10]	Large	Terra satellite (MODIS sensors)	X	O	X	O	X	X	O
Chi et al. (2017) [15]	Large	Satellite (passive microwave sensors)	X	O	X	X	X	X	O
Yan et al. (2019) [18]	Large	COMS satellite (GOCI)	X	X	O	O	X	X	O
Dumitru et al. (2019) [27]	Large	Sentinel-1 satellite (SAR)	X	X	O	O	X	X	O
Xiao et al. (2021) [16]	Large	CryoSat-2 satellite	X	X	X	O	X	X	O
Weissling et al. (2009) [20]	Small	Camera observation	X	X	O	O	X	X	O
Zhang and Skjetne. (2015) [21]	Small	Camera observation	X	X	O	O	X	X	O
Parmiggiani et al. (2019) [22]	Small	Camera observation	X	X	O	X	X	X	O
Renner et al. (2013) [23]	Medium	EM-bird sensor and camera observation	O	X	O	O	X	X	O
Hsieh et al. (2021) [26]	Medium	Marine radar	X	X	O	X	O	X	O
This study	Medium	Marine radar and ice radar	O	O	O	O	O	O	O

[1] Sea ice concentration; [2] Sea ice area; [3] Sea ice grayscale or thickness; [4] Distance between sea ice and the own-ship.
Note: O—considered, X—not considered.

be inconsistent; (2) as the selected range or interval of the radar distance scale ring may be different, the pixel distance corresponding to the same 1 nautical mile may be inconsistent; (3) as the radar display mode may be non-off-center or off-center, the pixel position of the center of the radar distance scale ring may be inconsistent.

In order to solve the above problems, this study proposed a radar image fusion process for near-field sea ice perception of polar autonomous ships. The process includes five steps, including radar image capture, basic information acquisition, redundant information elimination, spatial registration correction, and radar image fusion, as shown in Figure 1. These steps are explained as follows.

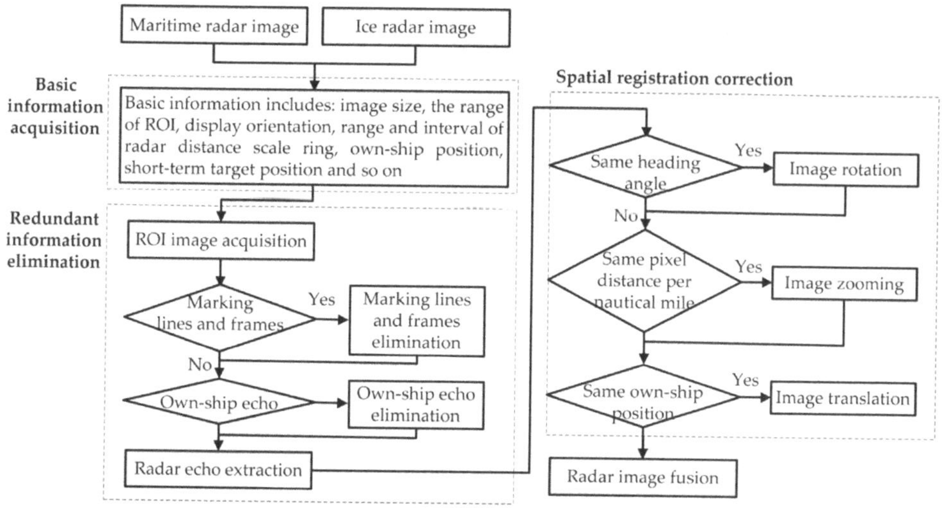

Figure 1. *Schematic diagram of radar image fusion process.*

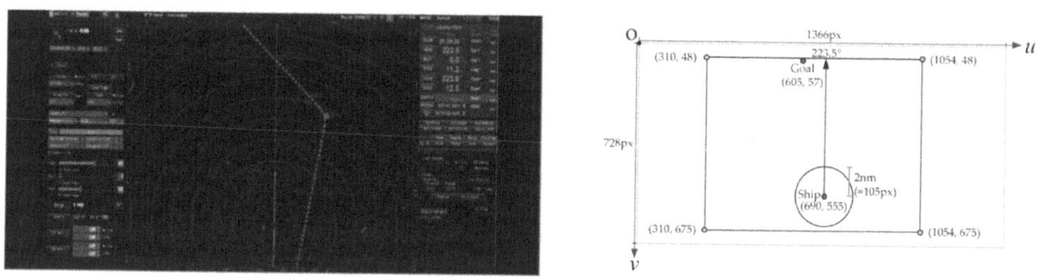

(a) (b)

Figure 2. *The marine radar image: (a) the original radar image; (b) the obtained basic information.*

2.1. Radar Image Capture and Basic Information Acquisition

The images of marine radar and ice radar at the same time point (note that there may be a time difference between the two radars or two sensors) were captured, respectively, as shown in Figures 2*a* and 3*a*; both radar images were saved in three-dimensional matrix form. Through image measurement software, the basic information such as image size, the range of region of interest (ROI), display orientation, range and interval of the radar distance scale ring, the position of the own-ship, and the short-term target position were obtained.

The basic information obtained from the marine radar image is shown in Figure 2b: the image size is 1366×728 pixels (px); the ROI is a rectangular region composed of four coordinate points $(310, 48)$, $(310, 675)$, $(1054, 48)$, and $(1054, 675)$; the radar display orientation is head-up and the heading angle is 223.5°; the radar display mode is off-center (that is, the center of the ring is located in the center of the screen); the range of the radar distance scale ring is set to 6 nautical miles (nm) and the ring diameter with a radius of 2 nm corresponds to 105px; the position of the own-ship is at the center of the distance scale ring, located at $(690, 555)$; the

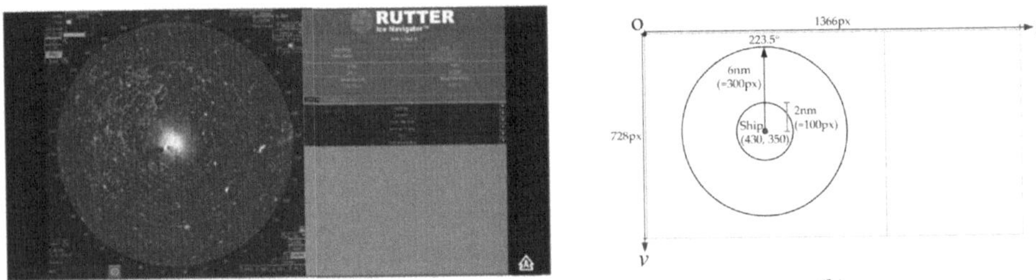

(a) (b)

Figure 3. *The ice radar image: (a) the original radar image; (b) the obtained basic information.*

short-term target position (i.e., the Goal in Figure 2a) is the intersection of the planning path (i.e., the red dashed line in Figure 2a) and the bearing scale, located at $(605, 57)$. The basic information obtained from the ice radar image is shown in Figure 3b: the image size is 1366×728px; the ROI is a circular area with $(430, 350)$ as the center and 300px as the radius; the radar display orientation is head-up and the heading angle is 223.5°; the radar display mode is non-off-center (that is, the center of the ring is located below the center of the screen); the range of the radar distance scale ring is set to 6 nm and the ring diameter with a radius of 2 nm corresponds to 100px; the position of the own-ship is at the center of the distance scale ring, located at $(430, 350)$. The acquisition of the above-mentioned basic information will be helpful for the processing and fusion of the radar image.

2.2. Redundant Information Elimination

In the original images of marine radar and ice radar, in addition to the target echo information that is most concerned by sea ice perception, there are also various redundant information such as words, numbers, and marking lines, which should be filtered and eliminated.

2.2.1. ROI Image Acquisition

According to the basic information such as image size and the range of the ROI obtained from the two original radar images, a mask matrix is created for each of them to display the inside of the ROI in white (with the value set to 1) and the outside of the ROI in black (with the value set to 0), as shown in Figure 4a,b. Then, the image of the ROI can be obtained through Equation (1), as shown in Figure 4c, d.

$$dst(I) = \text{src}\,1(I) \wedge \text{src}\,2(I) \text{ if } \text{mask}(I) \neq 0 \tag{1}$$

where *dst* represents the output image matrix; I represents the multi-dimensional index of elements in the image matrix; src1 and *src*2 represent two image matrices with the same size; mask represents the mask matrix.

2.2.2. Marking Lines and Frames Elimination

In addition to radar echo, there may also be occlusion and the interference of various marking lines and frames in the obtained image of the ROI. For example,

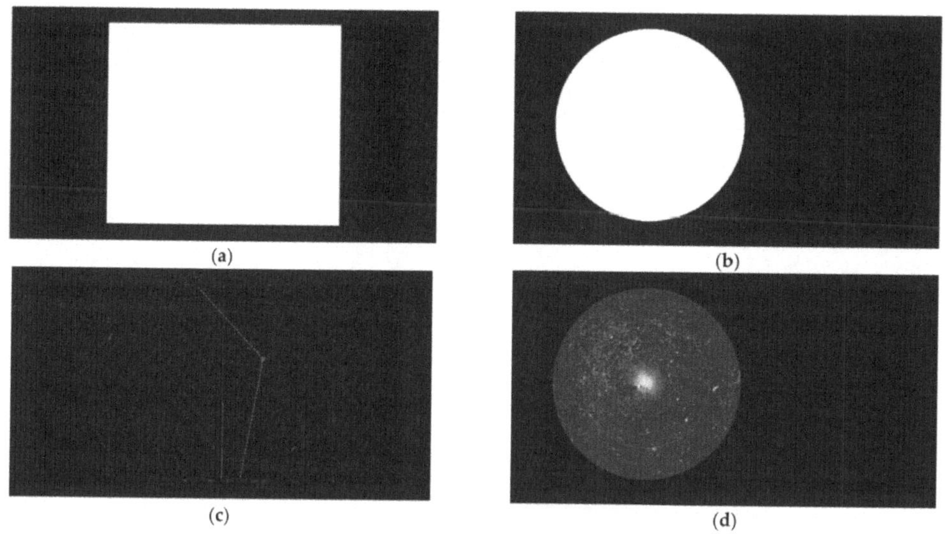

Figure 4. *The radar ROI image acquisition: (a) the marine radar mask image; (b) the obtained marine radar* ROI *image; (c) the marine radar mask image; (d) the obtained ice radar* ROI *image.*

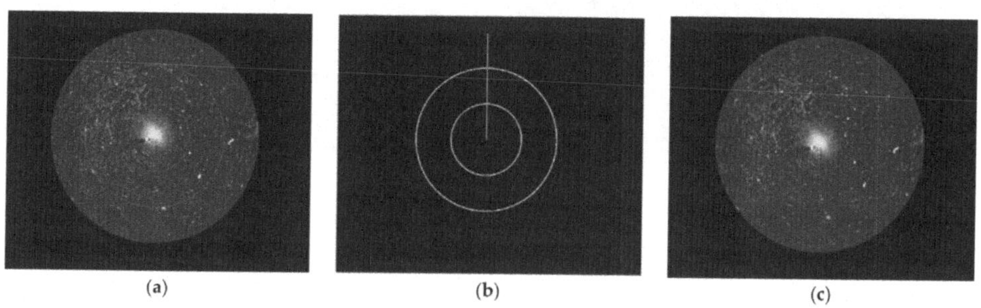

Figure 5. *The radar marking lines and frames elimination: (a) before elimination; (b) the elimination area; (c) after elimination.*

in the image of the ROI of the ice radar, there are red heading lines and yellow distance scale rings, as shown in Figure 5a. After creating the mask matrix to mark the areas to be eliminated, as shown in Figure 5b, the result of elimination can indeed improve the influence of the marking lines on radar echo, as shown in Figure 5c.

2.2.3. Own-Ship Echo Elimination

When a small and independent echo is displayed at the center of the radar distance scale ring, the echo may be the clutter generated by the steel hull of the own-ship rather than sea ice or other objects. The contour range information of all echoes in the image of the ROI are detected and compared with the position of the own-ship. After confirming the overlapping contour range information, the echo of the own-ship can be eliminated, as shown in Figure 6a,b.

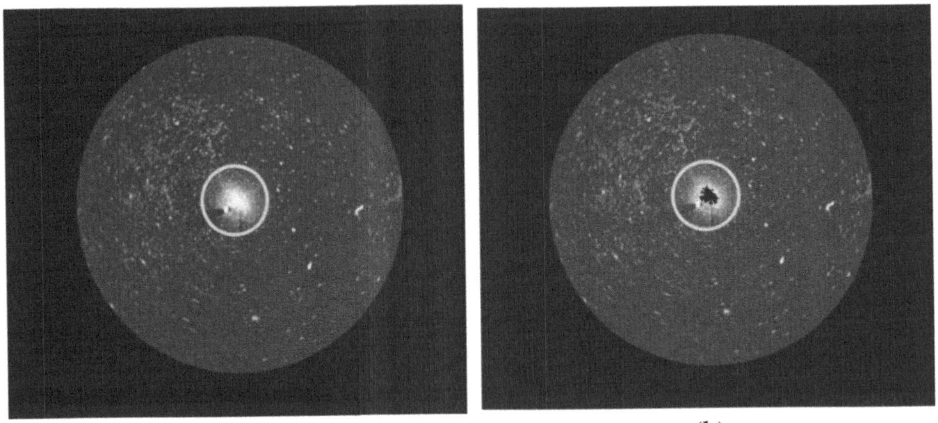

(a) (b)

Figure 6. *The radar own-ship echo elimination: (a) before elimination; (b) after elimination.*

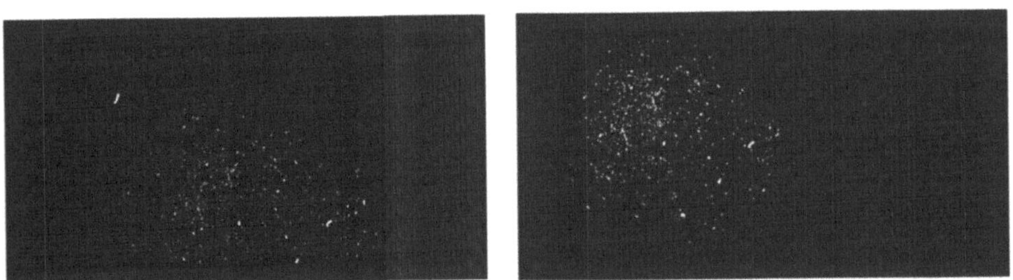

(a) (b)

Figure 7. *The radar echo extraction: (a) the obtained binary marine radar image; (b) the obtained binary ice radar image.*

2.2.4. Radar Echo Extraction

The marine radar echo is displayed in orange-yellow and the ice radar echo is displayed in grayish-white. In order to better extract the radar echoes with specific colors, the color space of image of the ROI is converted from RGB to HSV. Then, after setting the corresponding color threshold, the binary radar image can be obtained through Equation (2), as shown in Figure 7*a*, *b*.

$$dst(I) = \text{lowerb}(I) \leq \text{src}(I) \leq \text{upperb}(I) \tag{2}$$

where *dst* represents the output image matrix; I represents the multi-dimensional index of elements in the image matrix; lowerb represents the lower boundary threshold; src represents the input image matrix; upperb represents the upper boundary threshold.

2.3. Spatial Registration Correction

In order to carry out spatial registration correction, this study adopts the method of fixing the marine radar image and adjusting the ice radar image to make a geometric transformation of the ice radar image.

2.3.1. Image Rotation

When the heading angles of the marine radar and ice radar images are inconsistent, after confirming the rotation angle, the image can be rotated through affine transformation of Equation (3) or coordinate transformation of Equations (4) and (5).

For example, if the heading angle of the marine radar image is 0° and the heading angle of the ice radar image is 60°, after rotating the ice radar image counterclockwise by 60°, the same heading angle as the marine radar image can be obtained.

$$\begin{bmatrix} u' \\ v' \\ 1 \end{bmatrix} = A \begin{bmatrix} u \\ v \\ 1 \end{bmatrix} = \begin{bmatrix} \cos\theta & -\sin\theta & 0 \\ \sin\theta & \cos\theta & 0 \\ 0 & 0 & 1 \end{bmatrix} \begin{bmatrix} u \\ v \\ 1 \end{bmatrix} \tag{3}$$

$$u' = u\cos\theta - v\sin\theta \tag{4}$$

$$v' = u\sin\theta + v\cos\theta \tag{5}$$

where (u', v') represent the pixel coordinates in the transformed image; A represents the affine matrix; (u, v) represent the pixel coordinates in the original image; θ represents the rotation angle.

2.3.2. Image Zooming

When the pixel distances corresponding to the same 1 nm of the marine radar and ice radar images are inconsistent, after confirming the zooming ratio, the image can be zoomed through affine transformation of Equation (6) or coordinate transformation of Equations (7) and (8). For example, if 2 nm in the distance scale ring of the marine radar corresponds to 105px, and 2 nm in the distance scale ring of the ice radar corresponds to 100px, after magnifying the ice radar image by 1.05 times, the same corresponding pixel distance per nm as the marine radar image can be obtained.

$$\begin{bmatrix} u' \\ v' \\ 1 \end{bmatrix} = A \begin{bmatrix} u \\ v \\ 1 \end{bmatrix} = \begin{bmatrix} c_u & 0 & 0 \\ 0 & c_v & 0 \\ 0 & 0 & 1 \end{bmatrix} \begin{bmatrix} u \\ v \\ 1 \end{bmatrix} \tag{6}$$

$$u' = c_u u \tag{7}$$

$$v' = c_v v \tag{8}$$

where (u', v') represent the pixel coordinates in the transformed image; A represents the affine matrix; (u, v) represent the pixel coordinates in the original image; c_u and c_v represent the zooming ratio.

2.3.3. Image Translation

When the pixel positions of the own-ship of the marine radar and ice radar images are inconsistent, after confirming the vertical and horizontal movement values, the image can be translated through affine transformation of Equation (9) or coordinate transformation of Equations (10) and (11). For example, if the pixel position of the own-ship of the marine radar is located at $(690, 555)$, the pixel position of the own-ship of the ice radar is located at $(430, 350)$, and the corresponding pixel distance per nm of the two radar images is the same, after translating the ice radar image to the right by 260px and downward by 205px, the same pixel position of the own-ship as the marine radar image can be obtained.

$$\begin{bmatrix} u' \\ v' \\ 1 \end{bmatrix} = A \begin{bmatrix} u \\ v \\ 1 \end{bmatrix} = \begin{bmatrix} 1 \; 0 \; t_u \\ 0 \; 1 \; t_v \\ 0 \; 0 \; 1 \end{bmatrix} \begin{bmatrix} u \\ v \\ 1 \end{bmatrix} \tag{9}$$

$$u' = u + t_u \tag{10}$$

$$v' = v + t_v \tag{11}$$

where (u', v') represent the pixel coordinates in the transformed image; A represents the affine matrix; (u, v) represent the pixel coordinates in the original image; t_u represents the vertical movement value; t_v represents the horizontal movement value.

2.4. Radar Image Fusion

After eliminating redundant information and correcting spatial registration, the image fusion can be carried out through Equation (12) after confirming the respective weights of marine radar and ice radar images. The orange-yellow marine radar echo is less sensitive to sea ice perception, but it helps to outline the edge contours of the sea ice, while the grayish-white ice radar echo has a higher sensitivity to sea ice perception. The whiter and brighter the echo color is, the thicker the sea ice thickness is, which helps to master the structure and texture of the sea ice layer. The complementary and combination of the two kinds of radar image information will contribute to the subsequent development of sea ice perception application function.

$$dst(I) = \mathrm{src}\,1(I) \times alpha + \mathrm{src}\,2(I) \times beta + gamma \tag{12}$$

where dst represents the output image matrix; I represents the multi-dimensional index of elements in image matrix; src1 and src2 represent two image matrices with the same size; al pha represents the weight of image 1 ; beta represents the weight of image 2 ; gamma represents the add scalar value.

3. NEAR-FIELD SEA ICE RISK ASSESSMENT AND WARNING PROCESS

As the ship is sailing, in addition to considering the physical characteristics of nearfield sea ice (such as concentration, area, and grayscale), it is more important to consider the relationship between the own-ship and near-field sea ice (such as distance and relative bearing) from the perspective of the navigator. Therefore, this study proposes a set of nearfield sea ice risk assessment and warning processes that meet the needs of autonomous ships in the polar region and are suitable for two situations, as shown in Figure 8. By calculating the sea ice concentration of

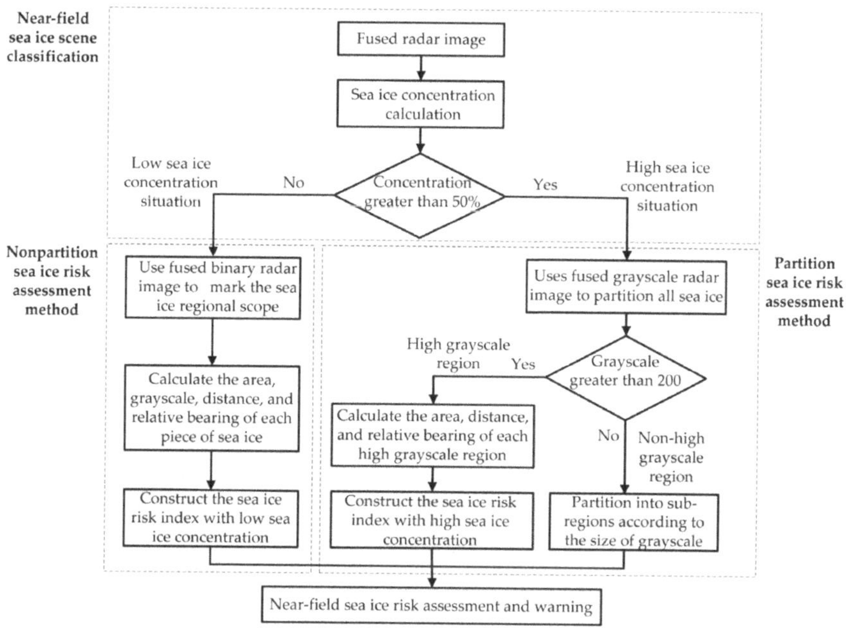

Figure 8. *Schematic diagram of near-field sea ice risk assessment and warning process.*

the fused radar image, the near-field sea ice scene with a sea ice concentration less than or equal to 50% is defined as the low-sea-ice-concentration situation, and the near-field sea ice scene with a sea ice concentration greater than 50% is defined as the high-sea-ice-concentration situation. Then, by using the risk variables such as sea ice area, sea ice grayscale, distance between sea ice and the own-ship, and relative bearing of sea ice and the own-ship, the sea ice risk index with low-sea-ice-concentration and the sea ice risk index with high-sea-ice-concentration are constructed. Finally, visualization processing is carried out according to the size of the risk index values of each piece of sea ice to achieve a better near-field sea ice risk assessment and warning effect.

3.1. Near-Field Sea Ice Risk Variables Calculation

3.1.1. Sea Ice Concentration

When calculating the sea ice concentration, the near-field area of the own-ship is defined as a circular area with the position of the own-ship at the center and a radius of 200px. Then, the sea ice concentration in the near-field area of the own-ship can be obtained through Equations (13) and (14).

$$\text{count} = \sum_{I:src(I)\neq 0} 1 \tag{13}$$

$$SIC = \frac{\text{count}}{\pi r^2} \times 100\% \tag{14}$$

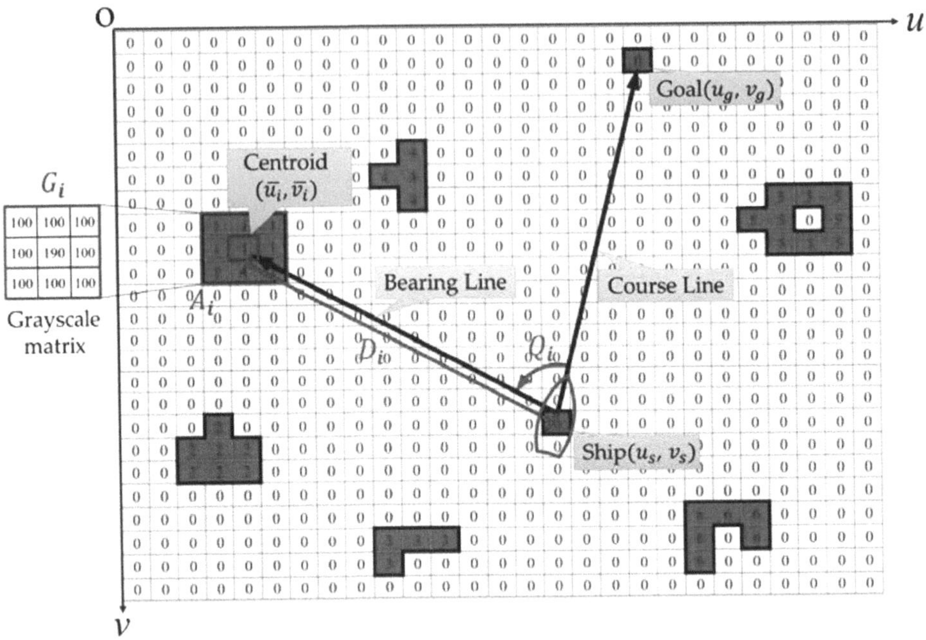

Figure 9. *Schematic diagram of sea ice marking in a low-sea-ice-concentration situation.*

where count represents the total number of sea ice pixels (i.e., the elements whose value is not 0 in the image matrix) in the near-field area of the own-ship; I represents the multidimensional index of elements in the image matrix; src represents the image matrix of the near-field area of the own-ship; SIC represents the value of sea ice concentration in the near-field area of the own-ship.

3.1.2. Sea Ice Area

When calculating the sea ice area, the total number of pixels contained in each piece of sea ice is the area value of the piece of sea ice, which can be obtained through Equation (15). For example, as shown in Figure 9, the sea ice image matrix numbered 1 contains 9px in total, so the area value of this piece of sea ice is 9.

$$A_i = \sum_{I:src(I)=i} 1 \tag{15}$$

where A_i represents the area value of the i-th piece of sea ice; I represents the multidimensional index of elements in the image matrix; src represents the sea ice image marker matrix.

3.1.3. Sea Ice Grayscale

When calculating the sea ice grayscale, the average grayscale value of each piece of sea ice is the grayscale value of the piece of sea ice, which can be obtained through Equations (16) and (17). For example, as shown in Figure 9, the sum of the grayscale values of all pixels in the sea ice image matrix numbered 1 is 990, and the area value is 9 , so the average grayscale value of this piece of sea ice is 110. Related studies have shown that there is an exponential relationship between sea ice albedo and sea ice thickness, and some scholars have proposed a retrieval model based on the exponential relationship between sea ice albedo and sea ice thickness [28,29]. Therefore, the higher the grayscale value of sea ice measured from the radar image, the thicker the sea ice may be.

$$G_{i-\text{ sum}} = \sum_{I:src(I)=i} G_{-}\text{ blend}(I) \tag{16}$$

$$G_i = \frac{G_{i-}\text{ sum}}{A_i} \tag{17}$$

where G_i_sum represents the sum of grayscale values of the i-th piece of sea ice; I represents the multi-dimensional index of elements in the image matrix; src represents the sea ice image marker matrix; G_blend represents the grayscale image fused according to a certain weight; G_i represents the average grayscale value of the i-th piece of sea ice; A_i represents the area value of the i-th piece of sea ice.

3.1.4. Distance between Sea Ice and the Own-Ship

When calculating the distance between sea ice and the own-ship, the distance refers to the Euclidean metric from the centroid of the sea ice to the position of the own-ship. Each piece of sea ice is regarded as an area with uniform mass distribution. The centroid of each piece of sea ice can be obtained through Equations (18) and (19), and then the distance between each piece of sea ice and the own-ship can be obtained through Equation (20). For example, as shown in Figure 9, the distance value between sea ice numbered 1 and the own-ship is 15px.

$$\overline{u_i} = \frac{\iint u_i d\theta}{\iint 1 d\theta} \tag{18}$$

$$\overline{v_i} = \frac{\iint v_i d\theta}{\iint 1 d\theta} \tag{19}$$

$$D_i = \sqrt{(u_s - \overline{u_i})^2 + (v_s - \bar{v}_i)^2} \tag{20}$$

where $(\overline{u_i}, \overline{v_i})$ represents the centroid coordinates of the i-th piece of sea ice; $(u\prime, v\prime)$ represents the pixel coordinates of the i-th piece of sea ice; $\theta \subset A_i$; D_i represents the distance value between the i-th piece of sea ice and the own-ship; (u_s, v_s) represents the position coordinates of the own-ship.

3.1.5. Relative Bearing of Sea Ice and the Own-Ship

When calculating the relative bearing of sea ice and the own-ship, the relative bearing refers to the included angle between the course line and the bearing line (i.e., the included angle between the line from the position of the own-ship to the short-term target position and the line from the position of the own-ship to the centroid of the sea ice, which is measured to the right or left based on the course line, and the range is 0° ∼ 180°), which can be obtained through Equations (21)-(24). For example, as shown in Figure 9, the relative bearing value of sea ice numbered 1 and the own-ship is 64.4°.

$$a_i = \sqrt{(u_s - \overline{u_i})^2 + (v_s - \overline{v_i})^2} \tag{21}$$

$$b_i = \sqrt{(u_g - \overline{u_i})^2 + (v_g - \overline{v_i})^2} \tag{22}$$

$$c_i = \sqrt{(u_s - u_g)^2 + (v_s - v_g)^2} \tag{23}$$

$$Q_i = \cos^{-1}\left(\frac{a_i^2 + c_i^2 - b_i^2}{2a_i c_i}\right) \tag{24}$$

where a_i represents the distance between the centroid of the i-th piece of sea ice and the position of the own-ship; b_i represents the distance between the short-term target position and the centroid of the i-th piece of sea ice; c_i represents the distance between the position of the own-ship and the short-term target position; Q_i represents the relative bearing value of the i-th piece of sea ice and the own-ship; (u_s, v_s) represents the position coordinates of the own-ship; $(\overline{u_i}, \overline{v_i})$ represents the centroid coordinates of the i-th piece of sea ice; (u_g, v_g) represents the coordinates of the short-term target position.

3.2. Nonpartition Sea Ice Risk Assessment Method for Low-Sea-Ice-Concentration Situation

In the low-sea-ice-concentration situation, as shown in the bottom left corner of Figure 8 , this study uses the fused binary radar image to mark the regional scope of each piece of sea ice, and calculates its area value, grayscale value, distance value from the own-ship, and the relative bearing value to the own-ship throughs Equations (15)-(24), to construct the near-field sea ice risk index with low-sea-ice-concentration.

The marking of the regional scope of each piece of sea ice is shown in Figure 9. The connected regions of each piece of sea ice are marked with positive integers starting from " 1 " in sequence, and the other background regions are marked with " 0 ". The regions marked with the same positive integer belong to the same sea ice region. Then, considering variables such as the area value, grayscale value, distance value from the own-ship, and the relative bearing value to the own-ship of each piece of sea ice, this study constructs the near-field sea ice risk index with low-sea-ice-concentration, as shown in Equation (25). Accordingly, the sea ice risk assessment and warning can be carried out based on the obtained risk index values of each piece of sea ice. The assessment method is to treat each piece of sea ice itself as a whole, without any further partition treatment.

$$\mathrm{LCRI}_i = \frac{\left[\frac{A_i}{A_{\max}} + \frac{G_i}{G_{\max}} + \frac{(D_{\max} - D_i)}{D_{\max}} + \frac{(Q_{\max} - Q_i)}{Q_{\max}}\right]}{4} \times 100\% \tag{25}$$

where A_i represents the area value of the i-th piece of sea ice; $A_{\max}$ represents the maximum area value in all sea ice regions; G_i represents the average grayscale value of the i-th piece of sea ice; $G_{\max}$ represents the maximum average grayscale value in all sea ice regions; D_i represents the distance value between the i-th piece of sea ice and the own-ship; $D_{\max}$ represents the maximum distance value from the own-ship in all sea ice regions; Q_i represents the relative bearing value of the i-th piece of sea ice and the own-ship; $Q_{\max}$ represents the maximum relative bearing value to the own-ship in all sea ice regions; $LCRI_i$ represents the risk index value of the i-th piece of sea ice.

3.3. Partition Sea Ice Risk Assessment Method for High-Sea-Ice-Concentration Situation

In the high-sea-ice-concentration situation, as shown in the bottom right corner of Figure 8 , if the risk index is still constructed for the whole sea ice, it is easy to cause the near-field area of the own-ship to be full of high-risk sea ices, and the effect of sea ice risk assessment and warning is poor. Thus, this study uses the fused grayscale radar image to partition all sea ice. The regions with a sea ice grayscale greater than 200 are defined as high grayscale regions, and the other regions with a sea ice grayscale less than or equal to 200 are defined as non-high grayscale regions. Then, for the high grayscale regions in each piece of sea ice, their area value, distance value from the own-ship, and the relative bearing value to the own-ship are calculated through Equations (15) and (18)-(24), to construct the near-field sea ice risk index with high-sea-ice-concentration.

The partition of each piece of sea ice is shown in Figure 10a. For example, there is a large intact sea ice in the front left of the own-ship. According to the size of the grayscale values, it can be partitioned into the high grayscale regions and the non-high grayscale regions. For the high grayscale regions, as shown in Figure 10b, considering variables such as the area value, distance value from the own-ship, and the relative bearing value to the own-ship, this study constructs the near-field sea ice risk index with high-sea-iceconcentration, as shown in Equation (26); although the other non-high grayscale regions do not involve the construction of the near-field sea ice risk index, they can still be partitioned into four sub-regions (e.g., $(1,50], (50,100], (100,150], (150,200]$) according to the size of the grayscale values, which is helpful for the visualization of subsequent risk warning, as shown in Figure 10c. Accordingly, the sea ice risk assessment and warning can be carried out based on the obtained risk index values of the high grayscale regions in each piece of sea ice. The assessment method is to partition each piece of sea ice to find out the priority regions that have the greatest impact on navigation safety in the whole sea ice.

$$HCRI_j = \frac{\left[\frac{A_j}{3A_\sigma} + \frac{(D_{\max} - D_j)}{D_{\max}} + \frac{(90 - Q_j)}{90}\right]}{3} \times 100\% \tag{26}$$

where A_j represents the area value of the j-th high grayscale region; A_σ represents the standard deviation of the area values of all high grayscale regions; D_j represents

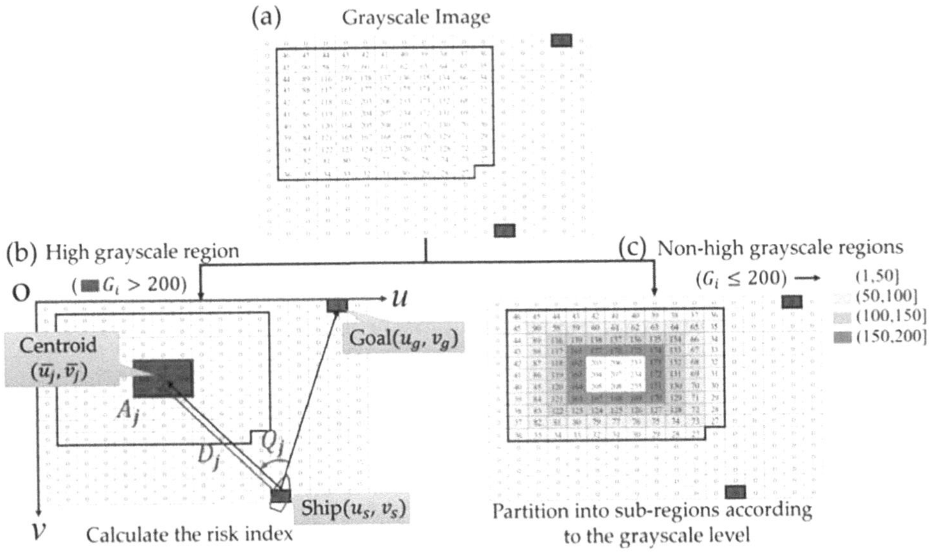

Figure 10. *Schematic diagram of sea ice partition in a high-sea-ice-concentration situation: (a) grayscale image; (b) high grayscale region; (c) non-high grayscale regions.*

the distance value between the j-th high grayscale region and the own-ship; $D_{\max}$ represents the maximum distance value from the own-ship in all high grayscale regions; Q_j represents the relative bearing value of the j-th high grayscale region and the own-ship; HCRI represents the risk index value of the j-th high grayscale region.

4. EXAMPLE DEMONSTRATION FOR NEAR-FIELD SEA ICE PERCEPTION AND WARNING

This study collected the replay videos of marine radar and ice radar of the polar research icebreaker Xuelong 2 from Shenzhen to Zhongshan Station for China's 36th Antarctic expedition in 2019. The radar images used in subsequent examples of low-seaice-concentration and high-sea-ice-concentration were all captured from these radar replay videos of this voyage.

4.1. Example 1

Example 1 used image software to capture the marine radar and ice radar images at 20:39:26 on 18 November 2019, as shown in Figure 11a,b. At this time, Xuelong 2 approached 63° south latitude, and the sea ice distribution was relatively scattered and sparse. Through the radar image fusion process in Section 2, after redundant information elimination, spatial registration correction, and image fusion, the fused radar image obtained is shown in Figure 11c.

In the fused radar image of example 1, through Equations (13) and (14), it can be calculated that the total number of sea ice pixels in the near-field area with the position of the own-ship at the center and a radius of 200px is 7883 , and the

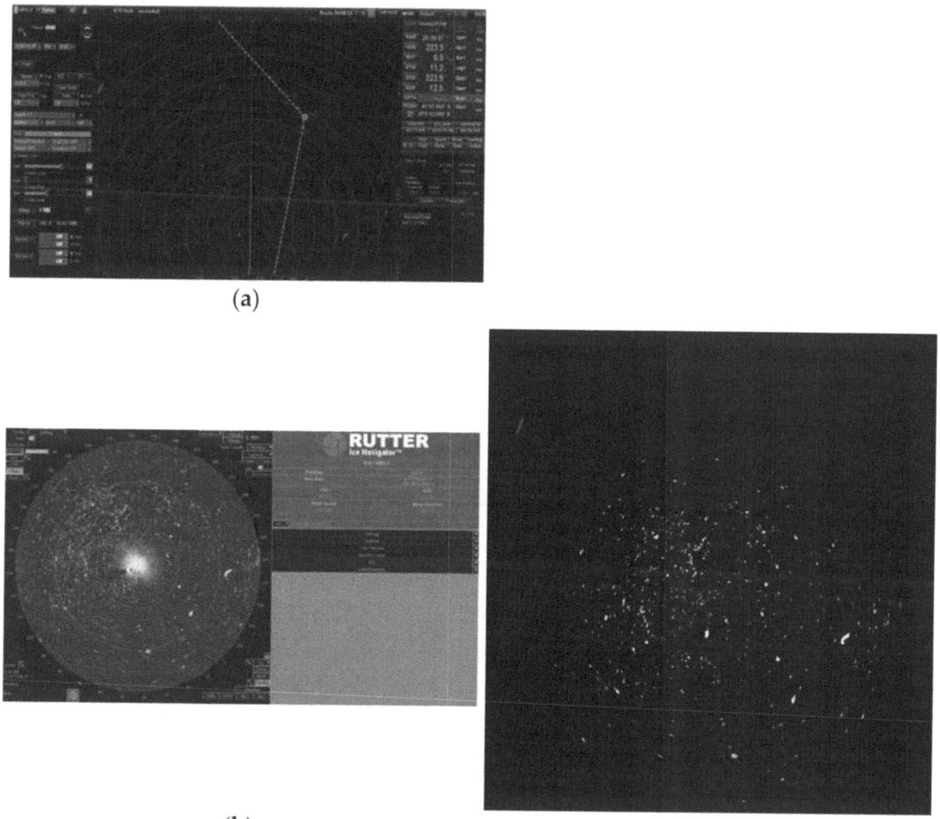

(a) (b) (c)

Figure 11. *The radar image of example 1: (a) the original marine radar image; (b) the original ice radar image; (c) the fused radar image.*

obtained sea ice concentration is 6.2% (less than 50%), which belongs to the low-sea-ice-concentration situation. Therefore, by adopting the nonpartition sea ice risk assessment method, the regional scope of each piece of sea ice in the fused binary radar image is marked, and a total of 824 pieces of sea ice can be obtained. The area value, grayscale value, distance value from the own-ship, and the relative bearing value to the own-ship of each piece of sea ice can be calculated through Equations (15)-(24) in Section 3.1. Then, the risk index value of each piece of sea ice can be obtained through Equation (25) in Section 3.2.

By drawing scatter plots, the correlation between the obtained sea ice risk index and each variable is analyzed separately, as shown in Figure 12a,d; the obtained sea ice risk index value ranges from 16.6% to 76.3%. It can be seen from Figure 12a that the minimum value of sea ice area is 1 and the maximum is 253 . The larger the sea ice area, the higher the risk, and the two are positively correlated. It can be seen from Figure 12*b* that the minimum value of sea ice grayscale is 0 and the maximum is 178.3. The greater the sea ice grayscale, the higher the risk, and

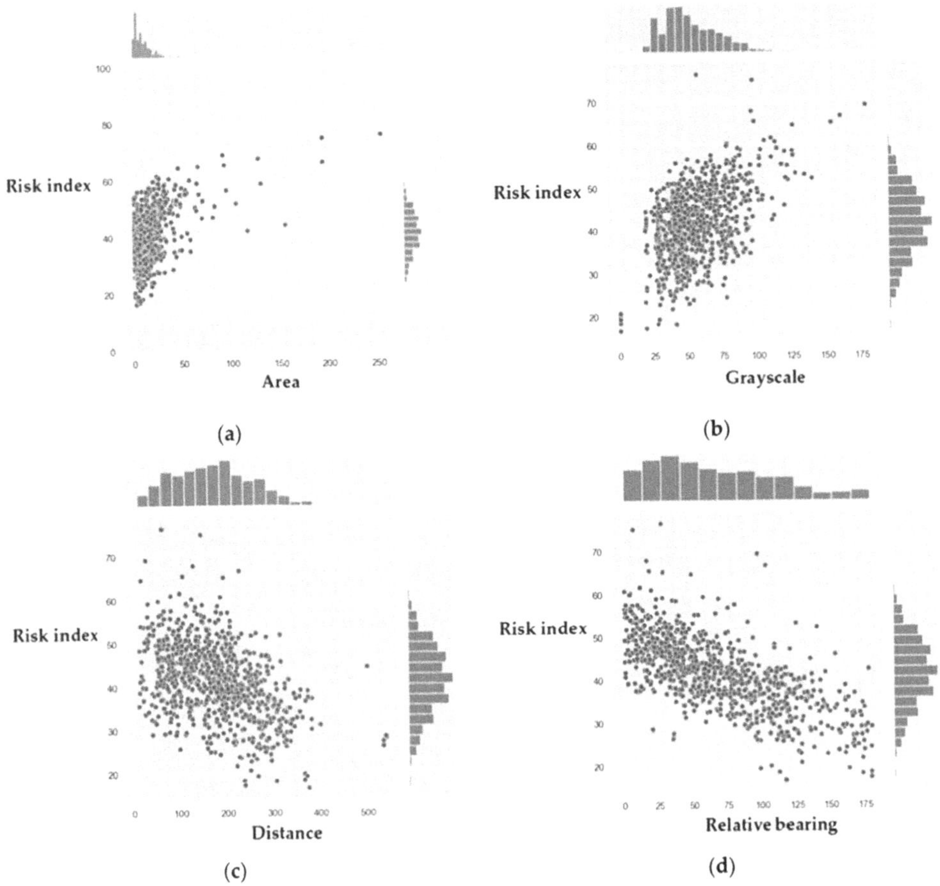

Figure 12. *The scatter plot of sea ice risk index vs. variables: (a) the risk index vs. sea ice area; (b) the risk index vs. sea ice grayscale; (c) the risk index vs. sea ice distance; (d) the risk index vs. sea ice relative bearing.*

the two are positively correlated. It can be seen from Figure 12c that the minimum distance value between sea ice and the own-ship is 11.7 and the maximum is 540 . The closer the sea ice, the higher the risk, and the two are negatively correlated. It can be seen from Figure 12d that the minimum relative bearing value of sea ice and the own-ship is 0.2°, and the maximum is 179.8°. The smaller the relative bearing (that is, the closer the position is to the front of the own-ship), the higher the risk, and the two are negatively correlated. Based on the above analysis results, the near-field sea ice risk index with low-sea-ice-concentration proposed in this study can indeed reasonably reflect the influence degree of each piece of sea ice on the navigation safety of the own-ship.

According to the size of the obtained risk index values, the regions of each piece of sea ice can be marked with different colors for sea ice risk assessment and visualization warning, as shown in Figure 13. When the sea ice risk index value is

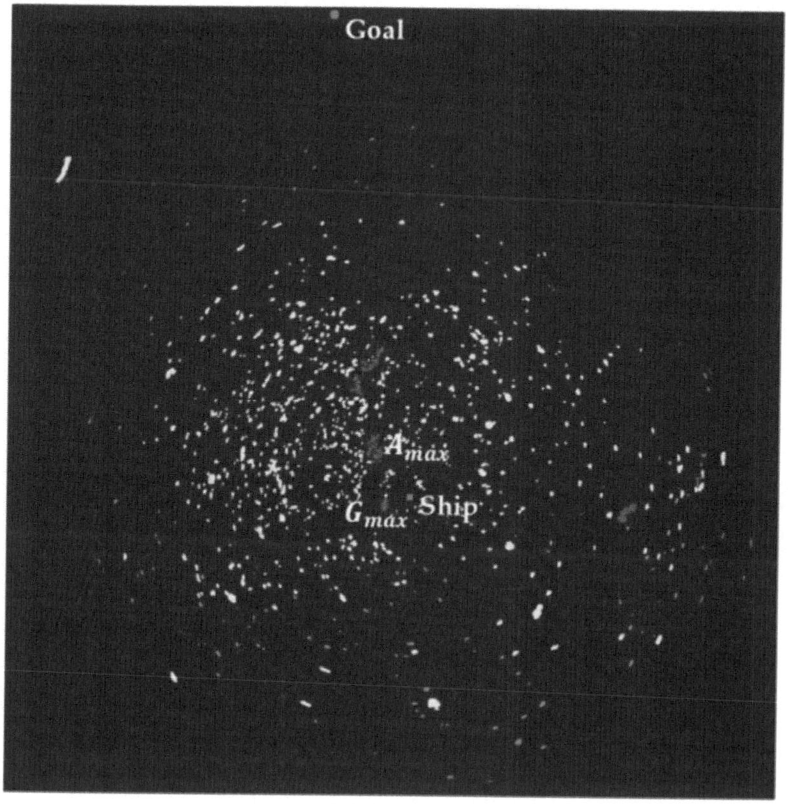

Figure 13. *Schematic diagram of sea ice risk assessment and visualization warning of example 1.*

greater than 66% $(HCRI_j > 66\%)$, the sea ice region is marked in red and belongs to high-risk sea ice that the own-ship should focus on. When the sea ice risk index value is less than or equal to 66% and greater than 33% $(33\% < HCRI_j \leq 66\%)$, the sea ice region is marked in yellow and belongs to medium-risk sea ice. When the sea ice risk index value is less than or equal to 33% $(HCRI_j \leq 33\%)$, the sea ice region is marked in green and belongs to low-risk sea ice. In addition, the A_{max} can be marked next to the sea ice with the largest area and the G_{max} can be marked next to the sea ice with the greatest grayscale to more intuitively display the sea ice information that threatens the navigation safety of the own-ship.

4.2. Example 2

Example 2 used image software to capture the marine radar and ice radar images at 13:45:53 on 19 November 2019, as shown in Figure 14a,b. At this time, Xuelong 2 approached 66° south latitude, and the sea ice distribution was relatively concentrated and significant. Through the radar image fusion process in Section 2, after redundant information elimination, spatial registration correction, and image fusion, the fused radar image obtained is shown in Figure 14c.

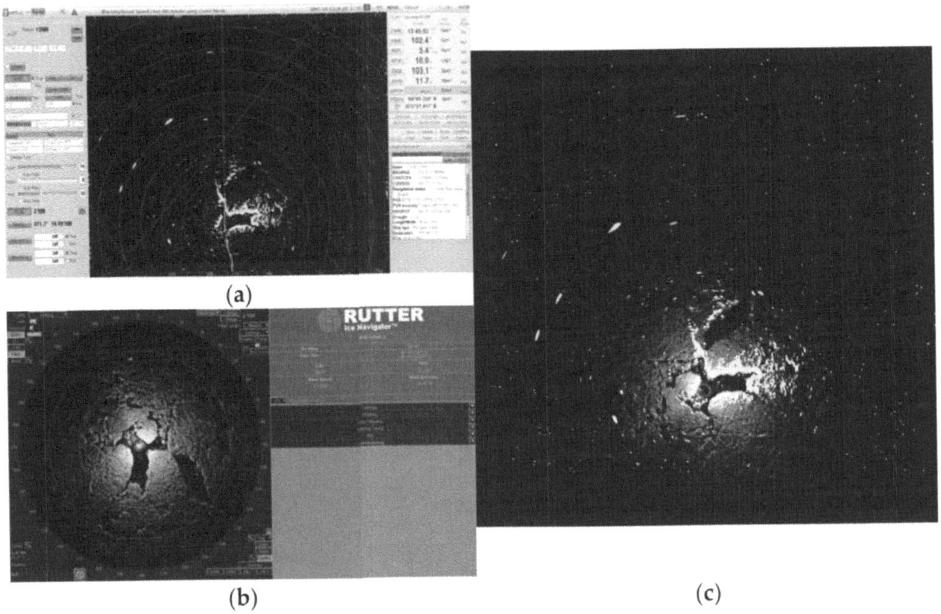

Figure 14. The radar image of example 2: (a) the original marine radar image; (b) the original ice radar image; (c) the fused radar image.

In the fused radar image of example 2, through Equations (13) and (14), it can be calculated that the total number of sea ice pixels in the near-field area with the position of the own-ship at the center and a radius of 200px is 96,489 , and the obtained sea ice concentration is 76.7% (greater than 50%), which belongs to the high-sea-ice-concentration situation. Therefore, by adopting the partition sea ice risk assessment method, all sea ice regions in the fused grayscale radar image are partitioned into the high grayscale regions with a sea ice grayscale greater than 200 and the non-high grayscale regions with a sea ice grayscale less than or equal to 200. For all high grayscale regions, the area value, distance value from the own-ship, and the relative bearing value to the own-ship can be calculated through Equations (15) and (18)-(24) in Section 3.1. Then, the risk index values of the high grayscale regions in each piece of sea ice can be obtained through Equation (26) in Section 3.2.

According to the size of the obtained risk index values of the high grayscale regions, the same criteria as in Section 4.1 (the risk index value is greater than 66%, less than or equal to 66% and greater than 33%, or less than or equal to 33%) can also be used to mark each high grayscale region in red, yellow, and green, respectively, for sea ice risk assessment and visualization warning, as shown in Figure 15a. Other non-high grayscale regions can also be partitioned into four sub-regions (e.g., $(1,50], (50,100], (100,150], (150,200]$) according to the size of the grayscale values, and then the high grayscale regions (also regard it as a sub-region) are added to draw the contour map of these five sub-regions and to mark them with

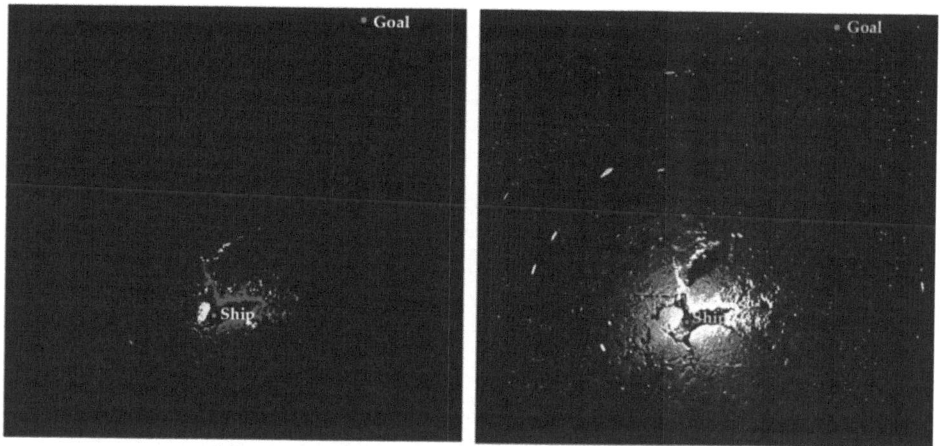

(a) (b)

Figure 15. *Schematic diagram of sea ice risk assessment and visualization warning of example 2: (a) the high grayscale region risk distribution; (b) the overall grayscale distribution.*

gradient colors from blue to white for sea ice risk assessment and visualization warning, as shown in Figure 15b.

Figure 15b shows the overall grayscale distribution of sea ice in the near-field of the own-ship. According to the relevant study results that there is an exponential relationship between sea ice albedo and sea ice thickness [28,29], the higher the grayscale value, the thicker the sea ice may be. Therefore, priority attention should be given to regions marked in white (i.e., the high grayscale regions with a sea ice grayscale greater than 200). Figure 15a shows the risk distribution of these high grayscale regions. These regions are partitioned into three levels according to the size of the risk index values, among which the region marked in red is the region that poses the greatest threat to the navigation safety of the own-ship and should be given more priority attention. It can be seen from the results of visualization warning that in the high-sea-ice-concentration situation, the sea ice situation in the near-field area of the own-ship can be fully mastered from the macro- and micro-perspective through the overall grayscale distribution map of sea ice and the risk distribution map of high grayscale regions.

In this example, if the nonpartition sea ice risk assessment method is still adopted, according to the size of the obtained risk index values, the same criteria as in Section 4.1 can also be used to mark the regions of each piece of sea ice in red, yellow, and green, respectively, for sea ice risk assessment and visualization warning, as shown in Figure 16. However, it can be seen from the results of visualization warning that in the near-field area of the own-ship, the own-ship is surrounded by high-risk sea ice marked in red, and it is impossible to distinguish the priority regions that have the greatest impact on navigation safety and should therefore be given more priority attention. Thus, the nonpartition sea ice risk assessment method adopted in the low-sea-ice-concentration situation is indeed not suitable for the high-sea-ice-concentration situation.

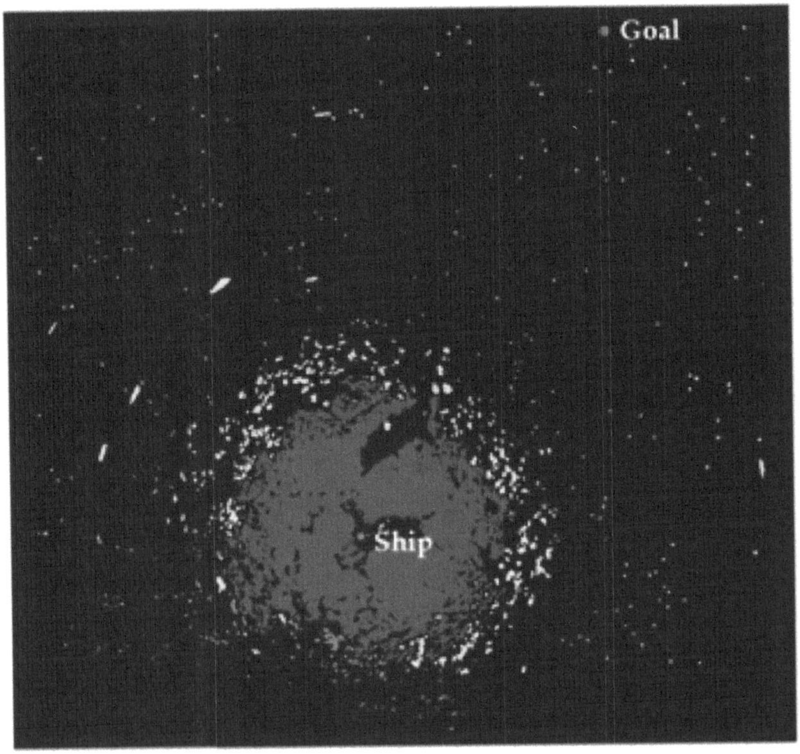

Figure 16. *Schematic diagram of sea ice risk assessment and visualization warning of example 2 using the nonpartition sea ice risk assessment method.*

4.3. Results and Discussion

From the above two examples, the radar image fusion process proposed in this study is demonstrated. In example 1, in addition to the occlusion and interference of the marking line and marking frame, the echo noise of the own-ship is particularly significant in the ice radar image. After eliminating redundant information, these problems have been improved. In example 2, the heading angle, the corresponding pixel distance per nautical mile, and the position of the own-ship are inconsistent between the marine radar and the ice radar images. The marine radar display orientation is head-up, the heading angle is $0°$, the distance scale ring with a radius of 2 nm corresponds to 53px, and the position of the own-ship is located at $(648, 548)$; the ice radar display orientation is north-up, the heading angle is $102.4°$, the distance scale ring with a radius of 2 nm corresponds to 75px, and the position of the own-ship is located at $(430, 350)$. After the correction of image rotation, zooming, and translation, these spatial registration problems have been solved obviously. It can be seen from Figures 11c and 14c that the radar image fusion effect in these two examples is good.

In addition, from the above two examples, it also demonstrates a set of near-field sea ice risk assessment and warning processes proposed in this study that meets the needs of autonomous ships in the polar region and is suitable for two situations. In the situation of low-sea-ice-concentration in example 1, because the sea ice distribution is relatively scattered and sparse, each piece of sea ice itself is regarded as a whole without any further partition treatment and the nonpartition sea ice risk assessment method proposed in this study is used to calculate the risk index. In the situation of high-sea-ice-concentration in example 2, because the sea ice distribution is relatively concentrated and significant, the partition sea ice risk assessment method proposed in this study is used to partition each piece of sea ice according to the size of the grayscale values, and then the risk index is calculated for the high grayscale regions (i.e., the priority regions that have the greatest impact on navigation safety). It can be seen from Figures 13 and 15 that the risk assessment and warning effects of near-field sea ice are good in both the low-sea-ice-concentration situation in example 1 and the high-sea-ice-concentration situation in example 2.

In fact, the key to near-field sea ice risk assessment and warning is to identify the "relatively" dangerous sea ice regions around the ship. Unlike obstacles of rigid structures such as islands, reefs, or sunken ships, sea ice belongs to flexible structures that ships can contact or break; and the stronger the icebreaking ability of the ship, the thicker the sea ice that can be contacted. For example, the hull structural strength of the Xuelong 2 meets the PC3 level, and it can continuously break ice at a speed of $2 \sim 3$ knots under sea conditions of 1.5 m of ice and 0.2 m of snow. Therefore, the ship can set the parameter values in the near-field sea ice risk index calculation formula proposed in this study according to its own conditions. With the support of two kinds of radar image information, the near-field sea ice risk index value obtained in this study is based on variable data such as sea ice concentration, sea ice area, sea ice grayscale, distance between sea ice and the own-ship, and relative bearing of sea ice and the own-ship. Unlike the partial and subjective judgment of the navigator, it can determine the sea ice risk situation in the near-field area of the ship more comprehensively and objectively, providing effective support for the assisted decision-making of autonomous navigation in the polar region.

5. CONCLUSIONS

For the navigation safety of autonomous ships in the polar region, this study chooses marine radar and ice radar as the near-field sea ice sensing equipment and proposes the radar image fusion process. The process can integrate the two kinds of radar image information well, solve the problems of redundant information and spatial registration during image fusion effectively, and perceive the sea ice information in the near-field area of the ship more comprehensively than a single sensor can.

Considering the relationship between the own-ship and near-field sea ice from the perspective of the navigator, this study proposes a set of near-field sea ice risk assessment and warning processes that meet the needs of autonomous ships in the polar region and are suitable for two situations. In the low-sea-ice-concentration situation, each piece of sea ice itself is regarded as a whole, and the nonpartition sea ice risk assessment method is adopted to calculate the risk index value of each piece of sea ice. In the high-sea-ice-concentration situation, the partition sea ice risk assessment method is used to find out the high grayscale regions that have the greatest impact on navigation safety in the whole sea ice, and calculate the risk index value of the high grayscale regions in each piece of sea ice.

According to the example demonstration results, through the integrated radar image fusion process and a set of near-field sea ice risk assessment and warning processes proposed in this study, the sea ice risk distribution in the near-field area of the ship can be well obtained in both low- and high-sea-ice-concentration situations. These processes and visualized warning maps can provide effective support for the assisted decision-making of autonomous navigation in the polar region.

On the basis of this study, follow-up research can carry out related work on how to maneuver ships to avoid sea ice at a lower risk. In fact, ship navigation is a dynamic process. The process proposed in this study can be used to master the sea ice risk distribution at each time point in the current navigation environment, and can also be used to evaluate the effect of avoiding sea ice.

Author Contributions

T.-H.H. and S.W. conceived and designed the experiments; B.L. performed the experiments; T.-H.H., B.L. and W.L. analyzed the data; S.W. contributed materials; T.-H.H. and B.L. wrote the paper. All authors have read and agreed to the published version of the manuscript. Funding: This research was funded by National Key Research and Development Program of China (Grant No. 2021YFC2801003, 2021YFC2801004), National Natural Science Foundation of China (Grant No. 52071199), and Shanghai Science and Technology Innovation Action Plan (Grant No. 21DZ1205803).

Acknowledgments

The authors would like to express their gratitude to Ning Xu, Yanping Zhao, and Xude Zhang of the China Polar Research Institute for their valuable suggestions on this paper.

REFERENCES

1. Stroeve, J.; Notz, D. Changing state of Arctic sea ice across all seasons. Environ. Res. Lett. 2018, 13, 10300110.
2. England, M.R.; Polvani, L.M.; Sun, L. Robust Arctic warming caused by projected Antarctic sea ice loss. Environ. Res. Lett. 2020, 15,104005 .
3. Yang, J.; Xiao, C.; Liu, J.; Li, S.; Qin, D. Variability of Antarctic sea ice extent over the past 200 years. Sci. Bull. 2021, 66, 2394-2404.
4. Guarino, M.V.; Sime, L.C.; Schröeder, D.; Irene, M.V.; Rosenblum, E.; Ringer, M.; Ridley, J.; Feltham, D.; Bitz, C.; Eric, J.S.; et al. Sea-ice-free Arctic during the last interglacial supports fast future loss. Nat. Clim. Chang. 2020, 10, 928-932.

5. Wright, R.G. Unmanned and Autonomous Ships-An Overview of Mass, 1st ed.; Routledge Taylor & Francis Group: New York, NY, USA, 2020; pp. 115-144. Available online: https://scholar.google.com/scholar?hl=zh-CN&as_sdt=0%2C5&q=Unmanned+and+autonomous+ships-An+Overview+of+MASS&btnG= (accessed on 9 March 2022).
6. Tsou, M.C. Multi-target collision avoidance route planning under an ECDIS framework. Ocean Eng. 2016, 121, 268-278.
7. Skauen, A.N. Quantifying the tracking capability of space-based AIS systems. Adv. Space Res. 2016, 57, 527-542.
8. Gu, W.; Liu, C.; Yuan, S.; Li, N.; Chao, J.; Li, L.; Xu, Y. Spatial distribution characteristics of sea-ice-hazard risk in Bohai, China. Ann. Glaciol. 2013, 54, 73-79.
9. Wang, L.; Xu, L.; Clausi, D.A. Sea ice concentration estimation during melt from dual-pol SAR scenes using deep convolutional neural networks: A case study. IEEE Trans. Geosci. Electron. 2016, 54, 4524-4533.
10. Zeng, T.; Shi, L.; Marko, M.; Cheng, B.; Zou, J.; Zhang, Z. Sea ice thickness analyses for the Bohai Sea using MODIS thermal infrared imagery. Acta Oceanol. Sin. 2016, 35, 96-104.
11. Spreen, G.; Kaleschke, L.; Heygster, G. Sea ice remote sensing using AMSR-E 89-GHz channels. J. Geophys. Res. 2008, 113, C02S03.
12. Beitsch, A.; Kaleschke, L.; Kern, S. Investigating high-resolution AMSR2 sea ice concentrations during the february 2013 fracture event in the Beaufort Sea. Remote Sens. 2014, 6, 3841-3856.
13. Posey, P.G.; Metzger, E.J.; Wallcraft, A.J.; Hebert, D.A.; Allard, R.A.; Smedstad, O.M.; Phelps, M.W.; Fetterer, F.; Stewart, J.S.; Meier, W.N.; et al. Improving Arctic sea ice edge forecasts by assimilating high horizontal resolution sea ice concentration data into the US Navy's ice forecast systems. Cryosphere 2015, 9, 1735-1745.
14. Hebert, D.A.; Allard, R.A.; Metzger, E.J.; Posey, P.G.; Preller, R.H.; Wallcraft, A.J.; Phelps, M.W.; Smedstad, O.M. Short-term sea ice forecasting: An assessment of ice concentration and ice drift forecasts using the U.S. Navy's Arctic Cap Nowcast/Forecast System. J. Geophys. Res. 2015, 120, 8327-8345.
15. Chi, J.; Kim, H. Prediction of Arctic sea ice concentration using a fully data driven deep neural network. Remote Sens. 2017, 9, 1305.
16. Xiao, F.; Zhang, S.; Li, J.; Geng, T.; Xuan, Y.; Li, F. Arctic sea ice thickness variations from CryoSat-2 satellite altimetry data. Sci. China Earth Sci. 2021, 64, 1080-1089.
17. Xu, Y.; Li, H.; Liu, B.; Xie, H.; Burcu, O.C. Deriving Antarctic sea-ice thickness from satellite altimetry and estimating consistency for NASA's ICESat/ICESat-2 missions. Geophys. Res. Lett. 2021, 48, e2021GL093425.
18. Yan, Y.; Huang, K.; Shao, D.; Xu, Y.; Gu, W. Monitoring the characteristics of the Bohai sea ice using high-resolution geostationary ocean color imager (GOCI) data. Sustainability 2019, 11, 777.
19. Haverkamp, D.; Soh, L.K.; Tsatsoulis, C. A comprehensive, automated approach to determining sea ice thickness from SAR data. IEEE Trans. Geosci. Remote. Sens. 1995, 33, 46-57.
20. Weissling, B.; Ackley, S.; Wagner, P.; Xie, H. EISCAM-Digital image acquisition and processing for sea ice parameters from ships. Cold Reg. Sci. Technol. 2009, 57,49-60.

21. Zhang, Q.; Skjetne, R. Image processing for identification of sea-ice floes and the floe size distributions. IEEE Trans. Geosci. Remote. Sens. 2015, 53, 2913-2924.
22. Parmiggiani, F.; Flores, M.M.; Wadhams, P.; Aulicino, G. Image processing for pancake ice detection and size distribution computation. Int. J. Remote Sens. 2019, 40, 3368-3383.
23. Renner, A.H.H.; Dumont, M.; Beckers, J.; Gerland, S.; Haas, H. Improved characterisation of sea ice using simultaneous aerial photography and sea ice thickness measurements. Cold Reg. Sci. Technol. 2013, 92, 37-47.
24. Xu, J.; Wang, H.; Cui, C.; Liu, P.; Zhao, Y.; Li, B. Oil spill segmentation in ship-borne radar images with an improved active contour model. Remote Sens. 2019, 11, 1698.
25. IMO. International Code for Ships Operating in Polar Waters (Polar Code); IMO: London, UK, 2015. Available online: https://wwwcdn. imo.org/localresources/en/MediaCentre/HotTopics/Documents/POLAR CODE TEXT AS ADOPTED.pdf (accessed on 9 March 2022).
26. Hsieh, T.H.; Wang, S.; Gong, H.; Liu, W.; Xu, N. Sea ice warning visualization and path planning for ice navigation based on radar image recognition. J. Mar. Sci. Technol. 2021, 29, 280-290.
27. Dumitru, C.O.; Andrei, V.; Schwarz, G.; Datcu, M. Machine Learning for Sea Ice Monitoring from Satellites. In Proceedings of the Photogrammetric Image Analysis & Munich Remote Sensing Symposium, Munich, Germany, 18-20 September 2019; pp. 83-89.
28. Allison, I.; Brandt, R.E.; Warren, S.G. East Antarctic sea ice: Albedo, thickness distribution, and snow cover. J. Geophys. Res. 1993 98, 12417 – 12429.
29. Su, H.; Wang, Y. Using MODIS data to estimate sea ice thickness in the Bohai Sea (China) in the 2009-2010 winter. J. Geophys. Res. 2012, 117, C10018.

CHAPTER 10

Adversarial AI Testcases for Maritime Autonomous Systems

Mathew J Walter, Aaron Barrett, David J Walker and Kimberly Tam

University of Plymouth, UK

ABSTRACT

Contemporary maritime operations such as shipping are a vital component constituting global trade and defence. The evolution towards maritime autonomous systems, often providing significant benefits (e.g., cost, physical safety), requires the utilisation of artificial intelligence (AI) to automate the functions of a conventional crew. However, unsecured AI systems can be plagued with vulnerabilities naturally inherent within complex AI models. The adversarial AI threat, primarily only evaluated in a laboratory environment, increases the likelihood of strategic adversarial exploitation and attacks on mission-critical AI, including maritime autonomous systems. This work evaluates AI threats to maritime autonomous systems in situ. The results show that multiple attacks can be used against real-world maritime autonomous systems with a range of lethality. However, the effects of AI attacks vary in a dynamic and complex environment from that proposed in lower entropy laboratory environments. We propose a set of adversarial test examples and demonstrate their use, specifically in the marine environment. The results of this paper highlight security risks and deliver a set of principles to mitigate threats to AI, throughout the AI lifecycle, in an evolving threat landscape.

Keywords: *maritime cyber security; adversarial AI; maritime autonomous systems*

1. INTRODUCTION

In recent years artificial intelligence (AI) has been utilised to automate many operations and processes abound in academia and industry. One globally crucial industry is maritime, which recognises the plethora of benefits automation could bring over contemporary vessels; these include reduced crew requirements, ease and optimisation of processes, increased crew safety, the possibility of significant operational cost reduction, and emission reduction [1-7]. Therefore, it is seemingly indisputable that greater automation, and hence the utilisation of maritime autonomous systems (MAS), will play a significant role in the maritime industry in future years. Furthermore, the development of these systems has already been initiated; for example, the work of [8] developed a reduced-crew autonomous ocean-travelling ship. Other work, such as the Mayflower autonomous ship, intends to be fully automated [9]. The Yara Birkeland project, based in Norway, has also seen successes with automated coastal hopping but with open questions around the cost of insurance, cyber security, and contingencies [7]. The Royal Navy is also making great use of uncrewed surface vessels (USV). [1]

Much of the advanced automation will be the product of AI, given its proven success, particularly in optimisation, clustering, classification and regression. However, whilst AI has great potential for significant benefits, the nature of subsymbolic AI makes the process of understanding the solution generation mechanism difficult to interpret, yielding a black-box nature, particularly so for deep neural networks relying on billions of parameters, in addition to the high-dimensional phenomena. Therefore, AI has been documented to be a security risk with the term adversarial AI (AAI) coined for the misuse of AI. The fields of AAI and eXplainable AI (XAI) have shown the dangers and safety risks poor development of mission-critical AI can exhibit, in some cases leading to the possibility of fatalities [10]. AI can be used to automate attacks on other technologies, as well as being a technology that can be vulnerable to attacks. AI can be attacked with multiple opportunities through the whole AI development process to the deployment of AI technology. What is more concerning is the lack of existing literature which considers AAI in the risk assessments and security of MAS, all whilst we are seeing increasing examples of adversarial AI [11, 12].

In this paper, the authors consider the threat of AI over a multi-domain literature analysis to parameterise AAI specific tests designed to strengthen MAS development. First, we consider the threats of today and the future that AI in the maritime environment may face. Given that AI for both land-based and air-based operations overlap but are not identical, this indicates that the challenges brought on by a marine environment (e.g., water distortion and reflections) will also have its own unique subset of challenges. By examining each class of MAS AI for vulnerabilities to AAI during its lifecycle, the authors are able to theorise a set of test cases. These can then be used to increase safe, reliable, and trustworthy AI solutions for maritime operations. Finally, we propose best practices to secure MAS within maritime environments. Ultimately, this research aims to highlight the fast-approaching dangers of ubiquitous AI / automation in MAS and motivate the inclusion of AAI in MAS risk assessment to mitigate against the dangers to all MAS stakeholders.

This paper offers the following novel contributions:

(1) A start-to-end lifecycle evaluation of AAI threats against maritime autonomous systems in situ;

(2) A comprehensive review and evaluation of AI threats to MAS considering the AAI and MAS literature;
(3) Case examples to support high-fidelity real-world tests over laboratory environments for AAI;
(4) Principles to better secure MAS against AAI and ways to enhance MAS AI security across its lifecycle.

The paper is structured as follows: Section 2 critically reviews existing literature across multiple domains to understand the current state of the art. We consider AI in autonomous maritime systems and then examine the threat of adversarial AI in that context. Section 3 considers the types of AI used in MAS and the existing threats to these systems. Section 4 presents an evaluation and analysis of general adversarial principles to MAS. Finally, we conclude and provide further work in Section 5.

2. BACKGROUND

2.1. AI in maritime autonomous systems

2.1.1. Sensors and instruments

Shipping is a crucial part of global trade, accounting for nearly 90% of international trade [13]. Waterborne vessels are also critical for human transport, naval defence, and scientific exploration and monitoring of the seas and inland bodies of water. The successful automation of shipping and other maritime operations and services could bring significant advantages over contemporary vessels. Many of these advantages include reducing costs and increasing safety. For example, having no crew aboard ameliorates human factor errors, safety from dangerous working conditions and adverse weather, captivity/attacks from pirates or criminals, more socially supportive conditions and even a reduction in the transmission of some pathogens. Other advantages include cost savings via not having to employ crew, more storage capacity, cheaper development of vessels (crew facilities and living spaces not required), and more economical and environmentally friendly vessels [14]. Some of these benefits, especially crew physical safety, can be obtained, to a reduced degree, with remote unmanned vessels. However, higher tiers of autonomy are needed to maximise these benefits. Furthermore, remote systems would be susceptible to many cyber security attacks, such as jamming and hacking the remote communication [15].

Maritime autonomy can be categorised into different levels, similar to the way autonomous cars and advanced air mobility (AAM), i.e., drones and aircraft, are defined. For example, the Maritime Safety Committee (MSC) of the International Maritime Organisation (IMO) categorises autonomy into four levels. Level 1 pertains to vessels with autonomous components which support the vessel's crew. Level 2 vessels also have crew aboard to support operations, but a remote control centre operates the vessel. Level 3 vessels are remotely controlled and unmanned.

Level 4 vessels are unmanned and fully autonomous. In this work, we will consider only level 4 systems. Other organisations also provide different autonomous level systems, e.g., [8].

As the work of [16] suggests, autonomous systems consist of perception and control elements. Perception elements can be considered the sensors or systems that collect information to be used by the control elements that control the vessel's actions. Contemporary vessels usually have a number of sensors on board to support crew decisions, therefore, leading to an existing framework for autonomous systems (although some sensors may require adaptation in MAS). These sensors and systems include; RADAR (radio detection and ranging) to find, usually large, objects with radio waves. The velocity of the object can be determined with doppler RADARs; object detection can be done with other sectors of the electromagnetic spectrum, such as light detection and ranging (LiDAR), which uses infrared light from lasers - these small wavelengths can detect smaller objects and more accurately detect features but at shorter ranges. Echosounders can be used in a similar way to RADAR and LiDAR; however, echosounders use sound pulses to detect underwater objects, such as the depth of the water. Echosounders can be forwards (or laterally) looking as well as vertical to assist in collision avoidance. Multibeam echosounders can give a 3D point cloud which can be geolocated with millimetric accuracy using RTK GNSS. Measuring echo return backscatter can give useful data about detected objects that go beyond purely the range and baring, giving details about the nature of the detected surface. CCTV/IR/multispectral cameras can be used to detect close-range objects, such as coastal landmarks or objects in the water, akin to LiDAR; furthermore, multiple cameras can be used to triangulate and detect the range of objects; objects can also be captured in colour at high resolution. An array of microphones can be used to detect audio cues on a vessel but may be disrupted by a lot of audio noise, such as the sound of waves, wind and other vessels. Directional microphone arrays are now available that can indicate the range and bearing of remote sounds. These will be essential in the future for autonomous vessel to perceive the direction of sound signals. Automatic identification system (AIS) uses very high frequency (VHF) radio to transmit and receive vessel locations and vessel data. Global navigation satellite systems (GNSS) (such as GPS or Galileo) can support dynamic positioning (DP) systems and location services. Electronic chart display and information system (ECDIS) renders charting information. Vessels can contain weather sensors (barometer, temperature, wind speed etc.). Vessels often contain systems for broadband and 3G/4G/5G, as well as VHF for communication. Cargo supervision systems often host an array of sensors such as internal temperature, humidity, smoke detectors. When considering shipping, sensors for monitoring cargo (e.g., food, gas, oil, passengers) are often specific to the specific maintenance needs of that cargo. Vessels often also contain fault diagnosis and voyage data recorder (VDR) systems that store sensor data for post-incident analyses.

Vessels may also contain specialist sensors unique to the vessel type, which determines its size (e.g., gross tonnage), area of operation (e.g., Arctic), and cargo type (e.g., fertilizer). Vessels could also use AI in airborne, surface or subsurface drones to extend sensor range. Studies also show the benefits of multi-sensor perception systems [16], which increase the accuracy of the available data by using multi-systems, e.g., a small object may not be detected by RADAR but instead detected by the camera through cross-validation of system data. Multiple sensors can also support an element of redundancy. Additionally, the challenges of detecting objects in their natural environment, for example, both on the surface of the water and subsurface, yields difficulties. The water itself can cause confusing distortions but also presents an arduous environment for the physical devices (e.g., salt corrosion).

2.1.2. Artificial intelligence

In fully autonomous maritime systems, AI is used to support, supplement, or replace crews in automating the operations of the vessel. The AI takes sensor data as input and makes decisions to automate the vessel's processes. Different AI types are required for different systems because of the range of tasks required by a fully autonomous marine system. The sensors previously discussed can be used as input features to support AI systems to safely navigate the ocean and maintain the functionality of the vessel. However, there exists an overlap in technology, e.g., in order to avoid objects, one requires a degree of situational awareness. We next consider AI for autonomous systems, as categorised in [8], which consists of several AI technologies connected to a DP system that controls the vessel. These technologies include:

- Situational awareness (SA) - the SA component is required to determine the vessel's real-time location and the vessel's environment (for example, the detection and range of objects). SA modules may also use natural language processing (NLP) to interpret incoming communications. This AI system can use a number of different types of methods and algorithms, such as convolutional neural networks (CNN), region proposal networks (RPN) and sensors to detect landmark objects and navigational cues such as coastal features or buoys [17]. In addition, the AI system could be supported by other non-AI systems, such as GNSS, to cross-validate the vessel's location.
- Collision avoidance-uses SA information and prevents the vessel from colliding with objects. These systems use computer vision object recognition to detect objects (SA module output) and feed into the local autonomous route planning modules to change the vessel's trajectory to avoid a collision. Some AI technologies used are CNNs [18] to locate objects and support vector machines (SVMs) which have previously been used to output a new trajectory to prevent collisions [19].
- Global autonomous optimal planning modules-ensure the vessel's movement along the optimal route; an optimal global route may depend on many objectives such as the quickest route, most fuel-efficient, economical and safest route (e.g., weather, global tensions, piracy). The common algorithms utilised are evolutionary algorithms (EAs) which can evolve optimal high-dimensional solutions, particle swarm optimisation (PSO) and ant colony optimisation

(ACO), which use the emergence properties of nature to find optimal solutions [20-22].

The vessel may also include AI to support specialist tasks such as auto berthing/mooring and engine condition maintenance which assist with the general functionality of the vessel. Other examples include Gaussian processes, neural networks, Bayesian modelling, and active learning that can be used for anomaly detection in autonomous vessels to detect deviations and unexpected events [17]

2.2. Adversarial AI

The advancement and ever-increasing size of neural networks increase the complexity of applications supported by AI. However, as the complexity of the model increases, explainability and hence the interpretability of the model decrease [23]. The lack of explainability for complex models, combined with high-dimensional phenomena and poor security principles, can give rise to adversarial AI. The work of [24] was one of the earliest to recognise that neural networks yield properties that can be vulnerable to adversarial attacks, and a 2023 survey paper identified 32 offensive AI capabilities [25]. Governments globally have begun to recognise the threat; notably, the 2021 U.S. National Security Commission on AI stated, "the U.S. government is not prepared to defend the United States in the coming artificial intelligence (AI) era". Many other countries are preparing for an Adversarial AI (AAI) wave by developing frameworks which attempt to secure AI systems [26, 27]. Furthermore, academic authors [11] have highlighted that "the number of adversarial attacks will continue to increase in the future as the economic benefits trend". As of now, adversarial AI has been demonstrated in a number of applications to support social engineering/spear phishing [28], biometric spoofing [29], computer vision object recognition [30-32], malware development avoiding network detection [33-36], NLP [37, 38], and attacks on cloud APIs [39] to name a few.

We recognise AI can be, and has been, used as an attacking tool, e.g., the automation of conventional cyber attacks, side-channel analysis, creation of deepfake media, OSINT collection and analysis. However, these more active AI threats are outside the scope of this paper. Instead, we focus on the inherent vulnerabilities within AI systems processes (in particular, threats to maritime autonomous systems), and how AAI tests can reveal those vulnerabilities to the developer. The primary adversarial goals of AAI are to attack the confidentiality, integrity and accessibility (CIA) triad for ML processes of AI systems.

(1) Confidentiality-sensitive data can be used during the training phase of the model and has been shown to be extractable from the model [39, 40]. This is of particular concern to a model which uses sensitive data (such as governments) and personal data (privacy concern). Furthermore, data is one of the most valuable resources in modern times [41] and developing AI can be a long and expensive process which could be bypassed with large financial gain by stealing the intellectual property (IP) of the model. As MAS is a new area of growth globally, competition is significantly high.

(2) Integrity-often involving the attacker aiming to get the AI system to misclassify an input to a specific target or any other false target, usually, so the system carries out an intended adversarial action such as allowing malicious traffic to pass through a network AI-based IDS [42]. This is a concern with physical object evasion in mission-critical AI, such as naval mine detection, for which an attack could damage the integrity of the AI.

(3) Accessibility-this adversarial goal is similar to denial of service type attacks where the attacker usually intends to cause high numbers of misclassification to deem the AI inoperable or cause a serious misclassification such as changing the interpretation of a perturbed stop sign; this should prevent the use and access of the AI [43].

These terms can overlap with the risk and threat commensurate with the application of the system, e.g., attacks on mission-critical AI posing the greatest threat. Before we consider the different types of existing attacks on AI, we introduce some key terms, namely, closed-box algorithms and open-box algorithms. We note particular confusion with the term black-box algorithm-in the general AI literature, black-box often refers to the poor interpretability of AI models-that is, given the model architecture, hyperparameters and even raw weight values, the combination of interpreting billions of weight values, makes the algorithm difficult to interpret the inner workings (how a prediction is being made by some instance passed through the model). However, the use of the term black-box in the adversarial AI literature appears to take a slightly different meaning which is that the attackers of the AI systems do not know any of the model's architecture, hyperparameters and weights but merely have access to only a model API input and the final result. In this and future sections, we instead use the term closed-box AI which will refer to the attacker only having access to the model inputs and outputs. In contrast, open-box will refer to having access to the inputs, outputs and also all the model's parameters and architecture. AAI survey papers use a range of nomenclature to classify attacks.

We now consider some of the most prominent adversarial AI methods disclosed in the AAI literature. We note that a large proportion of these methods are relevant to computer vision, and this is considered as one of the primary AI concerns in the near future [25]. Adversarial attacks are not limited to the post-deployment stage and can be attacked in earlier development stages of the ML pipeline. We categorise this literature into attacks that are performed in pre-deployment and -post-deployment stages.

In the pre-deployment stage, an attacker is concerned with altering the development of the AI model, known as poisoning attacks. These attacks can have a lasting effect on the model through the rest of the model's lifespan. Some key pre-deployment attacks include manipulating training data; this can be done by poisoning the training dataset. The motives for this attack could be the misclassification for evasion and misclassification to lower the classification rate. This can be done by changing the distribution of the training dataset by modifying feature values or injecting new training samples [44], as well as changing the training labels [45].

The post-deployment methods are more concerned with evasion and inversion attacks. Model extraction and model inversion attacks aim to acquire information about the model. Given the time-consuming and expensive training to collect data, preprocess and train a model, the information could include sensitive data points or information about the model's architecture to steal proprietary information or sensitive data [46, 47]. Furthermore, if one is able to recreate an accurate surrogate model, then one could use the model to create adversarial examples in an offline environment where one is more stealthy and able to avoid the actions being logged.

The most well-documented AAI literature is the evasion methods post-deployment, so we discuss these in detail.One of the earliest methods was the work of [24], which proposed perturbing samples to obtain a misclassification by the ML model during the deployment stage. The change to the sample was a minimisation optimisation problem which intended to make a minimal change (so that it was undetectable by the human eye) but enough to cause a misclassification. In order to optimise this problem, one needs to know the direction of sensitivity (e.g., positive/negative perturbations, perturbations of which features) and the magnitude to perturb (usually as small as possible). The work introduces the L-BFGS algorithm (the acronym is the author's initials), which was a method to solve the problem formulated as,

$$\min \|\delta X\| : f(X + \delta X) \neq Y \tag{1}$$

where f is the model's cost function, and Y is the true label of the instance X. The work also considers the disputed reasons for adversarial examples. They note that the cause of adversarial samples is the result of linear behaviour in high dimensional space [48]. The authors interestingly note a small perturbation to many features of an instance can be far greater than a larger single feature perturbation (such as the one-pixel attack [49]), which uses differential evolution to evolve solutions which intend to change one single pixel in the image. Although this method was fast to compute, the samples generated were often non-functional. Later, three variants of FGSM were developed; namely, one-step target class, Basic Iterative Method (BIM), and Iterative Least-Likely Class Method (ILCM) [50]. The one-step method changes the Y value in equation (1) from the true label to the desired class label and sets the equation equal to Y. Therefore the algorithm considered perturbing toward a specific class rather than just away from the true class. The BIM method considers equation (1), but instead iterates the algorithm over small step sizes which can produce numerous adversarial examples. Finally, ILCM also considers an iterative version of equation (1) but considers perturbing toward the class with the lowest recognition probability. DeepFool [51] is another tool which uses an iterative process to generate adversarial examples to create samples of an image which eventually crosses the class decision boundary.

Instead of gradient descent methods, Jacobian Based Saliency Maps (JSMA) [52] consider the Jacobian of a function matrix (forward derivative), i.e., how does a feature (pixel) change affect the change of a class probability? The saliency map is commonly used in the XAI literature to detect which pixels are making the most significant contributions to the model's prediction and hence which input features should be perturbed for a desired effect on model output—this map is usually utilised to craft adversarial samples or can be superimposed over the original image as a heatmap. JSMA allows for source-to-target class adversarial sample creation. Using the Jacobian of the model can allow one to determine the sensitivity of a model for specific inputs, i.e., greater sensitivity means perturbations have larger effects on the model's prediction of a class. To use JSMA, one must calculate the forward derivative matrix, also denoted the Jacobian j of the learned function F. This is defined in the original work as:

$$J_{\mathbf{F}}(\mathbf{X}) = \left[\frac{\partial \mathbf{F}(\mathbf{X})}{\partial x_1}, \frac{\partial \mathbf{F}(\mathbf{X})}{\partial x_2}\right], \tag{2}$$

for inputs x_1, x_2 and $\mathbf{F}(\mathbf{X})$ providing the model output.A salience map $S(\mathbf{X}, t)[i]$ can then be computed with the formulae:

$$S(\mathbf{X}, t)[i] = \begin{cases} 0 \quad \text{if } J_{it}(\mathbf{X}) < 0 \quad \text{or} \quad \sum_{j \neq t} J_{ij}(\mathbf{X}) > 0 \\ J_{it}(\mathbf{X}) \left| \sum_{j \neq t} J_{ij}(\mathbf{X}) \right| \quad \text{otherwise.} \end{cases} \tag{3}$$

This limits the Jacobian to be positive (positive effect of the pixel on classification) to decide if input feature i should be perturbed for an adversarial effect on the model. The work showed not all areas of the search space are equally difficult for crafting adversarial samples and that certain source-to-target class pairs are easier to craft than others. For this attack, only the model's output and inputs (closed-box) are required to calculate the Jacobian and create a saliency map. Other attacks include using EAs and PSO to optimise the problem [53] and generative adversarial networks GANs, such as AdvGAN [54], which is used to generate adversarial examples. Furthermore, there exists a range of open-source tools to enact these perturbation attacks.

In this work, we also consider patch-based attacks [55] which are a prevalent evasion attack within the literature.These attacks involve generating a highly salient digital or physical patch that can be applied to the image or physical environment to avoid object detection models [56-60] and classification models [55] by being more salient than other objects in the image and refocusing the attention of the model to the patch [55]. Cyber-physical patch attacks are often formulated as an optimisation problem akin to perturbation attacks. The problem can be formulated as [57],

$$\arg\max_{P} \mathbb{E}_{x,t,l}[\log \Pr(\hat{y} \mid A(P, x, l, t))] \tag{4}$$

where we aim to generate a patch P where $A(P, x, l, t)$ is the input taking a transformation function A which applies the patch using the original image x, location

l and rotation/scaling transformation t. We aim to maximise the loss function of the probability of classifying the input A to true classification label $\hat{y}$. The resulting optimisation formulates an adversarial patch which is superimposed onto the original image and inputted into the model. Common patch-based attacks include DPatch, which is a closed-box adversarial patch attack for object detection models, and the work of [61] considers dynamic patches (video), to name a few.

3. ADVERSARIAL AI IN MARITIME AUTONOMOUS SYSTEMS

We now consider evaluating the threat to ML systems utilised to operate autonomous systems in the maritime environment. Most of the adversarial attacks have been evaluated only in a limited laboratory environment; we aim to evaluate these attacks in the real world where the effects are unknown, yet have the greatest potential for impact. This is highlighted as a primary challenge of adversarial AI by adversarial AI authors, including in the literature survey of [62] "[the] need to verify the attack effect in real physical scenarios" and "the current defence technology research lacks the practice in the real world". We demonstrate these attacks in the MAS environment and provide the results and analysis to highlight the effects and practicability of these attacks in the real world. Where appropriate, we visually show some of these attacks in this work. While these threats consider adversarial attacks on AI, conventional cyber security attacks are just as pertinent. Furthermore, some AAI attacks require one to employ AAI and conventional cyber security attacks in unison. Also, conventional security, such as unpatched software and the jamming/spoofing of sensors, can affect both conventional cyber security and AAI-based security. The focus of this work only considers the nascent domain of AAI in MAS, which comprises very limited literature. Notably, in the literature review to date, we found a single publication [63] considering a few theoretical AAI attack possibilities against MAS.

In this publication, we will use Microsoft's failure modes in the ML framework [64] to comprehensively evaluate the type of threats to MAS and provide context to the maritime environment. Microsoft's failure modes in the ML framework categorise AAI attacks into a possible 11 classifications; we list these below. We provide several proof-of-concepts with this list. These are not intended to be an exhaustive set but to demonstrate the usefulness and feasibility of MAS AAI.

Class 1: Model inversion-Even if one is able to secure and protect the knowledge the ML uses to make an output, such as the features used during prediction, one may be able to query the model to determine the model's prerequisite features in a model inversion attack. Whilst this does not threaten the model's functionality, it could be used in the form of reconnaissance to support a future attack. Therefore, the attack is an abuse of the confidentiality of the system.

Class 2: Perturbation attack-In a perturbation attack, the attacker crafts a query which is submitted to the ML algorithm and the ML algorithm actions the attacker's desired response. For example, an attacker could evolve an adversarial data example with an EA, possibly in some underrepresented area/tail of the probability occurrence distribution or a near boundary instance, which causes the collision

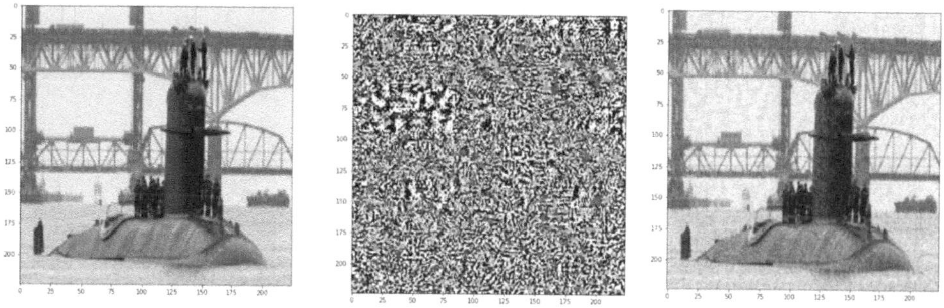

(a) Original Input (b) Perturb Map (c) Adversarial Sample

Figure 1. *Adversarial perturbation attack sample was generated for a pre-trained MobileNet* V_2 *image classification model. The input sample, a submarine image, is predicted as a submarine by the model, providing a confidence value of* 80.15%. *The adversarial sample is predicted as a llama by the model, providing a confidence value of* 10.74%. *The FGSM attack with* $\epsilon = 0.1$ *was set.*

(a) Original Input (b) Adversarial Image

Figure 2. *Adversarial perturbation attack sample was generated using real-world data on a pre-trained FasterRCNN object detection model. The input sample, a warship, is predicted as a vessel by the model, providing a confidence value of* 98.87%. *The adversarial sample is predicted as multiple incorrect objects by the model with the greatest prediction of person, providing a confidence value of* 99.33%. *The projected gradient descent attack with the maximum allowable pixel-wise difference between the original image and the adversarial image set at 32 . Both the FasterRCNN and YOLOV 3 model used in this work was trained on the COCO dataset labelling all ships, marine vessels and boats under the classification of 'boat'.*

avoidance system to output a false negative in a busy port and not make an appropriate collision avoidance maneuver. This threat has high consequences; it could be relatively simple to achieve the attack but requires access to modify (or create adversarial inputs and block the legitimate traffic) the input to the ML model-conventional cyber attacks could be leveraged to support this. An example of this attack can be seen in figure 1 and figure 2, it should be noted that the accuracy of the adversarial sample appears correlated with the quality of the original image. Therefore using low-resolution cameras may be more susceptible to attack as well as reducing the model classification accuracy.

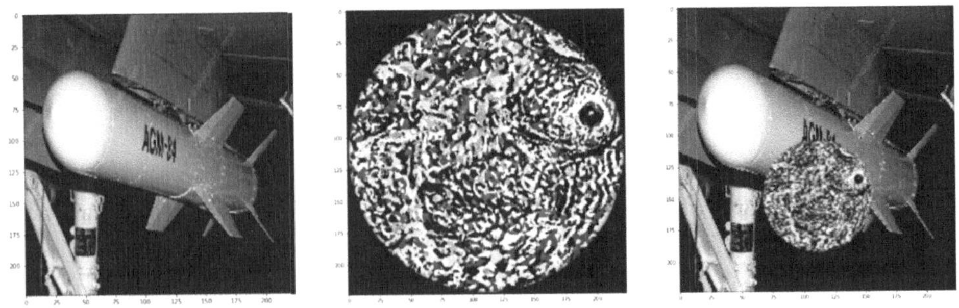

(a) Original Input (b) Patch (c) Patched Input

Figure 3. *Adversarial patch attack sample generated to attack the common pre-trained ResNet5o image classification model. The input sample, an image of an anti-shipping missile, is predicted as a missile by the model, providing a confidence value of* 61.12%. *An adversarial patch (b) is provided physically (e.g., as a sticker) or to the input image to change the classification prediction to a spotlight by the model, providing a confidence value of* 20.74%. *The patch attack [55] method was used. The patch size and shape can be optimised to increase the model classification (although out of scope for this research paper).*

Class 3: Membership inference-The attacker may be able to infer whether a data instance is a constituent of the training data used to train a model, potentially a breach of privacy.

Class 4: Model stealing-Through querying the model, the attacker may be able to determine information about the model parameters and architecture. With this information, the attack could recreate the model and essentially steal the model/IP. This could save the attacker time and money having to develop the model themselves; this also could be used to turn a black-box model into a white-box model for use with other attack methods. An adversary who steals a MAS model could recreate the model and perform offline attacks (non-logged events) for greater stealth and create more efficient and accurate attacks before applying it to the real online model.

Class 5: Adversarial example in the physical domain-An adversarial example in the physical domain is akin to a perturbation attack. It considers modifying physical properties; for example, an attacker could spoof certain sensor inputs to confuse the MAS vessel and cause a change in the vessel's trajectory or one could paint a signature on a hostile ship that the searching MAS CNN recognises as a benign object. Example patch attacks can be seen in figures 3 and 4 .

Class 6: Malicious ML provider recovering training data-Akin to a membership inference attack, the attacker may be able to infer the training data used to train a model. The difference from this attack is that the attacker can use queries to derive training data which could potentially be a breach of privacy. The data could contain sensitive information, breach confidentiality, and support model/IP theft.

Class 7: Attacking the ML supply chain-In this attack, the attacker could interfere with elements of the ML lifecycle. For example, capturing training/testing data and retraining new models can be resource-heavy (time and cost); therefore, engineers optimise time by reusing models (transfer learning) and existing datasets.

(a) Original Input (b) Patch (c) Patched Input

Figure 4. *A real-world adversarial patch attack sample [56] generated to attack the common pre-trained YOLOv3 object detection model. The input sample, an image of two vessels, is predicted as a vessel by the model, providing a confidence value of 99% and 92%. An adversarial patch (b) is provided digitally or physically (e.g., as a sticker) to change the detection prediction to a zebra by the model. Notice the patch can interfere with the prediction and hide nearby objects, such as the second 'boat', i.e. vessel, on its right in the photo. The patch does not need to be significantly large but can be closer to the camera than the evading objects to produce the desired effect.*

This creates a vector for attackers to manipulate data and models. For example, a model intended to be shared and reused for developing navigational AI could have neurons injected, which cause the vessel to change course given specific spoofed input signals.

Class 8: Backdoor ML-One could create a backdoor utilising the innate poor interpretability of an extensive neural network. For example, one could inject specific neurons or alter existing neuron weights to minimise the noise in the object detection CNN model but render a backdoor so that, given a specific input (a hostile vessel), a model predicts a desired output (misclassify). This could damage both the integrity and confidentiality of the ML model.

Class 9: Exploit software dependencies-This considers the conventional attack surface of software more generally. This could require an attacker to corrupt ML libraries or exploit buffer overflow attacks in the developing software (e.g., labelling application).

Class 10: Reprogramming the ML system-In this attack, the attacker takes the existing ML model and uses it to perform a nefarious task.

Class 11: Poisoning attack-Poisoning attacks involve manipulating the training data of the ML model. One could manipulate by injecting new values, new samples, or modifying the feature values or/and labels of the training data. This could be executed to reduce the integrity and availability of the ML model. For example, changing the distribution of the training data or injecting chaff data creates a high misclassification rate and hence reduces the integrity of the system; this could render a denial of service type attack. A MAS-related example could be that for a MAS search vessel, the images or acoustic signals of a certain hostile ship are incorrectly classified. This attack requires access to the training phase of ML development and so is more difficult to achieve than other attacks. It is also more likely that an attack to cause large misclassification for many classes would be noticed during the testing phase of the model's development. An example of AIS poisoning can be seen in Table 1 .

Table 1. *A sample of poisoned AIS data. The Maritime Mobile Service Identity (MMSI) number was altered (the last two values were modified), and the vessel velocity was modified using the velocity standard deviations. The GPS coordinates can be replaced with other vessels' coordinates. Poisoned data can be used to poison the model during training or spoof existing situational awareness AI.*

Time stamp	MMSI	Latitude	Longitude	Speed	IMO	Name	Destination
05/01/2021 01:34	209504011	27.09833	−79.88783	15.15792	9517411	CONTSHIP ICE	USMIA
05/01/2021 01:50	209504011	27.05333	−79.88583	15.15792	9517411	CONTSHIP ICE	USMIA
05/01/2021 02:10	209504011	26.99783	−79.88383	15.15792	9517411	CONTSHIP ICE	USMIA
05/01/2021 06:04	209504011	26.33928	−79.92229	14.75792	9517411	CONTSHIP ICE	USMIA
05/01/2021 06:10	209504011	26.32417	−79.92317	14.65792	9517411	CONTSHIP ICE	USMIA
05/01/2021 06:15	209504011	26.31133	−79.92383	14.85792	9517411	CONTSHIP ICE	USMIA
05/01/2021 06:20	209504011	26.29667	−79.92483	14.65792	9517411	CONTSHIP ICE	USMIA
05/01/2021 06:25	209504011	26.28433	−79.92533	14.75792	9517411	CONTSHIP ICE	USMIA
05/01/2021 06:30	209504011	26.269	−79.926	14.85792	9517411	CONTSHIP ICE	USMIA
05/01/2021 06:35	209504011	26.256	−79.92667	14.85792	9517411	CONTSHIP ICE	USMIA
05/01/2021 06:40	209504011	26.24117	−79.92717	14.85792	9517411	CONTSHIP ICE	USMIA
05/01/2021 06:45	209504011	26.227	−79.92767	14.65792	9517411	CONTSHIP ICE	USMIA
05/01/2021 06:50	209504011	26.21433	−79.92817	14.55792	9517411	CONTSHIP ICE	USMIA
05/01/2021 06:55	209504011	26.19967	−79.92883	14.75792	9517411	CONTSHIP ICE	USMIA
05/01/2021 07:00	209504011	26.09457	−79.99955	14.65792	9517411	CONTSHIP ICE	USMIA

3.1. Experimental setup and findings

In order to test the proof-of-concept adversarial perturbation and patch attacks, we examined and collected the relevant data from Plymouth sound (UK territorial waters). The vessel used four Omega 1080p cameras which can collect video media that can be fed into an object detection computer vision model at either the vessel or remote centre side. The camera array was mounted on a manned vessel, but in a way that it would have the same view of its immediate surroundings as a USV would. Refer to the figure captions for the specific models used and parameters to generate relevant attacks.

The primary findings of the study showed that the lab-developed attack methods worked well in a controlled environment. However, when performing these same attacks in a complex and dynamic environment like the sea, the effectiveness of the attacks varied more significantly. Just as the type of AI most effective depends on the environment and application of the model, so does the type of attack. The quality of the onboard cameras made object detection and hence evasion more difficult. The object's range from the camera influences the model's effectiveness and evasion. At sea, water distortion, water on the lens, and vibrations of the vessel also added to this effect. Lighting was another important variable where the position of the natural light could cause difficulties. One can observe in figure 2*a* and figure 4 different effects of light taken within an hour window on both the model accuracy and attack accuracy. We also consider the possible application of these attacks, for example, perturbation attacks, which would require the precise distortion of the input for misclassification-this would require access to the model input, which could be challenging in a marine environment. Furthermore, the generation time of the perturbation map for many attacks would generate significant delays to the input image-making it an unlikely vector of attack until more sophisticated and faster methods can be developed. This additional complication means one can not be certain of the effects and limitations of the AAI without evaluating the AI in the natural environment and application of the AI.

Other attacks, such as the patch attack seemed far more likely to be used in a real-world attack. For example, the patch could be physically placed on or near an object and surrounding objects (even objects not covered by the patch) would appear hidden to the model. This is potentially a way for attackers to evade and hide from object detection models. The patches also have a degree of transferability between models. However, from experimentation, the size and placement (camera-angle-distance relationship) alter the effect of the patch attack. The strength of the patch detection and distance from other objects also affect this hiding/evasion property of nearby objects. Patch attack of an image classifier is easier to achieve than of an object detection model but less practical as MAS are more likely to use object detection (for detecting multiple objects in frame) than image classification. For many of these reasons, we strongly advocate the testing of these methods and future novel methods in a more dynamic environment to gauge the real-world impacts of such attacks.

3.2. AI lifecycle

The ML development process has a whole lifecycle, denoted MLOps (analogous to DevOps shorting of development operations), commonly understood as defining the problem, data acquisition, data preprocessing, model training, model testing and post-deployment ops. Different attacks can occur at different stages in the development of the model, so each stage and transfer between stages are vulnerable. One could corrupt the environment during the early stages of data collection, manipulate the data preprocessing functions, exploit the hardware (particularly GPUs and CPUs) during training, obfuscate backdoors in transfer learning models and craft adversarial samples during deployment. In figure 5 , we illustrate the process of the ML pipeline and the possible known types of attacks discussed previously.

The attacks are considered at the various common stages of ML development; however, it is possible for other attacks to be performed in the interim stages of the ML cycle and for vulnerabilities to exist during the ML stage transition. For example, a data poisoning attack could occur when recalled by the training software immediately before model training. We can also see from the diagram that the majority of attacks are focused on the deployment stage of the model's lifecycle.

4. AI SECURITY PRINCIPLES FOR MAS

This section considers existing AAI attacks in maritime autonomous systems (Section 3) to create principles to secure maritime autonomous systems. Based on the eleven attack categories, we propose seven secure MAS principles, each with the objective of mitigating the respective AAI attack threat.

There is no one size fits all method of adversarial defence and risk assessments should be considered at the beginning of the ML development process to determine

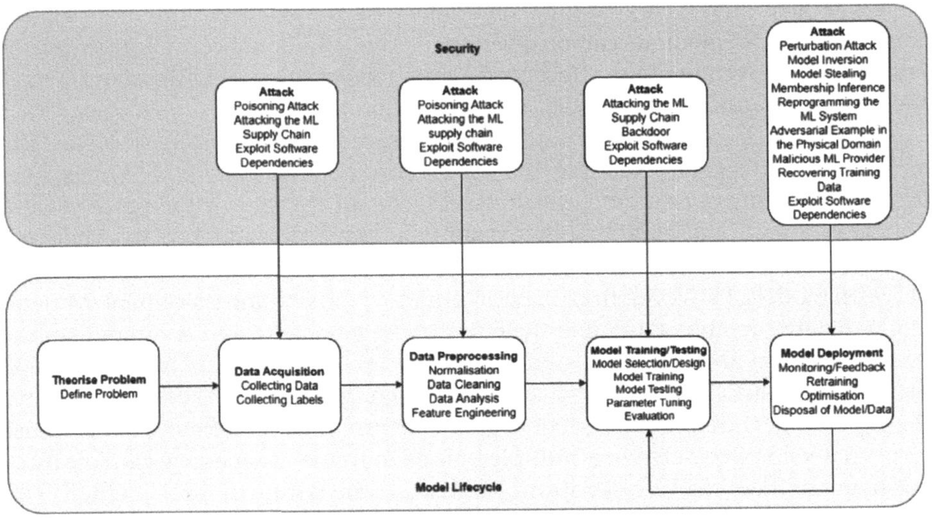

Figure 5. *The machine learning lifecycle and adversarial AI attack types*

the types of risks, threats, data and the use case of the AI system. It should also be noted that whilst following the suggested principles for maritime autonomous systems provides a degree of security, this will not completely secure the systems from all possible attacks; one reason for this is that a model would need to produce a safe mapping from output to all possible inputs which is an NP-hard problem. Furthermore, the adversarial AI threat is a fast-evolving landscape, and it is likely the case that novel threats will be detected over the coming years. These AAI principles have been generated by the authors for MAS AI based on the findings of this work and aim to reduce the real threat surface this industry faces. These principles for secure AI in maritime autonomous systems are as follows:

(1) **Enforce strong conventional cyber security principles**—In addition to strong adversarial AI defences, conventional security methods can complement AI security. This is why the first principle is to create strong conventional cybersecurity. At this early stage in AI development, attacks utilising poor conventional security practices (e.g., unpatched ML libraries) are the most likely vectors of attack. Good countermeasures include blocking bad internet protocol addresses, using CAPTCHA before inputs can be made and throttling/limiting queries to the model. Log inputs and events. Ensure ML libraries and systems are patched and up to date. Limit user access to the model and implement least privilege authority practices. Secure acquisition/storage/transport/decommissioning of model and data can prevent transfer learning/data/model poisoning-based attacks. These strong conventional security measures can protect against Reprogramming ML, attacking the supply chain, and exploiting software dependencies threats.

(2) **Develop risk assessment/security assessment before starting ML projects**—Whilst this is a property of enforcing strong conventional cyber security, it was too important not to have its own principle. Consider the application of the ML, the types of ML and their associated vulnerabilities and the generation process to develop a risk assessment. Consider who, why and how an attack might benefit from attacking the application. Mission-critical and security-sensitive AI would likely require a more secure approach to AI development. Furthermore, some applications may further increase risk by utilising processes such as continual learning, which prove additional attack vectors for attackers [27]. A real-world example of this attack was the Microsoft Tay Twitter-trained bot in the first few hours of deployment [65]. Moreover, whilst convenient, the reuse of models (transfer learning / AI repurposing) and data increases the opportunities for exploitation.

(3) **Maximise the model's robustness**—Maximising model robustness reduces the attacking space available to prevent perturbation and adversarial examples. The exploitation of model robustness is one of the simplest of attacks to implement given the scalability of the attack (many models are not robust against all possible adversarial space). It is also possible to use the attack to cause a failure (sometimes not so obvious) of mission-critical systems. Maximising the model's robustness should also provide the additional benefits of protecting against accidental adversarial attacks / errors. Furthermore, the work of [27] also considers forcing model architecture and capacity proportionate to training data to improve robustness and reduce unnecessary feature space whilst covering the distribution of training data.

(4) **Maximise explainability and insight for trusted developers and minimise for untrusted users**—Explainability should play a significant role in supporting the development of adversarial AI defences in the coming years. Having greater explainability provides many benefits, but in the context of security, having a better understanding of the ML system's decision processes can support; locating poisoned models, the system limitations, transferability, robustness, and trustworthiness to enhance the security of the system. However, this knowledge could also be used to find and exploit weaknesses, such as locating adversarial space. Therefore one should limit the explainability outputted by the model to untrusted users as well as the technical details of the model, e.g., parameter values and model architecture.

(5) **Regulate the input and output of the model**—This principle ties in with revealing too much information about the model, which could be used for nefarious activities against the model, e.g., one could avoid revealing exact probabilities of detection for a classification model to prevent some gradient-based attacks. Regulating the input of the model can prevent adversarial queries from successfully triggering backdoors or exploiting the model [27].

(6) **Recognise the exploitation of the model and understand the risks of exploitation**—Having indicators of compromise for the model will not stop

adversarial attacks from happening but could allow one to identify and isolate threats. Understanding the effects of a compromised system will allow one to understand the risks and develop effective tailored security approaches in depth.

(7) **Sensor redundancy/harmonisation and data correlation**—Utilising multiple fused sensor inputs can be used to bring assurance to situational awareness modules. For example, relying exclusively on a single sensor to deduce the presence or absence of objects is likely to increase the ease of attack. However, sensor fusion of the CV, LiDAR, forward-looking sonar, AIS, and RADAR, all feeding into a navigational AI system, would increase the overall robustness of the system. The requirement of fooling multiple sensors would be exponentially more challenging to mount a successful attack than that of a single sensor, and anomalous behaviour becomes more apparent if not all sensors are fooled, which could lead to a lower confidence weighting provided to data which significantly differs / appears adversarial from other sensors' results.

4.1. Countermeasures

This section details possible countermeasures which could be used to implement the six principles proposed above to protect MAS AI against AAI; these are summarised in Table 2. We first consider adversarial training. Adversarial training requires synthesising adversarial samples from a model and using those adversarial samples as training data to train the model to produce an element of model robustness against adversarial attacks [24, 48]. The samples can be iteratively generated by retraining the model and regenerating adversarial training data [66]. By creating new samples, we increase the distribution of the dataset for the robustness and accuracy of the model. The first instance of adversarial training was [24], which used FSGM to create and then inject samples into the training data. Many variations and improvements of adversarial training exist, such as using GANs [67-69]. DNN verification tools can be used to locate adversarial samples [70]. It is worth noting to search all the sample space is an NP-hard problem. The three-step null label method blocks the transferability of attacks between models. The method adds a new null label to a one-hot encoding, and the network is trained with some adversarial samples which are labelled as null, therefore if the model input is classified highly as null, this indicates an adversarial input [71] and reduces adversarial attacks happening between models.

Further countermeasures include regularisation. Regularisation is used in ML to prevent overfitting of a model during training (adding a penalty, i.e. regular term, to a cost function); this can reduce the possible adversarial attack space by making the model more robust to small perturbations. Methods of regularisation include feature pruning to prune activations and neurons from a network [72]. Neuron dropout can be used during model training to stochastically remove neurons which can prevent overfitting on small datasets and potentially remove a backdoor [73]. Other methods include adding a layer of random noise to the model after the input layer so that during forward propagation, the noise creates slightly different outcomes to make the model more robust against small permutations [74].

Table 2. *The associated defensive countermeasures to prevent exploitation from each adversarial attack. There is much overlap between the defences resulting in the effect of the sum of multiple defensive measures being greater than its individual constituents.*

AAI Attack	Defences
Perturbation Attack	Adversarial training, Regularisation, Ensemble methods, Input validation and manipulation/preprocessing, Gradient masking, Model distillation, Adversarial sample detection, Explainability
Poisoning Attack	Regularisation, Ensemble methods, Input validation and manipulation/preprocessing, Explainability
Model Inversion	Input validation and manipulation/preprocessing, Adversarial sample detection, Explainability, Preventing information loss
Membership Inference	Input validation and manipulation/preprocessing, Adversarial sample detection, Explainability, Preventing information loss
Model stealing	Preventing information loss
Reprogramming the ML	Regularisation, Ensemble methods, Gradient masking, Model distillation, Explainability, Preventing information loss
Adversarial Example in Physical Domain	Adversarial training, Regularisation, Ensemble methods, Input validation and manipulation/preprocessing, Gradient masking, Model distillation, Adversarial sample detection, Explainability
ML Provider Recovering Training Data	Regularisation, Ensemble methods, Input validation and manipulation/preprocessing, Gradient masking, Model distillation, Explainability, Preventing information loss
Attacking the ML Supply Chain	Strong conventional cybersecurity practices
Backdoor ML	Regularisation, Ensemble methods, Input validation and manipulation/preprocessing, Model distillation, Explainability, Preventing
Exploit Software Dependencies	Strong conventional cybersecurity practices

Ensemble methods is a term used to represent the combination of multiple ML models constituting an overall model. Common methods include using gradient boosting such as XGBoost [75]. This method can reduce the likelihood of training data poisoning as the individual models are trained on different datasets; therefore, when combining the models, the good models can reduce the effect of the poisoned models [76]. It is also possible that adversarial samples are fewer with a greater distribution of training data.

Input validation (or sanitation) and manipulation/preprocessing, which can control the data going into the model can help prevent attacks. Input reconstruction can be used to remove the adversarial effect from input data analogous to input sanitisations to prevent Structured Query Language (SQL) attacks. Input reconstruction was suggested in the work of [77], which proposed transforms applied to input images before making model predictions (clipping, JPEG compression, rescaling depth, etc.). Feature compression/data compression, as in ComCNN [78], can be used to reduce the feature depth of the input, e.g., reduce the colour depth of pixels and increase robustness at the cost of reduced input accuracy. Inputs could also be filtered, smoothed, and have random noise applied on input to sanitise the data. Regression analysis can be used to locate data outliers during input [79]. Image preprocessing (such as random image padding) can be used to prevent a backdoor attack from being triggered. Input denoising works by attempting to remove noise from the input. Tools include (high-level representation guided denoiser HGD [80]). GANs can be used to clean data by recreating a similar image to the input. MagNet [81] and defence-GAN [82] are tools that can also be used to recreate input images with reduced adversarial noise on a more similar manifold to the benign data.

A countermeasure that would be useful against the perturbation attacks shown in this paper, and others, is gradient masking. Gradient masking can reduce the likelihood of an attacker acquiring the model's gradient and hence reduce against gradient-based adversarial attacks [83]. There is no reduction in the adversarial sample size. However, this attack aims to make it more difficult for open-box probing, by masking the useful gradient, at finding these samples. The effect works by creating less smooth (sharper) boundaries for classification. Gradient masking methods include model distillation and dropout.

Model distillation requires training a smaller, less complex model based on the original complex model. This reduction/compression in model complexity, whilst maintaining similar model accuracy, can help prevent adversarial attacks by creating a model with smoother loss and hence less sensitive to small perturbations [84]. The original output is used as a soft label, and the original label is used as a hard label.

Adversarial sample detection can help protect MAS AI against adversarial AI via input monitoring. Instead of sterilising the input, one could attempt to detect if the input is an adversarial sample before being accepted or rejected by the model. The input can then be determined as adversarial or benign before being decided whether to be entered into the model. For example, these detection models can be created as a binary classifier and determine if the input follows a similar distribution to the training data [85].

Explainability covers a number of terms, such as trustworthiness, causality, transferability, and informativeness which all support the understanding and hence the security of the ML models. Explainability is a hot topic, and many methods exist to support model explainability. Improving AI explainability is critical as this sector develops AI for mission-critical operations.

Preventing information loss considers the threat of model stealing in a few ways. To protect against data stealing, one could use PATE [86], which splits the training data into subsets and trains multiple models on the subsets before the models are combined, and the systems vote on the predicted outcome. Watermarking can also be used to place a unique watermark in the model, which can be evaluated to determine if it was stolen [87].

5. CONCLUSION

This work has provided an evaluation of AI security in maritime autonomous systems. A literature review revealed the potential vulnerabilities in MAS AI that could be exposed through a set of adversarial AI test cases strategically designed to test AI used in MAS operations. However, this study of the current state-of-the-art MAS security has also highlighted the inherent vulnerabilities of only testing adversarial AI in limited laboratory environments. Given the extreme differences in marine environments based on location, weather, and time of day, it is also clear that any AAI dataset must also be evaluated in a real-world environment to be truly useful and cyber-resilient maritime AI. After the evaluation of these results in situ, we developed a series of secure AI in MAS principles which can be used to mitigate these threats across the AI's lifecycle.

In further work, we recognise the limited preparation and understating of AAI in MAS technologies by developers, security professionals, and marine regulators. Therefore, knowledge could be disseminated by the secure AI principles and AAI employee training. We would also consider the evaluation of other attacks and their associated defences in the maritime environment (in a range of conditions); we would then consider the effects of underwater distortion etc. Further, we aim to evaluate a range of real-world AI (existing commercial and military AI systems) against AAI - this will allow one to gauge the secondary effects of the attack too, e.g., if one interferes with the CV object detection, how would that impact the collision avoidance module of a vessel? As well as evaluate the effectiveness of AAI and defence in a complex and dynamic environment. Furthermore, we would like to consider the probability of each attack in a maritime autonomous environment with some attacks more effective and likely than others in the real-world environment. As we see greater accessibility of AI, we are also likely to see an increase in the misuse of AI (e.g., AI to support clandestine / smuggling operations) as well an increase in the exploitation of AI systems. The importance of the security of AI is increasing with its use in mission-critical systems (e.g., we are seeing increasing use of maritime autonomous systems by militaries [88, 89]). Whilst many of these AAI attacks have not been utilised in the real world, as the potential financial gain of these attacks increase and the increased use of AI in mission-critical systems, adversaries will look to exploit these methods and the requirement to prepare for the fast-evolving AAI threat landscape is today.

Acknowledgments

This work was supported by the Turing's Defence and Security programme through a partnership with the UK government in accordance with the framework agreement between GCHQ & The Alan Turing Institute. The authors would also like to thank the University of Plymouth for their use of their autonomous fleet in order to collect real world data.

REFERENCES

1. Felski A, Zwolak K. The ocean-going autonomous ship_challenges and threats. J Mar Sci Eng. 2020; 8(1) : 41.
2. Kretschmann L, Burmeister H-C, Jahn C. Analyzing the economic benefit of unmanned autonomous ships: an exploratory cost-comparison between an autonomous and a conventional bulk carrier. Res Transp Bus Manag. 2017;25: 76-86.
3. Morris D. Worlds first autonomous ship to launch in 2018. Fortune [Internet]. 2017 [cited 2017 Jul 22]. Available fromhttps://fortune.com/2017/07/22/first-autonomous-ship-yara-birkeland/.
4. Munim ZH. Autonomous ships: a review, innovative applications and future maritime business models. Supply Chain Forum: Int J. 2019;20: 266-279.
5. Porathe T, Prison J, Man Y. Situation awareness in remote control centres for unmanned ships. In: Proceedings of Human Factors in Ship Design & Operation, 26-27 February 2014, London, UK. Buckinghamshire, UK: CORE; 2014. 93p.
6. Tsvetkova A, Hellström M. Creating value through autonomous shipping: an ecosystem perspective. Marit Econ Logist. 2022;24: 255-277.
7. Ziajka-Poznańska E, Montewka J. Costs and benefits of autonomous shipping - a literature review. Appl Sci. 2021;11(10):4553.
8. Royce R. Remote and autonomous ships. In: AAWA position paper. Oslo, Norway: DNV; 2016. 9 Anderson M. Bon voyage for the autonomous ship mayflower. IEEE Spectr. 2019;57(1):36-39.
9. Caruana R, Lou Y, Gehrke J, Koch P, Sturm M, Elhadad N. Intelligible models for healthcare: predicting pneumonia risk and hospital 30-day readmission. In: Proceedings of the 21th ACM SIGKDD International Conference on Knowledge Discovery and Data Mining. New York, USA: ACM; 2015. p. 1721-1730.
10. Kong Z, Xue J, Wang Y, Huang L, Niu Z, Li F. A survey on adversarial attack in the age of artificial intelligence. Wirel Commun Mob Comput. 2021;2021: 4907754.
11. Qiu S, Liu Q Zhou S, Wu C. Review of artificial intelligence adversarial attack and defense technologies. Appl Sci. 2019;9(5):909.
12. Kaluza P, Kölzsch A, Gastner MT, Blasius B. The complex network of global cargo ship movements. J R Soc Interface. 2010;7(48):1093-1103.
13. Askari HR, Hossain MN. Towards utilising autonomous ships: a viable advance in industry 4.0. J Int Marit Saf Environ Aff Shipp. 2022;6(1):39-49.

14. Fan C, Wróbel K, Montewka J, Gil M, Wan C, Zhang D. A framework to identify factors influencing navigational risk for maritime autonomous surface ships. Ocean Eng. 2020;202: 107188.

15. Thombre S, Zhao Z, Ramm-Schmidt H, García JMV, Malkamäki T, Nikolskiy S, et al. Sensors and AI techniques for situational awareness in autonomous ships: a review. In: IEEE Transactions on Intelligent Transportation Systems. Piscataway, NJ: IEEE; 2020.

16. Noel A, Shreyanka K, Gowtham K, Satya K. Autonomous ship navigation methods: a review. In: International Conference on Marine Engineering and Technology Oman 2019 (ICMET Oman) [Internet]; 2019 Nov 5-7; Muscat, Oman. Military Technological College Oman; 2019. Available from https://doi.org/10.24868/icmet.oman.2019.028.

17. Bentes C, Frost A, Velotto D, Tings B. Ship-iceberg discrimination with convolutional neural networks in high resolution SAR images. In: Proceedings of EUSAR 2016: 11th European Conference on Synthetic Aperture Radar. Hamburg, Germany: VDE; 2016. p. 1-4.

18. Wang J, Xiao Y, Li T, Chen CLP. A survey of technologies for unmanned merchant ships. IEEE Access. 2020;8: 224461-224486.

19. Kim H, Kim S-H, Jeon M, Kim JH, Song S, Paik K-J. A study on path optimisation method of an unmanned surface vehicle under environmental loads using genetic algorithm. Ocean Eng. 2017;142: 616-624.

20. Song CH. Global path planning method for USV system based on improved ant colony algorithm. In: Applied mechanics and materials. vol. 568, Switzerland: Trans Tech Publications; 2014. p. 785-788.

21. Zhang Y, Gong D-w, Zhang J-h. Robot path planning in uncertain environment using multi-objective particle swarm optimisation. Neurocomputing. 2013;103: 172-185.

22. Arrieta AB, Díaz-Rodríguez N, Del Ser J, Bennetot A, Tabik S, Barbado A, et al. Explainable artificial intelligence (XAI): concepts,.taxonomies, opportunities and challenges toward responsible AI. Inf Fusion . 2020;58: 82-115.

23. Szegedy C, Zaremba W, Sutskever I, Bruna J, Erhan D, Goodfellow I, et al. Intriguing properties of neural networks [Internet]. arXiv; 2013. Available from:https://arxiv.org/abs/1312.6199.

24. Mirsky Y, Demontis A, Kotak J, Shankar R, Gelei D, Yang L, et al. The threat of offensive AI to organizations. Comput Secur. 2022;124: 103006.

25. Caroline B, Christian B, Stephan B, Luis B, Giuseppe D, Damiani E, et al. Securing machine learning algorithms. Athens, Greece: ENISA; 2021.

26. Kate S. Introducing our new machine learning security principles. Boca Raton, FL: CRC Press; 2022.

27. Seymour J, Tully P. Weaponizing data science for social engineering: automated E2E spear phishing on twitter. Black Hat USA. 2016;37: 1-39. 29 Dang H, Liu F, Stehouwer J, Liu X, Jain AK. On the detection of digital face manipulation. In:

Proceedings of the IEEE/CVF Conference on Computer Vision and Pattern Recognition. Seattle, WA, USA: IEEE; 2020. p. 5781-5790.

28. Akhtar N, Mian A. Threat of adversarial attacks on deep learning in computer vision: a survey. IEEE Access. 2018;6: 14410-14430.

29. Elsayed G, Shankar S, Cheung B, Papernot N, Kurakin A, Goodfellow I, et al. Adversarial examples that fool both computer vision and time-limited humans. In: 32nd Conference on Neural Information Processing Systems (NeurIPS 2018), Montréal, Canada. Red Hook, NY: Curran Associates Inc.; 2018. 31 p.

30. Wang Z, She Q Ward TE. Generative adversarial networks in computer vision: a survey and taxonomy. ACM Comput Surv (CSUR). 2021;54(2):1-38.

31. Al-Dujaili A, Huang A, Hemberg E, OReilly U-M. Adversarial deep learning for robust detection of binary encoded malware. In: 2018 IEEE Security and Privacy Workshops (SPW). San Francisco, USA: IEEE; 2018. p. 76 – 82.

32. Kolosnjaji B, Demontis A, Biggio B, Maiorca D, Giacinto G, Eckert C, et al. Adversarial malware binaries: evading deep learning for malware detection in executables. In: 2018 26th European Signal Processing Conference (EUSIPCO). San Francisco, USA: IEEE; 2018. p. 533-537.

33. Li D, Li Q Ye Y, Xu S. Arms race in adversarial malware detection: a survey. ACM Comput Surv (CSUR). 2021; 55(1) : 1 – 35.

34. Maiorca D, Biggio B, Giacinto G. Towards adversarial malware detection: lessons learned from pdf-based attacks. ACM Comput Surv (CSUR). 2019;52(4):1-36.

35. Morris JX, Lifland E, Yoo JY, Grigsby J, Jin D, Qi Y. Textattack: a framework for adversarial attacks, data augmentation, and adversarial training in NLP [Internet]. arXiv; 2020. Available from:https://arxiv.org/abs/2005.05909.

36. Wallace E, Feng S, Kandpal N, Gardner M, Singh S. Universal adversarial triggers for attacking and analyzing NLP [Internet]. arXiv; 2019. Available from:https://arxiv.org/abs/1908.07125.

37. Juuti M, Szyller S, Marchal S, Asokan N. PRADA: protecting against DNN model stealing attacks. In: 2019 IEEE European Symposium on Security and Privacy (EuroS&P). Stockholm, Sweden: IEEE; 2019. p. 512-527.

38. Wang B, Gong NZ. Stealing hyperparameters in machine learning. In: 2018 IEEE Symposium on Security and Privacy (SP). San Francisco, USA: IEEE; 2018. p. 36-52.

39. Kessler J. Data protection in the wake of the GDPR: California's solution for protecting "the world's most valuable resource". South Calif Law Rev. 2019;93: 99.

40. Sewak M, Sahay SK, Rathore H. Adversarialuscator: an adversarial-DRL based obfuscator and metamorphic malware swarm generator. In: 2021 International Joint Conference on Neural Networks (IJCNN). Piscataway, NJ: IEEE; 2021. p. 1-9.

41. Gu T, Dolan-Gavitt B, Garg S. Badnets: identifying vulnerabilities in the machine learning model supply chain [Internet]. arXiv; 2017. Available from:https://arxiv.org/abs/1708.06733.

42. Barreno M, Nelson B, Sears R, Joseph AD, Tygar JD. Can machine learning be secure? In: Proceedings of the 2006 ACM Symposium on Information, Computer and Communications Security. New York, USA: ACM; 2006. p. 16 – 25.

43. Biggio B, Nelson B, Laskov P. Support vector machines under adversarial label noise. In: Asian Conference on Machine Learning. Cambridge, MA: PMLR; 2011. p. 97-112.

44. Fredrikson M, Jha S, Ristenpart T. Model inversion attacks that exploit confidence information and basic countermeasures. In: Proceedings of the 22nd ACM SIGSAC Conference on Computer and Communications Security. New York: ACM; 2015. p. 1322-1333. 47 Orekondy T, Schiele B, Fritz M. Knockoff nets: stealing functionality of black-box models. In: Proceedings of the IEEE/CVF Conference on Computer Vision and Pattern Recognition. Piscataway, NJ: IEEE; 2019.p. 4954-4963.

45. Goodfellow IJ, Shlens J, Szegedy C. Explaining and harnessing adversarial examples [Internet]. arXiv; 2014. Available from:https://arxiv.org/abs/1412.6572.

46. Su J, Vargas DV, Sakurai K. One pixel attack for fooling deep neural networks. IEEE Trans Evol Comput. 2019;23(5) : 828 – 841.

47. Kurakin A, Goodfellow I, Bengio S. Adversarial machine learning at scale [Internet]. arXiv; 2016. Available from:https://arxiv.org/abs/1611.01236.

48. Moosavi-Dezfooli S-M, Fawzi A, Frossard P. Deepfool: a simple and accurate method to fool deep neural networks. In: Proceedings of the IEEE Conference on Computer Vision and Pattern Recognition. Las Vegas, NV, USA: IEEE; 2016. p. 2574-2582.

49. Papernot N, McDaniel P, Jha S, Fredrikson M, Celik ZB, Swami A. The limitations of deep learning in adversarial settings. In: IEEE European Symposium on Security and Privacy (EuroS@P). Piscataway, NJ: IEEE; 2016. p. 372-387.

50. Chen J, Su M, Shen S, Xiong H, Zheng H. Poba-ga: perturbation optimized black-box adversarial attacks via genetic algorithm. Comput Secur. 2019;85: 89-106.

51. Xiao C, Li B, Zhu J-Y, He W, Liu M, Song D. Generating adversarial examples with adversarial networks [Internet]. arXiv; 2018. Available from: https://arxiv.org/abs/1801.02610.

52. Brown TB, Mané D, Roy A, Abadi M, Gilmer J. Adversarial patch [Internet]. arXiv; 2017. Available from: https://arxiv.org/abs/1712.09665.

53. Lee M, Kolter Z. On physical adversarial patches for object detection [Internet]. arXiv; 2019. Available from:https://arxiv.org/abs/1906.11897.

54. Liu X, Yang H, Liu Z, Song L, Li H, Chen Y. Dpatch: an adversarial patch attack on object detectors [Internet]. arXiv; 2018. Available from:https://arxiv.org/abs/1806.02299.

55. Song D, Eykholt K, Evtimov I, Fernandes E, Li B, Rahmati A, et al. Physical adversarial examples for object detectors. In: 12th USENIX Workshop on Offensive Technologies (WOOT 18). Berkeley, CA, USA: USENIX; 2018.

56. Wu H, Yunas S, Rowlands S, Ruan W, Wahlstrom J. Adversarial detection: attacking object detection in real time [Internet]. arXiv; 2022. Available from:https://arxiv.org/abs/2209.01962.

57. Yang C, Kortylewski A, Xie C, Cao Y, Yuille A. Patchattack: a black-box texture-based attack with reinforcement learning. In: Computer Vision-ECCV 2020: 16th European Conference, Glasgow, UK, August 23-28, 2020, Proceedings, Part XXVI. Berlin: Springer; 2020. p. 681-698.

58. Hoory S, Shapira T, Shabtai A, Elovici Y. Dynamic adversarial patch for evading object detection models [Internet]. arXiv; 2020. Available from: https://arxiv.org/abs/2010.13070.

59. Liang H, He E, Zhao Y, Jia Z, Li H. Adversarial attack and defense: a survey. Electronics. 2022;11(8):1283.

60. Yoo J-W, Jo Y-H, Cha Y-K. Artificial intelligence for autonomous ship: potential cyber threats and security. J Korea Inst Inf Secur Cryptol. 2022;32(2):447-463.

61. Kumar RSS, Brien DO, Albert K, Viljöen S, Snover J. Failure modes in machine learning systems [Internet]. arXiv; 2019. Available from: https://arxiv.org/abs/1911.11034.

62. Wolf MJ, Miller K, Grodzinsky FS. Why we should have seen that coming: comments on microsoft's tay "experiment", and wider implications. ACM SIGCAS Comput Soc. 2017;47(3):54-64.

63. Kim S, Kim H. Zero-centered fixed-point quantization with iterative retraining for deep convolutional neural network-based object detectors. IEEE Access. 2021;9: 20828-20839. 67 Kannan H, Kurakin A, Goodfellow I. Adversarial logit pairing [Internet]. arXiv; 2018. Available from:https://arxiv.org/abs/1803.06373.

64. Lee H, Han S, Lee J. Generative adversarial trainer: defense to adversarial perturbations with gan [Internet]. arXiv; 2017. Available from: https://arxiv.org/abs/1705.03387.

65. Madry A, Makelov A, Schmidt L, Tsipras D, Vladu A. Towards deep learning models resistant to adversarial attacks [Internet]. arXiv; 2017. Available from:https://arxiv.org/abs/1706.06083.

66. Qian YG, Zhang XM, Wang B, Li W, Chen JH, Zhou WJ, et al. Towards robust DNNs: a Taylor expansion-based method for generating powerful adversarial examples [Internet]. arXiv; 2020. Available from:. https://arxiv.org/abs/2001.08389.

67. Hosseini H, Chen Y, Kannan S, Zhang B, Poovendran R. Blocking transferability of adversarial examples in black-box learning systems [Internet]. arXiv; 2017. Available from:https://arxiv.org/abs/1703.04318.

68. Dhillon GS, Azizzadenesheli K, Lipton ZC, Bernstein J, Kossaif J, Khanna A, et al. Stochastic activation pruning for robust adversarial defense [Internet]. arXiv; 2018. Available from:https://arxiv.org/abs/1803.01442.

69. Liu K, Dolan-Gavitt B, Garg S. Fine-pruning: defending against backdooring attacks on deep neural networks. In: Research in Attacks, Intrusions, and Defenses: 21st International Symposium, RAID 2018, Proceedings 21; 2018 Sep 10-12; Heraklion, Crete, Greece. Cham: Springer; 2018. p. 273-294.

70. Liu X, Cheng M, Zhang H, Hsieh C-J. Towards robust neural networks via random self-ensemble. In: Proceedings of the European Conference on Computer Vision (ECCV). Cham: Springer; 2018. p. 369-385.

71. Chen T, He T, Benesty M, Khotilovich V, Tang Y, Cho H. 2015. Xgboost: extreme gradient boosting, R package version 0.4 – 2.1 – 4.

72. Li D, Li Q. Adversarial deep ensemble: evasion attacks and defenses for malware detection. IEEE Trans Inf Forensics Secur. 2020;15: 3886-3900.

73. Guo C, Rana M, Cisse M, Van Der Maaten L. Countering adversarial images using input transformations [Internet]. arXiv; 2017. Available from:https: / / arxiv.org / abs / 1711.00117.

74. Jia X, Wei X, Cao X, Foroosh H. ComDefend: an efficient image compression model to defend adversarial examples. In: Proceedings of the IEEE / CVF Conference on Computer Vision and Pattern Recognition. Long Beach, CA, USA: IEEE; 2019. p. 6084-6092.

75. Jagielski M, Oprea A, Biggio B, Liu C, Nita-Rotaru C, Li B. Manipulating machine learning: poisoning attacks and countermeasures for regression learning. In: 2018 IEEE Symposium on Security and Privacy (SP). Piscataway, NJ: IEEE; 2018. p. 19-35.

76. Liao F, Liang M, Dong Y, Pang T, Hu X, Zhu J. Defense against adversarial attacks using high-level representation guided denoiser. In: Proceedings of the IEEE Conference on Computer Vision and Pattern Recognition. Bellingham, WA: SPIE; 2018. p. 1778-1787.

77. Meng D, Chen H. Magnet: a two-pronged defense against adversarial examples. In: Proceedings of the 2017 ACM SIGSAC Conference on Computer and Communications Security. New York, USA: ACM; 2017. p. 135 – 147.

78. Samangouei P, Kabkab M, Chellappa R. Defense-gan: protecting classifiers against adversarial attacks using generative models [Internet]. arXiv; 2018. Available from:https: / / arxiv.org / abs / 1805.06605.

79. Folz J, Palacio S, Hees J, Dengel A. Adversarial defense based on structure-to-signal autoencoders. In: 2020 IEEE Winter Conference on Applications of Computer Vision (WACV). Piscataway, NJ: IEEE; 2020. p. 3568 – 3577.

80. Papernot N, McDaniel P, Wu X, Jha S, Swami A. Distillation as a defense to adversarial perturbations against deep neural networks. In: 2016 IEEE Symposium on Security and Privacy (SP). Piscataway, NJ: IEEE; 2016. p. 582 – 597. 85 Tanay T, Griffin L. A boundary tilting persepective on the phenomenon of adversarial examples [Internet]. arXiv; 2016. Available from:https: / / arxiv.org / abs / 1608.07690.

81. Papernot N, Abadi M, Erlingsson U, Goodfellow I, Talwar K. Semi-supervised knowledge transfer for deep learning from private training data [Internet]. arXiv; 2016. Available from:https: / / arxiv.org / abs / 1610.05755

82. Adi Y, Baum C, Cisse M, Pinkas B, Keshet J. Turning your weakness into a strength: watermarking deep neural networks by backdooring. In: 27th USENIX Security Symposium (USENIX Security 18). Berkeley, CA, USA: USENIX; 2018. p. 1615-1631.

83. Hall A. Autonomous minehunter to trial uncrewed operations in the gulf. Navy News [Internet]; 2023 [cited 2023 Feb 13]. Available from: https://www.royalnavy.mod.uk/news-and-latest-activity/news/2023/february/13/20230213 autonomous-minehunter-to-trial-uncrewed-operations-in-the-gulf.

84. Hall A. Dstl and DASA research underpins royal navy maritime autonomy. Navy News [Internet]; 2023 [cited 2023 Jan 26]. Available from: https://www.gov.uk/government/news/dstl-and-dasa-researchunderpins-royal-navy-maritime-autonomy.

INDEX